DIGITAL SIGNAL PROCESSING TECHNIQUES AND APPLICATIONS IN RADAR IMAGE PROCESSING

WILEY SERIES ON INFORMATION AND COMMUNICATIONS TECHNOLOGIES

Series Editors: Russell Hsing and Vincent K. N. Lau

The Information and Communications Technologies (ICT) book series focuses on creating useful connections between advanced communication theories, practical designs, and end-user applications in various next generation networks and broadband access systems, including fiber, cable, satellite, and wireless. The ICT book series examines the difficulties of applying various advanced communication technologies to practical systems such as WiFi, WiMax, B3G, etc., and considers how technologies are designed in conjunction with standards, theories, and applications.

The ICT book series also addresses application-oriented topics such as service management and creation and end-user devices, as well as the coupling between end devices and infrastructure.

T. Russell Hsing, PhD, is the Executive Director of Emerging Technologies and Services Research at Telcordia Technologies. He manages and leads the applied research and development of information and wireless sensor networking solutions for numerous applications and systems. Email: thsing@telcordia.com

Vincent K.N. Lau, PhD, is Associate Professor in the Department of Electrical Engineering at the Hong Kong University of Science and Technology. His current research interest is on delay-sensitive cross-layer optimization with imperfect system state information. Email: eeknlau@ee.ust.hk

Wireless Internet and Mobile Computing: Interoperability and Performance
Yu-Kwong Ricky Kwok and Vincent K. N. Lau

RF Circuit Design
Richard C. Li

Digital Signal Processing Techniques and Applications in Radar Image Processing
Bu-Chin Wang

DIGITAL SIGNAL PROCESSING TECHNIQUES AND APPLICATIONS IN RADAR IMAGE PROCESSING

Bu-Chin Wang

WILEY

A JOHN WILEY & SONS, INC., PUBLICATION

Published by John Wiley & Sons, Inc., Hoboken, New Jersey
Published simultaneously in Canada

Library of Congress Cataloging-in-Publication Data

Wang, Bu-Chin.

Digital signal processing techniques and applications in radar image processing / Bu-Chin Wang.

p. cm.

ISBN 978-0-470-18092-1

1. Signal processing—Digital techniques. 2. Remote sensing. I. Title.

TK5102.9.W36 2008

621.36′78–dc22

2008004941

Printed in the United States of America

10 9 8 7 6 5 4 3 2 1

To my mother Chun-Ying Wang,
in memory of my father Lan-Din Wang, and
to my wife Rhoda and our children,
Anna and David

CONTENTS

PREFACE

In the past few decades, the principles and techniques of digital signal processing (DSP) have been used in applications such as data and wireless communication, voice and speech analysis and synthesis, and video and image compression and expansion. Radar image processing is considered the primary application of the remote sensing field and is a new and emerging area for DSP applications. Although the primary application of satellite-based radar imaging is military surveillance, the low cost and real-time processing capability of radar imaging, together with its capability to operate under any environmental conditions (e.g., night or day, rain or snow, fog or clear sky) have opened up many commercial applications. Sea ice monitoring and disaster monitoring of events such as forest fires, floods, volcano eruptions, earthquakes, and oil spills are examples of satellite-based radar imaging applications. Airborne-based radar systems also have made radar imaging more affordable and popular. Furthermore, exploration of underground natural resources is an example of a new application.

The processing of radar images, in general, consists of three major fields: DSP principles and communication theory, knowledge of antenna and radar operation, and algorithms used to process the radar images. The purpose of this book is to include the material in these fields in one publication, to provide the reader with a thorough understanding of how radar images are processed. To further familiarize the reader with the theories and techniques used in processing radar images, MATLAB*-based programs are utilized extensively in this book in both the synthesis and analysis of the radar image. In this way, the signal waveforms are therefore made visible at various stages during computer simulation, and the capability of three-dimensional (3D) graphical displays makes many abstract results easier to understand. This book is aimed at engineers or students who have some knowledge of DSP theory and limited knowledge of communication theory and/or antenna theory, but are interested in advanced DSP applications, especially in the remote sensing field.

This book consists of three major groups of chapters. Chapters 1 and 2 provide an overview of DSP principles, reviewing signal characteristics in both analog and digital domains and describing some DSP techniques that serve as key tools in radar images processing. Chapters 3–5 discuss the basics of antenna theory, radar

*MATLAB is a registered trademark of math Works, Inc., Natick, MA 01760.

operation principles, modulation/demodulation, and radar target detection techniques. Chapters 6–9 discuss the properties and formation of radar images and then try to model the processing of radar images. The principles of radar image data synthesis are presented and demonstrated with computer-simulated examples. Both the range–Doppler and the Stolt interpolation algorithms are described and applied to the simulated image data and satellite radar-based image data. The results are analyzed and compared. MATLAB* programs are used extensively during the generation of various waveforms of signal processing, radar detection, and synthesis/simulation of radar image processing.

The first two chapters briefly review the DSP principles. Chapter 1 describes the characteristics of signals, followed by Fourier series representation of periodic signals. Fourier transform is then introduced to represent a signal, whether in periodic or nonperiodic form. Sampling theory and interpolation filter are derived, and some advanced sampling and interpolation techniques are reviewed. Resampling from unevenly spaced data to obtain evenly spaced data is briefly discussed at the end of the chapter. Chapter 2 addresses the discrete signal transformation in both time and frequency domains. Discrete Fourier transform (DFT), together with some of its characteristics, are reviewed. Windowing functions and the well-known fast Fourier transform (FFT) technique are covered. The discrete cosine transform (DCT), which is the byproduct of DFT, is introduced. A graphical representation of DFT provides an overview of the relationship between a continuous signal and a discrete signal. It also provides signal variations in both time and frequency domains. The chapter ends with an example of resampling with fractional interpolation based on DFT technique.

Chapters 3–5 provide a background review on antenna theory and radar operation principles. Chapter 3 starts the review of the electromagnetic field with the Maxwell equation, followed by the electromagnetic (EM) fields generated from the infinitesimal dipole. Finite-length dipole- and half-wavelength dipole-based linear antenna arrays are described. Some commonly used antennas, including the microstrip antenna, are also covered. Chapter 4 deals with the basic theory of radar signal processing. The radar range equations and other related parameters are reviewed. The Doppler frequency due to relative movement between radar and target is briefly discussed with respect to the wavefront. Some target range and motion direction detection techniques are also revealed at the end of chapter. Chapter 5 provides broad coverage of modulation/demodulation and target detection techniques used by radar systems. Amplitude modulation (AM)-based pulse Doppler frequency radar is first reviewed, followed by discussion of target detection techniques. Frequency modulation (FM)-based radars, which include pulsed linear FM (LFM), continuous-wave LFM and stepped LFM signals, are then briefly discussed. Also covered in this chapter are in-phase–quadrature-phase (I–Q) demodulator and pulse compression (or matched filtering), which serve as important tools in radar signal processing

Chapters 6–9 discuss the main topic of this book: radar image formation and processing. Chapter 6 starts with a survey of some popular imaging radars and possible applications, followed by the description of the geometry of stripmap synthetic aperture radar (SAR), which consists of broadside SAR and squint SAR. The role of Doppler frequency in radar image formation is analyzed. Also covered are the range migraion, geometric distortion, and resolution of image radar. Chapter 7 discusses the ideal system model of radar imaging. The reconstruction of 2D target function is modeled by two independent 1D functions. The model of 1D range imaging is first described, followed by discussion of the 1D cross-range imaging. Data acquisition and the frequency spectrum of radar image are also reviewed. Chapter 8 discusses the principles of radar image generation and how to synthesize the radar image. Examples of synthesizing radar image data for broadside SAR and squint SAR are presented, which include single and multiple targets. The range–Doppler algorithm on processing radar images is then reviewed and applied to the synthesized data. Chapter 9 reviews some radar image processing techniques in the wavenumber domain. The Stolt interpolation technique on radar image processing is briefly reviewed and applied to some simulated image data. The real satellite radar signal is then processed by both range–Doppler and Stolt interpolation algorithms. A comparison on these two algorithms is also provided.

Some of the material in this book was presented to graduate students in Su-Zhou University in China, and the feedback from the students was incorporated into this book. It is my hope that this book can provide enough knowledge for readers to become familiar with radar image processing. Although I have made every effort to make this a thorough and accurate book, errors and mistakes are inevitable. Any comments or feedback from readers will be welcomed and appreciated.

Acknowledgment

I would like to thank Dr. Russell Hsing of Telcordia for his support and inspiration throughout the process of writing this book. His advice made the publication of this book possible and is greatly appreciated.

Finally, I owe a lot to my family for their patience and understanding as I worked on this book. My wife, Rhoda, my children, Anna and David, my son-in-law, Scott Chong, and granddaughter, Jocelyn, all helped make this book possible in numerous ways, and I am grateful to them.

BU-CHIN WANG, PHD

LIST OF SYMBOLS

A_s	Azimuth sample spacing along the azimuth or y axis, meters
α	Frequency change rate for LFM signal, hertz per second (Hz/s)
β	Frequency change rate for LFM signal along azimuth direction, Hz/s
B, B_t	Frequency bandwidth, Hz
B_D, B_{Dop}	Doppler frequency bandwidth, Hz
B_{Ds}	Doppler frequency bandwidth in terms of slow time s, Hz
B_{Du}	Doppler frequency bandwidth in terms of radar position u, Hz
B_{ku}	Spatial frequency bandwidth in the k_u domain, Hz
c	Speed of light, 3×10^8 meters per second (m/s)
f	Frequency, Hz
f_b	Baseband frequency, Hz
f_0	Reference frequency, Hz
f_c	Carrier frequency, Hz
f_s	Sampling frequency, Hz
f_m	Maximum frequency, Hz
f_{PRF}	Pulse repetition frequency, Hz
f_D	Doppler frequency, Hz
f_{Dr}	Doppler frequency in terms of slant range r, Hz
f_{Ds}	Doppler frequency in terms of slow time s, Hz
f_{Du}	Doppler frequency in terms of radar position u, Hz
f_{Dc}	Centroid of Doppler frequency, Hz
f_{Dcr}	f_{Dc} with respect to slant range r, Hz
f_{Dcs}	f_{Dc} with respect to slow time s, Hz
f_{Dcu}	f_{Dc} with respect to radar position u, Hz
$\dot{f}_{Dcs}$	Derivative of f_{Dcs}, or slope of f_{Dcs}, Hz/s
$\dot{f}_{Dc}$	Derivative of f_{Dc}, or slope of f_{Dc}, Hz/s
$\dot{f}_D$	Derivative of f_D, or slope of f_D, Hz/s
f_{DU}	Upper bound of Doppler frequency bandwidth, Hz
f_{DL}	Lower bound of Doppler frequency bandwidth, Hz
f_{DUr}	f_{DU} with respect to slant range r, Hz
f_{DLr}	f_{DL} with respect to slant range r, Hz
f_{DUs}	f_{DU} with respect to slow time s, Hz

f_{DLs}	f_{DL} with respect to slow time s, Hz
$f_0(x)$	Ideal target function along range or x axis
$f_0(y)$	Ideal target function along azimuth or y axis
$f(x)$	Target function along range or x axis
$f(y)$	Target function along azimuth or y axis
$F_0(k)$	Fourier transform of $f_0(x)$
$F_0(k_y)$, $F_0(k_u)$	Spatial Fourier transform of $f_0(y)$
$F(k)$	Fourrier transform of $f(x)$
$F(k_y)$, $F(k_u)$	Spatial Fourier transform of $f(y)$
$h_r(t)$	Range matched filter
$h_{az}(s)$	Azimuth matched filter
$h_{az}(t,s)$	2D azimuth matched filter
$H_{az}(\omega, \omega_D)$	2D spatiotemporal Fourier transform of $h_{az}(t,s)$
$I_m(k_u)$	Gating function in wavenumber k_u domain due to mth target
$i_m(u)$	Inverse spatial Fourier transform of $I_m(k_u)$
k_x	Spatial wavenumber, corresponding to spatial Fourier transform of x, 1/meter (m^{-1}; reciprocal meter)
k_u	Spatial wavenumber, corresponding to spatial Fourier transform of u, m^{-1}
k_{um}	Spatial wavenumber, corresponding to mth target
k'_u	Spatial wavenumber changing rate
k_y	Same as k_u, m^{-1}
k	Wavenumber, corresponding to f, m^{-1}
k_b	Wavenumber in baseband corresponding to f_b, m^{-1}
k_0	Wavenumber, corresponding to f_0, m^{-1}
k_c	Wavenumber, corresponding to f_c, m^{-1}
λ	Wavelength of electromagnetic wave, m
L	Antenna length, m
Ls	Synthetic aperture length, m
Lsa	Ls for target located at range R_{0a}, m
$Lsa2$	Half size of Lsa, m
Lsx	Ls for target located at range R_{0x} for $x = b, c, d \ldots$, m
$Lsx2$	Half size of Lsx, m
$Lsai$	Lsa for ith target
$Lsxi$	Lsx for ith target
L_{total}	Sum of Lsx from all targets, m
M	Number of rows in a 2D data array
M_{amb}	Doppler ambiguity
N	Number of columns in a 2D data array
Naz	Number of azimuth lines within the synthetic aperture length Ls
$Nazi$	Number of azimuth lines within Ls for target located at range R_{0i}
N_r	Number of samples within the transmitter pulse duration
N_{rx}	Number of sample difference between range R_x, for $x = b, c \ldots$, and range reference R_a
$p(t)$	Transmitted radar signal

$P(\omega)$	Fourier transform of $p(t)$
$p_b(t)$	Baseband transmitted radar signal
psf(t)	Pont spread function in time domain
psf(ω)	Fourier transform of psf(t)
R_0	Shortest distance between target and radar, m
R_{0a}	R_0 with target located at range $x = a$, m
R_{0i}	R_0 with target located at range $x = b, c\ \ldots$, m
$R_{u_i n}$	Slant range between target n and radar located at u_i, m
R_c	Distance between target and radar when target is under illunination of radar center beam, m
R_1	Distance between target and radar when radar starts to illuminate the target, m
R_3	Distance between target and radar when radar stops to illuminate the target, m
R_s	Slant range sample spacing, m
R_S	shortest distance from radar to ground along the range (x-axis) direction, m
R_L	Longest distance from radar to ground along the range (x-axis) direction, m
$R(s)$	Slant range in terms of slow time s, m
$R(u)$	Slant range in terms of radar position u, m
ΔR	Slant range difference with respect to R_0 (broadside SAR case) or R_3 (squint SAR case), m
ΔR_r	Slant range resolution, m
ΔR_{gr}	Ground range resolution, m
ΔR_a	Angular resolution, m
s	Slow time variable along radar moving direction, seconds (s)
s_c	Slow time when target is illuminated by center beam of radar, s
$s(t)$	Target reflected signal
$s_b(t)$	Baseband signal of $s(t)$
$S(\omega)$	Fourier transform of $s(t)$
$S_b(\omega)$	Fourier transform of $s_b(t)$
$S(t,u)$	Target reflected signal when radar is at location $y = u$
$S(\omega, \omega_D)$	2D Spatiotemporal Fourier transform of $s(t,u)$
$s_m(t,u)$	mth target reflected signal when radar is at location $y = u$
$S_m(\omega, \omega_D)$	2D spatiotemporal transform of $s_m(t,u)$
$s_{bm}(t,u)$	Baseband version of $s_m(t,u)$
$S_{bm}(\omega, \omega_D)$	Baseband version of $S_m(\omega, \omega_D)$
$s_{r0}(u)$	Range reference fouction for radar located at $(0, u)$
$S_{r0}(k_u)$	Spatial Fourier transform of $s_{r0}(u)$
τ	Pulse duration time or echo delay time, s
τ_{u_i}	Echo delay time when radar is located at u_i, s
$\tau_{u_i n}$	Echo delay time due to target n and when radar is located at u_i, s
T, T_0	Period of periodic signal, s
T_a	Time duration when target is under illumination of radar beam, s

T_p	Pulse duration time, s
u	Radar position variable along the azimuth or y axis, m
u_i	Radar position at $u = u_i$, m
u_2	Radar location when target is under illumination of radar center beam, m
u_1	Radar location when radar begins to illuminate target, m
u_3	Radar location when radar ceases to illuminate target, m
$\overline{V}$	Radar velocity, a vector along azimuth direction m/s
V_r	Radar's radial velocity, a scalar along the target direction, m/s
W	Antenna width, m
X_c	Centerpoint of target area along x axis, m
X_0	Half size of target area along range (x-axis) direction, m
Y_0	Half size of target area along azimuth (y-axis) direction, m
Y_c	Centerpoint of target area along y axis, m
y_i	Target location along y axis, m
$<rn>$	2D signal array corresponding to target n
ω	Angular frequency, Hz
ω_b	Angular baseband frequency, Hz
ω_0	Angular reference frequency, Hz
ω_c	Angular carrier frequency, Hz
σ_i	Reflection coefficient from ith target
θ	Angle, radians, or degrees
θ_H	Horizontal beamwidth $= \lambda/L$, radians or degrees
θ_V	Vertical beamwidth $= \lambda/W$, radians or degrees
$\theta_{3\text{dB}}$	Antenna 3-dB beamwidth, radians or degrees
$\theta_m(u)$	Aspect angle with respect to mth target when radar is at location $y = u$, radians or degrees
θ_u	Equal to $\theta_m(u)$ for single target
θ_q	Radar squint angle, radians or degrees
$\delta(t)$	Delta or impulse function
$\text{sinc}(t)$	Sampling or interpolation filter function
$\text{Rect}(t)$	Time-domain rectangular pulse with duration $\lvert t\rvert \leq \frac{1}{2}$
$\lvert\chi(\tau, f_D)\rvert$	Radar ambiguity function
$\mathfrak{F}$	Fourier transform operator
$\mathfrak{F}^{-1}$	Inverse Fourier transform operator
$J(u)$	Jacobian of transformation

LIST OF ILLUSTRATIONS

Figures

Table

1

SIGNAL THEORY AND ANALYSIS

A *signal*, in general, refers to an electrical waveform whose amplitude varies with time. Signals can be fully described in either the time or frequency domain. This chapter discusses the characteristics of signals and identifies the main tools used for signal processing. Some functions widely used in signal processing are described in Section 1.1. A quick review of the linear system and convolution theory is covered in Section 1.2. Fourier series representation of periodic signals is discussed in Section 1.3. Fourier transform of nonperiodic signals and periodic signals are covered in Sections 1.4 and 1.5, respectively. Section 1.6 describes sampling theory together with signal interpolation. Some advanced sampling and interpolation techniques are reviewed in Section 1.7.

1.1 SPECIAL FUNCTIONS USED IN SIGNAL PROCESSING

1.1.1 Delta or Impulse Function $\delta(t)$

The delta function or impulse function $\delta(t)$ is defined as

$$\begin{aligned} \delta(t) &= \infty \quad \text{for} \quad t = 0, \\ &= 0 \quad \text{for} \quad t \neq 0. \end{aligned}$$

and

$$\int_{-\infty}^{\infty} \delta(t)dt = 1. \tag{1.1}$$

Digital Signal Processing Techniques and Applications in Radar Image Processing, by Bu-Chin Wang.

On the basis of this definition, one can obtain

$$\int_{-\infty}^{\infty} f(t)\delta(t)dt = f(0)$$

and

$$\int_{-\infty}^{\infty} f(t)\delta(t - t_0)dt = f(t_0).$$

1.1.2 Sampling or Interpolation Function sinc (*t*)

The function sinc (t) is defined as

$$\begin{aligned} \operatorname{sinc}(t) &= 1 \quad \text{for} \quad t = 0 \\ &= \frac{\sin(\pi t)}{\pi t} \quad \text{otherwise} \end{aligned}$$

and

$$\int_{-\infty}^{\infty} \operatorname{sinc}(t)dt = 1. \tag{1.2}$$

A sinc (t) function for $t = -4$ to 4 is depicted in Fig. 1.1. Notice that sinc $(t) = 0$ for all integers of t, and its local maxima corresponds to its intersection with the $\cos(\pi t)$.

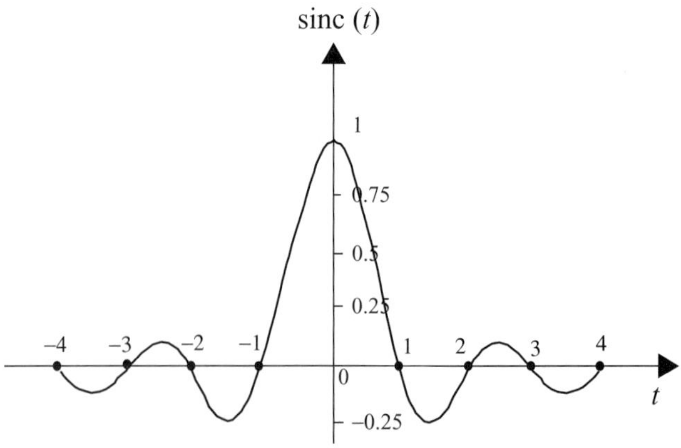

FIGURE 1.1 A sinc (*t*) function.

1.2 LINEAR SYSTEM AND CONVOLUTION

A linear system, as shown in Fig. 1.2, can be represented as a box with input $\underline{\mathbf{x}}$, output $\underline{\mathbf{y}}$ and a system operator $\mathbf{H}$ that defines the relationship between $\underline{\mathbf{x}}$ and $\underline{\mathbf{y}}$. Both $\underline{\mathbf{x}}$ and $\underline{\mathbf{y}}$ can be a set of components.

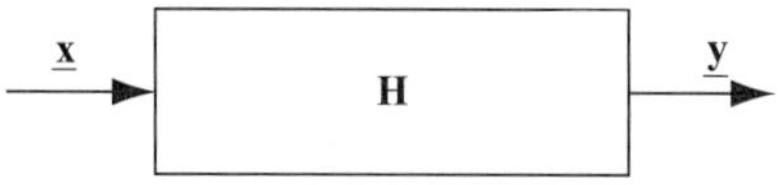

FIGURE 1.2 A linear system.

A system is linear if and only if

$$\mathbf{H}(a\underline{\mathbf{x}} + b\underline{\mathbf{y}}) = a\mathbf{H}\underline{\mathbf{x}} + b\mathbf{H}\underline{\mathbf{y}}. \tag{1.3}$$

where a and b are constants, $\underline{\mathbf{x}}$ is the system's input signal, and $\underline{\mathbf{y}}$ is the output signal.

In addition, a linear system having the fixed input–output relation

$$\mathbf{H}\underline{\mathbf{x}}(t) = \underline{\mathbf{y}}(t)$$

is time-invariant if and only if

$$\mathbf{H}\underline{\mathbf{x}}(t - \tau) = \underline{\mathbf{y}}(t - \tau)$$

for any $\underline{\mathbf{x}}(t)$ and any τ. In the following discussion, only the linear and time-invariant system is considered.

Let $p_\tau(t)$ be a pulse with amplitude $1/\Delta\tau$ and duration $\Delta\tau$; then any function $f(t)$ can be represented as

$$f(t) \approx \sum_{n=-\infty}^{\infty} f(n\Delta\tau)p_\tau(t - n\Delta\tau)\Delta\tau. \tag{1.4a}$$

Figure 1.3 illustrates the relationship between $p_\tau(t)$ and the function $f(t)$. Figure 1.3a shows a rectangular polygon with amplitude $1/\Delta\tau$ and duration$\Delta\tau$; Fig. 1.3b

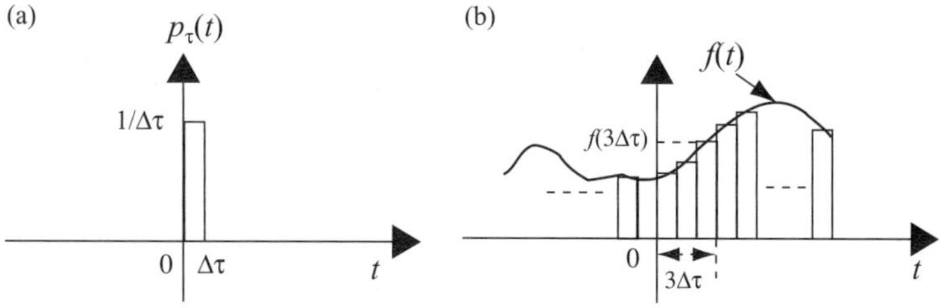

FIGURE 1.3 Graphical representations of a function in terms of pulses.

displays how a function $f(t)$ can be approximated by a series of delayed rectangular polygon $p_\tau(t- n\Delta\tau)$ with amplitude $f(n\Delta\tau)\Delta\tau$.

As $\Delta\tau \to 0, n\Delta\tau \to \tau$. Therefore

$$p_\tau(t) \to \delta(t)$$

and

$$p_\tau(t - n\Delta\tau) \to \delta(t - \tau).$$

The summation of Eq. (1.4a) then becomes

$$f(t) = \int_{-\infty}^{\infty} f(\tau)\delta(t - \tau)d\tau. \tag{1.4b}$$

Let $h(t)$ be the impulse response of a system:

$$\mathbf{H}\,\delta(t) = h(t).$$

Then, for any input function $x(t)$, the output function $y(t)$ can be expressed as

$$\begin{aligned} y(t) &= \mathbf{H}\,x(t) \\ &= \mathbf{H}\int_{-\infty}^{\infty} x(\tau)\delta(t - \tau)d\tau \\ &= \int_{-\infty}^{\infty} x(\tau)h(t - \tau)d\tau \\ &= x(t) * h(t). \end{aligned} \tag{1.5}$$

where the asterisk (symbol $*$) refers to convolution. If $x(t) = \delta(t)$, then

$$\begin{aligned} y(t) &= x(t) * h(t) \\ &= \delta(t) * h(t) \\ &= h(t). \end{aligned}$$

Equation (1.5) states the relationship between the input function $x(t)$, the impulse response or system function $h(t)$, and the output function $y(t)$. It serves as a fundamental equation and is widely used in linear and time-invariant systems. A simple block diagram that illustrates this relationship is shown in Fig. 1.4

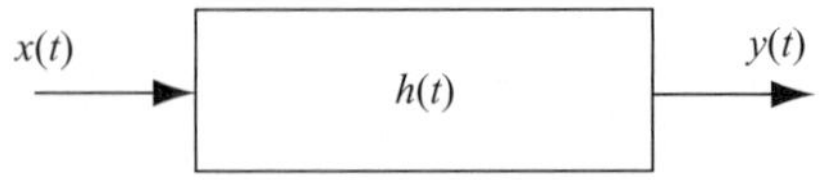

FIGURE 1.4 A time-invariant linear system.

1.2.1 Key Properties of Convolution

1.2.1.1 Commutative

By letting $\lambda = t - \tau$, Eq. (1.5) becomes

$$\begin{aligned} y(t) &= \int_{-\infty}^{\infty} x(\tau)h(t-\tau)d\tau \\ &= \int_{-\infty}^{\infty} x(t-\lambda)h(\lambda)d\lambda \\ &= h(t) * x(t). \end{aligned}$$

Therefore

$$\begin{aligned} y(t) &= x(t) * h(t) \\ &= h(t) * x(t). \end{aligned} \tag{1.6}$$

1.2.1.2 Associative

If

$$y(t) = [x(t) * h(t)] * z(t),$$

then

$$\begin{aligned} y(t) &= x(t) * [h(t) * z(t)] \\ &= [x(t) * z(t)] * h(t). \end{aligned} \tag{1.7}$$

1.2.1.3 Distributive

If

$$y(t) = x(t) * h(t) + x(t) * z(t),$$

then

$$y(t) = x(t) * [h(t) + z(t)]. \tag{1.8}$$

1.2.1.4 Timeshift

If

$$y(t) = x(t) * h(t),$$

then

$$\begin{aligned} y(t-\tau) &= x(t-\tau) * h(t) \\ &= x(t) * h(t-\tau). \end{aligned} \tag{1.9}$$

1.3 FOURIER SERIES REPRESENTATION OF PERIODIC SIGNALS

A signal $g_p(t)$ is called a *periodic signal* with period T_0 if it remains unchanged after it has been shifted forward or backward by T_0, that is

$$g_p(t) = g_p(t +/- T_0),$$

where $T_0 = 2\pi/\omega_0$.

There are three different Fourier series representations for a periodic signal. The first two representations are in terms of trigonometric functions, while the third is in exponential form. The three Fourier series representations of a periodic signal $g_p(t)$ are described below.

1.3.1 Trigonometric Fourier Series

A periodic signal $g_p(t)$ can be represented as

$$g_p(t) = a_0 + \sum_{n=1}^{\infty} a_n \cos(n\omega_0 t) + \sum_{n=1}^{\infty} b_n \sin(n\omega_0 t), \tag{1.10a}$$

where a_n and b_n for $n = 1,2,\ldots$ can be computed as

$$a_0 = \frac{1}{T_0}\int_0^{T_0} g_p(t)\,dt, \tag{1.10b}$$

$$a_n = \frac{2}{T_0}\int_0^{T_0} g_p(t)\cos(n\omega_0 t)\,dt, \tag{1.10c}$$

$$b_n = \frac{2}{T_0}\int_0^{T_0} g_p(t)\sin(n\omega_0 t)\,dt. \tag{1.10d}$$

1.3.2 Compact Trigonometric Fourier Series

Alternatively, a periodic signal $g_p(t)$ can be represented as

$$g_p(t) = c_0 + \sum_{n=1}^{\infty} c_n \cos(n\omega_0 + \theta_n), \tag{1.11a}$$

where

$$c_0 = a_0, \tag{1.11b}$$

$$c_n = \sqrt{a_n^2 + b_n^2}, \tag{1.11c}$$

$$\theta_n = \tan^{-1}\frac{b_n}{a_n}. \tag{1.11d}$$

1.3.3 Exponential Fourier Series

A periodic signal $g_p(t)$ can also be represented as

$$g_p(t) = \sum_{n=-\infty}^{\infty} G_n e^{jn\omega_0 t}, \tag{1.12a}$$

where

$$G_n = \frac{1}{T_0}\int_0^{T_0} g_p(t) e^{-jn\omega_0 t}\, dt,$$

or

$$G_0 = a_0, \tag{1.12b}$$

$$G_n = \frac{a_n - j\,b_n}{2}, \tag{1.12c}$$

$$G_{-n} = \frac{a_n + j\,b_n}{2}. \tag{1.12d}$$

Example 1.1 Figure 1.5 shows a periodic signal $g_p(t)$, which is expressed as

$$g_p(t) = \sum_{n=-\infty}^{\infty} g(t - nT_0),$$

with

$$\begin{aligned} g(t) &= A \quad \text{for } -\frac{\tau}{2} \leq t \leq \frac{\tau}{2}, \\ &= 0 \quad \text{otherwise.} \end{aligned}$$

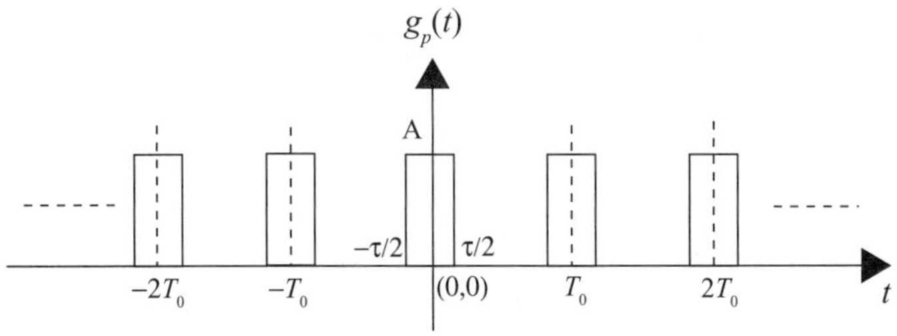

FIGURE 1.5 A periodic signal $g_p(t)$.

The Fourier series coefficients of $g_p(t)$ in terms of these three representations can be computed as the following equations show:

1. From Eq. (1.10)

$$g_p(t) = a_0 + \sum_{n=1}^{\infty} a_n \cos(n\omega_0 t) + \sum_{n=1}^{\infty} b_n \sin(n\omega_0 t).$$

$$\begin{aligned}
a_0 &= \frac{1}{T_0}\int_0^{T_0} g_p(t)dt \\
&= \frac{1}{T_0}\int_{-\tau/2}^{\tau/2} A\,dt \\
&= \frac{A\tau}{T_0}. \\
a_n &= \frac{2}{T_0}\int_0^{T_0} g_p(t)\cos(n\omega_0 t)\,dt \\
&= \frac{2}{T_0}\int_{-\tau/2}^{\tau/2} A\cos n\omega_0 t\,dt \\
&= \frac{2A}{n\pi}\sin\frac{n\pi\tau}{T_0}. \\
b_n &= \frac{2}{T_0}\int_0^{T_0} g_p(t)\sin(n\omega_0 t)\,dt \\
&= \frac{2}{T_0}\int_{-\tau/2}^{\tau/2} A\sin(n\omega_0 t)\,dt \\
&= 0.
\end{aligned}$$

2. From Eq. (1.11)

$$g_p(t) = c_0 + \sum_{n=1}^{\infty} c_n \cos(n\omega_0 + \theta_n).$$

$$\begin{aligned}
c_0 &= a_0 \\
&= \frac{A\tau}{T_0}. \\
c_n &= \sqrt{a_n^2 + b_n^2} \\
&= a_n \\
&= \frac{2A}{n\pi}\sin\frac{n\pi\tau}{T_0}. \\
\theta_n &= \tan^{-1}\frac{b_n}{a_n} \\
&= 0.
\end{aligned}$$

3. From Eq. (1.12)

$$
\begin{aligned}
g_p(t) &= \sum_{n=-\infty}^{\infty} G_n e^{jn\omega_0 t}. \\
G_0 &= a_0. \\
G_n &= G_{-n} \\
&= \frac{a_n}{2} \\
&= \frac{A}{n\pi} \sin \frac{n\pi\tau}{T_0} \\
&= \frac{A\tau}{T_0} \frac{\sin(n\pi\tau/T_0)}{n\pi\tau/T_0} \\
&= \frac{A\tau}{T_0} \operatorname{sinc}\left(\frac{n\tau}{T_0}\right).
\end{aligned}
$$

Figure 1.6 displays the G_n for the case when $A = 1$ and $\tau = T_0 / 2$. Notice that the Fourier series coefficients of $\{G_n\}$ are discrete and the dashed line represents the envelope of $\{G_n\}$.

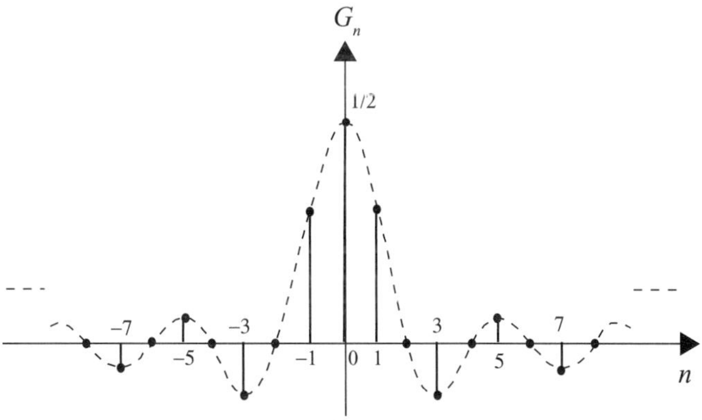

FIGURE 1.6 Fourier series coefficients of a periodic pulse.

Example 1.2 Let the periodic signal $g_p(t)$ in Example 1.1 be modified with $\tau = 0$ and $A = \infty$, such that $A\tau = 1$:

$$
\begin{aligned}
g_p(t) &= \sum_{n=-\infty}^{\infty} g(t - nT_0) \\
&= \sum_{n=-\infty}^{\infty} \delta(t - nT_0).
\end{aligned}
$$

The three Fourier series representations of a periodic impulse train can be computed as follows:

1. From Eq. (1.10)

$$\begin{aligned} a_0 &= \frac{1}{T_0}\int_0^{T_0} g_p(t)dt \\ &= \lim_{\substack{\tau\to 0\\ A\to\infty}} \frac{1}{T_0}\int_{-\tau/2}^{\tau/2} \delta(t)dt \\ &= \frac{1}{T_0}. \\ a_n &= \lim_{\substack{\tau\to 0\\ A\to\infty}} \frac{2}{T_0}\int_{-\tau/2}^{\tau/2} \delta(t)\cos(n\omega_0 t)\,dt \\ &= \frac{2}{T_0}. \\ b_n &= 0. \end{aligned}$$

Therefore

$$\begin{aligned} g_p(t) &= \sum_{n=-\infty}^{\infty} \delta(t - nT_0) \\ &= \frac{1}{T_0} + \sum_{n=1}^{\infty} \frac{2}{T_0}\cos(n\omega_0 t). \end{aligned}$$

2. From Eq. (1.11)

$$\begin{aligned} c_0 &= a_0 \\ &= \frac{1}{T_0}. \\ c_n &= \sqrt{a_n^2 + b_n^2} \\ &= a_n \\ &= \frac{2}{T_0}. \\ \theta_n &= \tan^{-1}\frac{b_n}{a_n} \\ &= 0. \end{aligned}$$

Therefore

$$g_p(t) = \sum_{n=-\infty}^{\infty} \delta(t - nT_0)$$
$$= \frac{1}{T_0} + \sum_{n=1}^{\infty} \frac{2}{T_0} \cos(n\omega_0 t).$$

3. From Eq. (1.12)

$$G_0 = a_0$$
$$= \frac{1}{T_0}.$$
$$G_n = G_{-n}$$
$$= \frac{a_n}{2}$$
$$= \frac{1}{T_0}.$$

Therefore

$$g_p(t) = \sum_{n=-\infty}^{\infty} \delta(t - nT_0)$$
$$= \sum_{n=-\infty}^{\infty} \frac{1}{T_0} e^{jn\omega_0 t}.$$

1.4 NONPERIODIC SIGNAL REPRESENTATION BY FOURIER TRANSFORM

A periodic signal $g_p(t)$ can always be represented in one of the three Fourier series forms described in the previous section. Consider the signal based on exponential representation as shown in Eq. (1.12a), that is

$$g_p(t) = \sum_{n=-\infty}^{\infty} G_n e^{jn\omega_0 t},$$

where G_n can be derived as

$$G_n = \frac{1}{T_0} \int_0^{T_0} g_p(t) e^{-jn\omega_0 t} dt$$

or

$$G_n = \frac{1}{T_0}\int_{-T_0/2}^{T_0/2} g_p(t)e^{-jn\omega_0 t}\,dt.$$

A nonperiodic signal $g(t)$ can be considered as a periodic signal $g_p(t)$ with the period $T_0 \to \infty$:

$$g(t) = \lim_{T_0\to\infty} g_p(t).$$

Notice that $T_0 \to \infty$ implies $\omega_0 = (2\pi/T_0) \to 0$.

Let $\omega_0 \approx \Delta\omega$; then

$$T_0 G_n = \int_{-T_0/2}^{T_0/2} g_p(t)e^{-jn\,\Delta\omega t}\,dt.$$

From the integration shown above, it can be seen that $T_0 G_n$ is a function of $n\Delta\omega$; therefore, one can define

$$T_0 G_n = G(n\Delta\omega),$$

and

$$\begin{aligned} g_p(t) &= \sum_{n=-\infty}^{\infty} \frac{G(n\Delta\omega)}{T_0} e^{jn\,\Delta\omega t} \\ &= \sum_{n=-\infty}^{\infty} \frac{G(n\Delta\omega)\Delta\omega}{2\pi} e^{jn\,\Delta\omega t}. \end{aligned}$$

Now

$$\begin{aligned} g(t) &= \lim_{T_0\to\infty} g_p(t) \\ &= \lim_{T_0\to\infty} \frac{1}{2\pi} \sum_{n=-\infty}^{\infty} G(n\Delta\omega)e^{jn\,\Delta\omega t}\,\Delta\omega \\ &= \frac{1}{2\pi}\int_{-\infty}^{\infty} G(\omega)e^{j\omega t}\,d\omega. \end{aligned} \tag{1.13}$$

Similarly

$$\begin{aligned} G(\omega) &= \lim_{\Delta\omega\to 0} G(n\Delta\omega) \\ &= \lim_{T_0\to\infty} T_0 G_n \\ &= \int_{-\infty}^{\infty} g(t)e^{-j\omega t}\,dt. \end{aligned} \tag{1.14}$$

Equations (1.13) and (1.14) are referred to as the *Fourier transform pair*. $G(\omega)$ in Eq. (1.14) is considered as the direct Fourier transform of $g(t)$, while $g(t)$ in Eq. (1.13) is the inverse Fourier transform of $G(\omega)$. The transform pair can also be expressed as

$$G(\omega) = \mathfrak{F}[g(t)]$$

and

$$g(t) = \mathfrak{F}^{-1}[G(\omega)].$$

The Fourier transform pair can also be expressed symbolically as

$$g(t) \leftrightarrow G(\omega).$$

Some key properties of the Fourier transform are listed below:

1. $$G(-\omega) = G^*(\omega) \tag{1.15}$$

   where the asterisk (*) denotes *complex conjugate of*.
2. If $g(t) \leftrightarrow G(\omega)$, then

   $$G(t) \leftrightarrow 2\pi g(-\omega). \tag{1.16}$$

3. $$g(at) \leftrightarrow \frac{1}{|a|} G\left(\frac{\omega}{a}\right). \tag{1.17}$$

4. $$g(t - t_0) \leftrightarrow G(\omega)e^{-j\omega t_0}. \tag{1.18}$$

5. $$g(t)e^{j\omega_0 t} \leftrightarrow G(\omega - \omega_0). \tag{1.19}$$

6. If $g_1(t) \leftrightarrow G_1(\omega)$ and $g_2(t) \leftrightarrow G_2(\omega)$, then

   $$g_1(t) * g_2(t) \leftrightarrow G_1(\omega)G_2(\omega), \tag{1.20}$$

   where the asterisk denotes convolution.
7. If $g_1(t) \leftrightarrow G_1(\omega)$ and $g_2(t) \leftrightarrow G_2(\omega)$, then

   $$g_1(t)g_2(t) \leftrightarrow \frac{1}{2\pi} G_1(\omega) * G_2(\omega). \tag{1.21}$$

   where the asterisk denotes convolution.

Example 1.3 Let $g(t)$ be defined as

$$g(t) = 1 \quad \text{for } -\frac{\tau}{2} \le t \le \frac{\tau}{2},$$
$$= 0 \quad \text{otherwise.}$$

The Fourier transform of $g(t)$ can be computed as

$$\begin{aligned} G(\omega) &= \mathfrak{F}[g(t)] \\ &= \int_{-\infty}^{\infty} g(t)e^{-j\omega t}\,dt \\ &= \int_{-\tau/2}^{\tau/2} e^{-j\omega t}\,dt \\ &= \tau \frac{\sin \omega\tau/2}{\omega\tau/2} \\ &= \tau \operatorname{sinc}(\omega\tau/2\pi). \end{aligned}$$

Figure 1.7 displays the time domain function $g(t)$ and its Fourier transform $G(\omega)$.

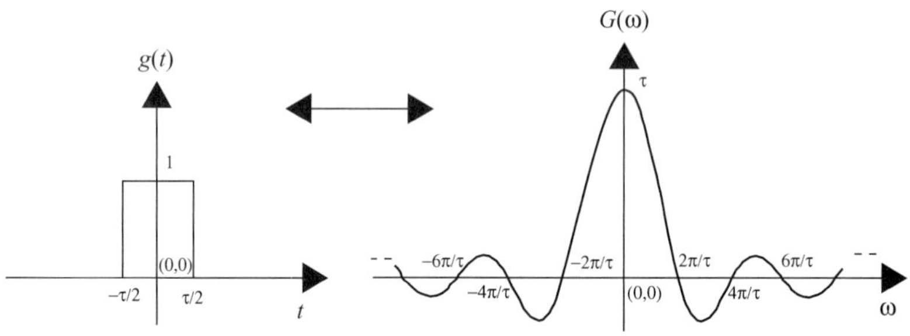

FIGURE 1.7 A single pulse $g(t)$ and its Fourier transform $G(\omega)$.

By comparing Figs. 1.6 and 1.7, one can see that the Fourier series coefficients of a periodic pulse train is the discrete version of the Fourier transform of a single pulse.

Example 1.4 Let $G(\omega)$ be defined as

$$G(\omega) = 1 \quad \text{for } -2\pi f_m \le \omega \le 2\pi f_m,$$
$$= 0 \quad \text{otherwise.}$$

The inverse Fourier transform of $G(\omega)$ can be computed as

$$
\begin{aligned}
g(t) &= \mathfrak{F}^{-1}[G(\omega)] \\
&= \frac{1}{2\pi}\int_{-\infty}^{\infty} G(\omega)e^{j\omega t}\,d\omega \\
&= \frac{1}{2\pi}\int_{-2\pi f_m}^{2\pi f_m} e^{j\omega t}\,d\omega \\
&= 2f_m \frac{\sin 2\pi f_m t}{2\pi f_m t} \\
&= 2f_m \,\mathrm{sinc}\,(2f_m t).
\end{aligned}
$$

Figure 1.8 displays the frequency domain function $G(\omega)$ and its inverse Fourier transform $g(t)$.

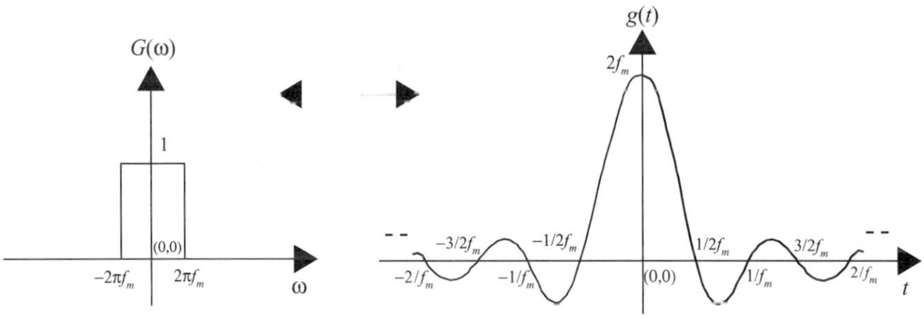

FIGURE 1.8 A single-pulse frequency spectrum $G(\omega)$ and its inverse Fourier transform $g(t)$.

Example 1.5 Let $g(t) = \delta(t)$, the Fourier transform of $g(t)$ can be computed as

$$
\begin{aligned}
G(\omega) &= \mathfrak{F}[g(t)] \\
&= \int_{-\infty}^{\infty} g(t)e^{-j\omega t}\,dt \\
&= \int_{-\infty}^{\infty} \delta(t)e^{-j\omega t}\,dt \\
&= 1.
\end{aligned}
$$

Therefore, $\delta(t)$ and 1 are a Fourier transform pair:

$$\delta(t) \leftrightarrow 1.$$

Similarly, if $G(\omega) = \delta(\omega)$, then

$$\begin{aligned} g(t) &= \mathcal{F}^{-1}[G(\omega)] \\ &= \frac{1}{2\pi}\int_{-\infty}^{\infty} \delta(\omega) e^{j\omega t}\, d\omega \\ &= \frac{1}{2\pi}. \end{aligned}$$

Therefore, 1 and $2\pi\delta(\omega)$ are a Fourier transform pair:

$$1 \leftrightarrow 2\pi\delta(\omega).$$

1.5 FOURIER TRANSFORM OF A PERIODIC SIGNAL

Although the Fourier transform was derived from the nonperiodic signal, it can also be used to represent the periodic signal. The Fourier transform of a periodic signal can be computed by first representing the periodic signal in terms of a Fourier series expression, then transforming each Fourier series coefficient (represented in exponential form) into the frequency domain. The following examples illustrate the Fourier transform of periodic signals.

Example 1.6 Let $g_p(t)$ be a periodic impulse train, expressed as

$$g_p(t) = \sum_{n=-\infty}^{\infty} \delta(t - nT_0).$$

The Fourier transform of $g_p(t)$ can be computed by first expressing the periodic signal $g_p(t)$ in terms of the Fourier series in the exponential form. Thus, from Example 1.2, we obtain

$$\begin{aligned} g_p(t) &= \sum_{n=-\infty}^{\infty} G_n e^{jn\omega_0 t} \\ &= \frac{1}{T_0} \sum_{n=-\infty}^{\infty} e^{jn\omega_0 t}, \end{aligned}$$

where $\omega_0 = 2\pi/T_0$. From Example 1.5, the Fourier transform of $g_p(t)$ can then be computed as

$$G(\omega) = \mathfrak{F}[g_p(t)]$$
$$= \frac{1}{T_0}\sum_{n=-\infty}^{\infty}\int_{-\infty}^{\infty} e^{-j(\omega-n\omega_0)t}\,dt$$
$$= \frac{2\pi}{T_0}\sum_{n=-\infty}^{\infty}\delta(\omega-n\omega_0).$$

Figure 1.9 displays the time-domain impulse train $g_p(t)$ and its Fourier transform $G(\omega)$. Both $g_p(t)$ and $G(\omega)$ appear to be impulse trains. Notice that the amplitude of the impulse train in the frequency domain is ω_0 and its spectrum repeated at $\pm n\omega_0$ with $n = 1,2,\ldots$ and $\omega_0 = 2\pi/T_0$.

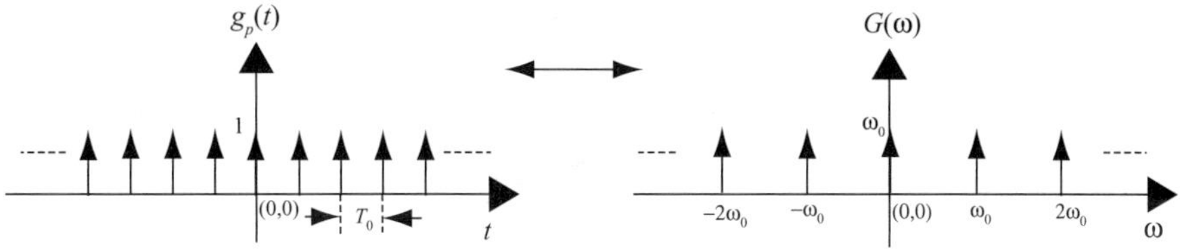

FIGURE 1.9 A periodic impulse train and its Fourier transform.

Example 1.7 Consider the signal $g_p(t)$ shown in Example 1.1, expressed as

$$g_p(t) = \sum_{n=-\infty}^{\infty} g(t - nT_0),$$

where

$$g(t) = A \quad \text{for } -\frac{\tau}{2} \le t \le \frac{\tau}{2},$$
$$= 0 \quad \text{otherwise.}$$

The Fourier transform of the periodic pulse train $g_p(t)$ can be computed in two steps:

Step 1 A periodic signal should first be expressed in terms of the Fourier series in exponential form:

$$g_p(t) = \sum_{n=-\infty}^{\infty} G_n e^{jn\omega_0 t}.$$

The Fourier series coefficients G_n can be computed as

$$\begin{aligned} G_0 &= \frac{A\tau}{T_0}. \\ G_n &= G_{-n} \\ &= \frac{a_n}{2} \\ &= \frac{A}{n\pi}\sin\frac{n\pi\tau}{T_0}. \end{aligned}$$

Therefore, $g_p(t)$ can be expressed as

$$g_p(t) = \sum_{n=-\infty}^{\infty} \frac{A}{n\pi}\sin\left(\frac{n\pi\tau}{T_0}\right)e^{jn\omega_0 t}.$$

Step 2 The Fourier transform of $g_p(t)$ can then be computed as

$$\begin{aligned} G(\omega) &= \mathfrak{F}[g_p(t)] \\ &= \int_{-\infty}^{\infty}\sum_{n=-\infty}^{\infty} \frac{A}{n\pi}\sin\left(\frac{n\pi\tau}{T_0}\right)e^{jn\omega_0 t}e^{-j\omega t}dt \\ &= \sum_{n=-\infty}^{\infty} \frac{A}{n\pi}\sin\left(\frac{n\pi\tau}{T_0}\right)\int_{-\infty}^{\infty} e^{-j(\omega-n\omega_0)t}dt. \end{aligned}$$

Example 1.5 has shown that

$$1 \leftrightarrow 2\pi\delta(\omega).$$

Therefore

$$\begin{aligned} G(\omega) &= 2\pi\sum_{n=-\infty}^{\infty} \frac{A}{n\pi}\sin\frac{n\pi\tau}{T_0}\delta(\omega - n\omega_0) \\ &= A\frac{2\pi\tau}{T_0}\sum_{n=-\infty}^{\infty}\frac{\sin(n\pi\tau/T_0)}{n\pi\tau/T_0}\delta(\omega - n\omega_0). \end{aligned}$$

If $\tau = T_0/2$ and $A = 1$, then

$$G(\omega) = \pi\sum_{n=-\infty}^{\infty} \text{sinc}\,(n/2)\,\delta(\omega - n\omega_0).$$

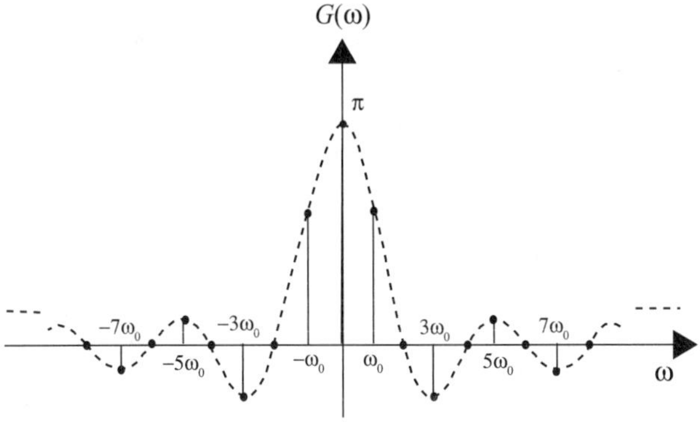

FIGURE 1.10 Fourier transform of a periodic pulse train.

Notice that $G(\omega)$ is a discrete signal that exists when $\omega = n\omega_0$ with n equal to an integer. Figure 1.10 demonstrates the frequency spectrum of $G(\omega)$ for $n = -8$ to 8, and its envelope is shown as sinc $(n/2)$.

Note that from Figs. 1.6 and 1.10, for the same periodic sequence $g_p(t)$, the amplitude of Fourier series coefficients G_n and the amplitude of Fourier transform $G(\omega)$ are the same with a scaling difference of only 2π.

1.6 SAMPLING THEORY AND INTERPOLATION

The sampling theory states that any signal that is frequency band-limited to f_m can be reconstructed from samples taken at a uniform time interval of $T_s \leq 1/(2f_m)$. The time interval $T_s = 1/(2f_m)$ is called the *Nyquist interval*, and the corresponding sampling rate is known as the *Nyquist rate*. The sampling theory can be derived as explained below.

Consider a signal $x(t)$ with its Fourier transform as $X(\omega)$ and its frequency spectrum band-limited to f_m. Let $g_p(t)$ be a unit impulse train as described in Example 1.6. Multiplication of $x(t)$ with $g_p(t)$ yields the sampled signal $x_s(t)$:

$$\begin{aligned} x_s(t) &= x(t)g_p(t) \\ &= x(t)\sum_{n=-\infty}^{\infty} \delta(t - nT_s). \end{aligned} \tag{1.22}$$

As shown in Example 1.6, a periodic pulse train can be expressed in terms of the Fourier series; that is, with $\omega_s = 2\pi/T_s$, are obtains

$$\begin{aligned} x_s(t) &= x(t)\frac{1}{T_s}\sum_{n=-\infty}^{\infty} e^{jn\omega_s t} \\ &= \frac{1}{T_s}\sum_{n=-\infty}^{\infty} x(t)e^{jn\omega_s t}. \end{aligned}$$

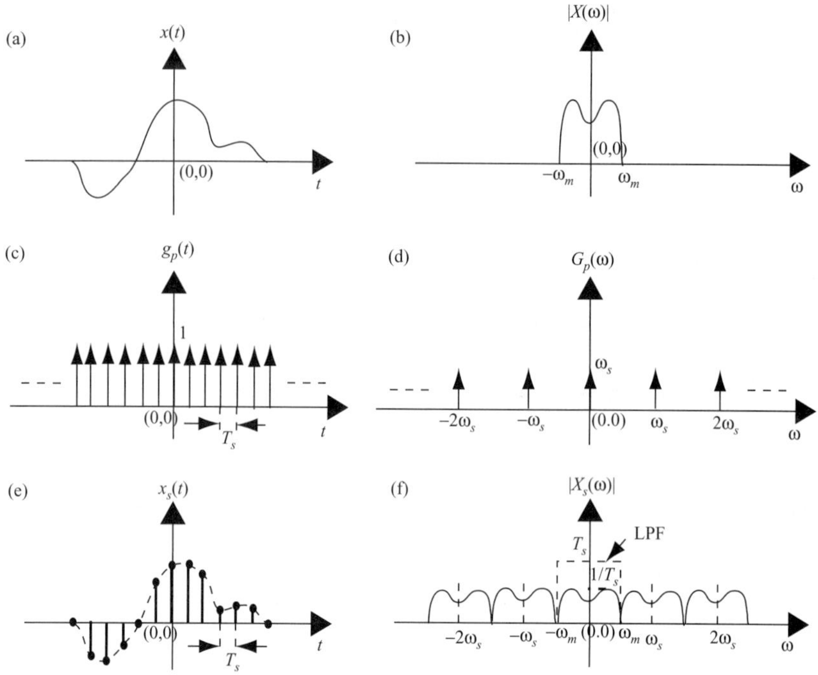

FIGURE 1.11 Graphical representations of the sampling theory.

By taking the Fourier transform of $x_s(t)$, one obtains

$$X_s(\omega) = \frac{1}{T_s} \sum_{n=-\infty}^{\infty} X(\omega - n\omega_s). \tag{1.23}$$

This equation states that after multiplication of $x(t)$ by the unit impulse train $g_p(t)$, the new frequency spectra $X_s(\omega)$ consists of $X(\omega)$, plus replica located at $\omega = \pm\, n\omega_s$, for $n = 0,1,2,\ldots$. The amplitude of $X_s(\omega)$ is attenuated by a factor of $1/T_s$.

Figure 1.11 illustrates the sampling theory. The original signal $x(t)$ and its analog frequency spectrum $|X(\omega)|$ are shown in Figs. 1.11a and 1.11b. A periodic impulse train $g_p(t)$ and its spectra are shown in Figs. 1.11c and 1.11d. By multiplying $x(t)$ with $g_p(t)$, one can then display the resultant $x_s(t)$ in Fig. 1.11e with the corresponding spectra shown in Fig. 1.11f.

From Fig. 1.11f, one can see that to prevent overlap between the neighboring spectra, the sampling frequency $\omega_s = 2\pi/T_s$ must satisfy the requirement that $\omega_s \geq (2 \times 2\pi f_m) = 2\omega_m$.

To reconstruct the original signal $x(t)$ from the digitized signal $x_s(t)$, one needs to filter out the spectrum $X(\omega)$, as shown in Fig. 1.11b, from the spectra $X_s(\omega)$, as shown in Fig. 2.11f. A lowpass filter (LPF) with a cutoff frequency at $2\pi f_m$ and a gain equal to T_s meets the filtering requirement. By passing the $x_s(t)$ through this lowpass filter, one can reconstruct the original signal $x(t)$.

Letting $H(\omega)$ be such a rectangular LPF, one can compute the time-domain function $h(t)$ as

$$h(t) = \frac{1}{2\pi} \int_{-2\pi f_m}^{2\pi f_m} T_s e^{j\omega t} d\omega$$
$$= 2 f_m T_s \frac{\sin(2\pi f_m t)}{2\pi f_m t}.$$

Let the sampling frequency $f_s = 2f_m$; then

$$h(t) = \mathrm{sinc}\,(2 f_m t). \tag{1.24}$$

In the time domain, passing the signal $x_s(t)$ through the filter $h(t)$ is equivalent to having $x_s(t)$ convolved with $h(t)$:

$$x(t) = x_s(t) * h(t)$$
$$= \left[x(t) \sum_{n=-\infty}^{\infty} \delta(t - nT_s) \right] * h(t)$$
$$= \sum_{n=-\infty}^{\infty} x(nT_s)\delta(t - nT_s) * \mathrm{sinc}\,(2 f_m t)$$
$$= \sum_{n=-\infty}^{\infty} x(nT_s)\,\mathrm{sinc}\,(2 f_m t - 2 f_m n T_s)$$
$$= \sum_{n=-\infty}^{\infty} x(nT_s)\,\mathrm{sinc}\,(2 f_m t - n). \tag{1.25}$$

Equation (1.25) states that $x(t)$ can be reconstructed from its discrete samples $x(nT_s)$ and the interpolation function sinc $(2f_m t - n)$. Notice that $x(nT_s)$ are all equally spaced with time interval T_s for integers $n = -\infty$ to ∞, and also that $x(t)$ is a continuous function in the time domain. Thus, Eq. (1.25) can be used to find $x(t_1)$, for $t_1 = nT_s + \Delta T_s$ with $\Delta < 1$, based on the discrete values of $x(nT_s)$. The new discrete time sequence $x(t_1)$ can be considered as a resampling of $x(t)$, or interpolated from $x(nT_s)$.

In practical applications, the resampling process that utilizes the interpolation filter sinc $(2f_m t)$ is simplified by two approximations. First, the interpolation filter is chosen with a finite number of sidelobes. For radar image processing, an 8-tap sinc filter is normally used to generate a new interpolated sample. The second approximation involves choosing a finite number of interpolation intervals (or fractional shifts). This means that the value of $\Delta < 1$ is discrete and finite. A set of 16 sinc filters with a minimum of $\frac{1}{16}$ interpolation sample intervals, or $\Delta = \frac{1}{16}$, serves well for radar

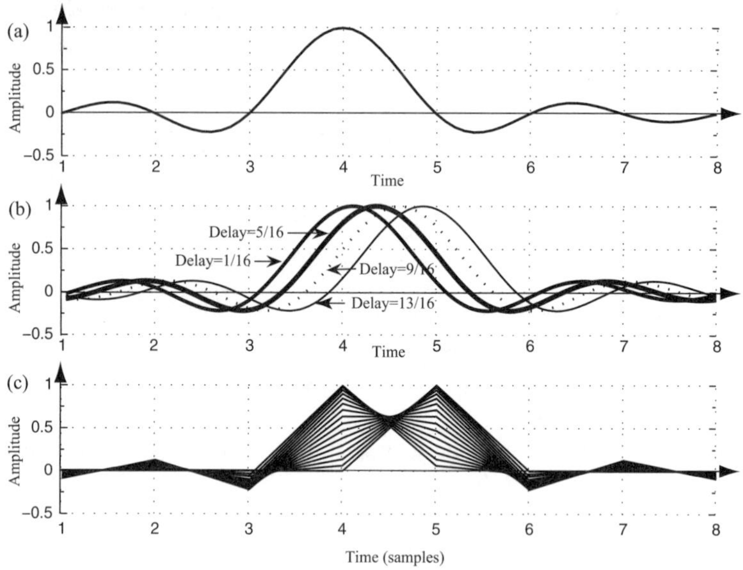

FIGURE 1.12 Interpolation filters.

image processing. This set of 16 sinc filters provides a fractional sample shift from $\frac{1}{16}, \frac{2}{16}, \ldots, \frac{15}{16}$. The $\frac{16}{16} = 1$ is not considered a sample shift.

Figure 1.12 displays the waveforms of 16 sets of interpolation filters with each having 8-tap coefficients. Figure 1.12a shows a reference sinc (t) function with 5 sidelobes around the mainlobe. The window function is normally applied to the interpolation filter, yet no window or rectangular window is used here for simplification. Figure 1.12b displays four interpolation filters with delays equal to $\frac{1}{16}, \frac{5}{16}, \frac{9}{16}$, and $\frac{13}{16}$ sample intervals with respect to the top reference function, respectively. Figure 1.12c displays the digitized version of all 16 interpolation filters, and each one corresponds to a $\frac{1}{16}$ sample delay from each other.

Table 1.1 lists the coefficients of the 16 interpolation filters, with each one having 8 coefficients. The first row of filters has a shift of $\frac{1}{16}$ sample interval, while the last one has $\frac{16}{16} = 1$ or no sample shift. From Table 1.1 and Fig. 1.12c, one can see that the interpolation filters are symmetric; that is, filter coefficients of row 1 with $\frac{1}{16}$ sample shift are the same as that of row 15 with $\frac{15}{16}$ sample shift, except that they are time-reversed. The filter coefficients of rows 1, 5, 9, and 13 of Table1.1 correspond to the four sinc filters shown in Fig. 1.12b.

To illustrate the principle of resampling based on an interpolation filter, consider a signal $x(n)$ that consists of three normalized frequencies, $f_1 = 0.15, f_2 = 0.25$, and $f_3 = 0.45$, and is expressed as

$$x(n) = 0.35 \cos(2\pi n f_1) + 0.2 \sin(2\pi n f_2) - 0.4 \cos(2\pi n f_3),$$

where $n = 1,2, \ldots, 32$.

TABLE 1.1 16 sets of 8-tap interpolation filters

−0.0203	0.0301	−0.0584	0.9936	0.0662	−0.0321	0.0211	−0.0158
−0.0390	0.0573	−0.1083	0.9745	0.1392	−0.0650	0.0424	−0.0314
−0.0555	0.0808	−0.1489	0.9432	0.2177	−0.0976	0.0629	−0.0464
−0.0693	0.1000	−0.1801	0.9003	0.3001	−0.1286	0.0818	−0.0600
−0.0799	0.1144	−0.2016	0.8469	0.3850	−0.1568	0.0985	−0.0718
−0.0871	0.1238	−0.2139	0.7842	0.4705	−0.1810	0.1120	−0.0811
−0.0908	0.1281	−0.2172	0.7136	0.5550	−0.1998	0.1218	−0.0876
−0.0909	0.1273	−0.2122	0.6366	0.6366	−0.2122	0.1273	−0.0909
−0.0876	0.1218	−0.1998	0.5550	0.7136	−0.2172	0.1281	−0.0908
−0.0811	0.1120	−0.1810	0.4705	0.7842	−0.2139	0.1238	−0.0871
−0.0718	0.0985	−0.1568	0.3850	0.8469	−0.2016	0.1144	−0.0799
−0.0600	0.0818	−0.1286	0.3001	0.9003	−0.1801	0.1000	−0.0693
−0.0464	0.0629	−0.0976	0.2177	0.9432	−0.1489	0.0808	−0.0555
−0.0314	0.0424	−0.0650	0.1392	0.9745	−0.1083	0.0573	−0.0390
−0.0158	0.0211	−0.0321	0.0662	0.9936	−0.0584	0.0301	−0.0203
−0.0000	0.0000	−0.0000	0.0000	1.0000	−0.0000	0.0000	−0.0000

To resample $x(n)$ to obtain a new set of digitized samples with a 0.5 sample shift, row 8 of the interpolation filter from Table 1.1 is chosen. The new set of data $x(n + 0.5)$ is computed from Eq. (1.25) by convolving the eight old data samples of $x(n)$ with the interpolation filter. Figure 1.13 displays the results of the resampling process. Only the effective samples from samples 4–28 of the convolution output are displayed for comparison.

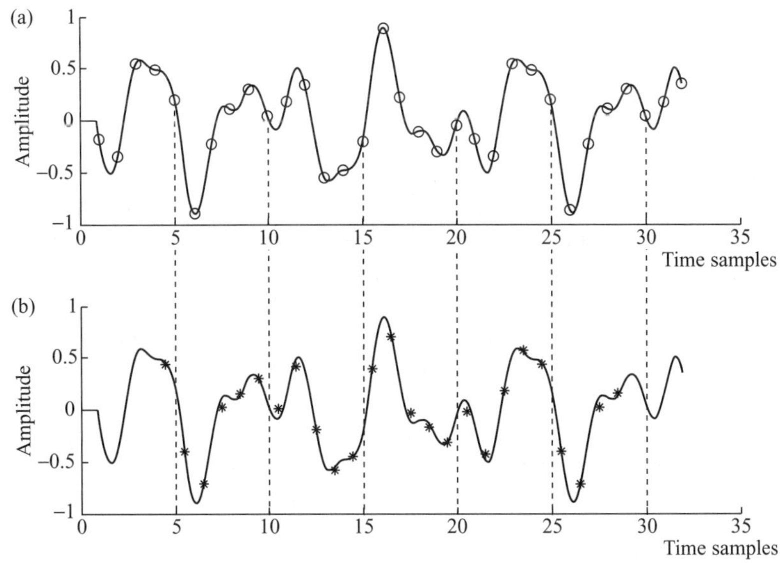

FIGURE 1.13 Original and half-sample-shifted digital signals.

Figure 1.13a shows the digitized signal $x(n)$, marked by hollow circles (∘), together with the original analog signal $x(t)$. Figure 1.13b shows the interpolated (or half-sample-shifted) signal $x'(n)$, marked by asterisks (*), together with the original analog signal $x(t)$. As can be seen, the interpolated and digitized signal closely matches the original signal.

The sampling theory discussed so far is based on a unit impulse train. In practice, the impulse train is replaced with a finite-time-duration pulse train. Assuming, that all conditions are the same except that the narrow pulse train with duration τ is used, the only difference will be that the amplitude of the Fourier series coefficient $1/T_s$ is replaced with $\sin(n\pi\tau/T_s)/n\pi$. This, in turn, causes Eq. (1.23) to become

$$X_s(\omega) = \sum_{n=-\infty}^{\infty} \frac{\sin(n\pi\tau/T_s)}{n\pi} X(\omega - n\omega_s) \tag{1.26}$$

The lowpass filter defined in Eq. (1.24) can then be applied to the digitized signal $x_s(t)$. Accordingly, the $x(t)$ can again be reconstructed from its digitized samples $x_s(t)$ by passing the $x_s(t)$ through a lowpass filter $h(t)$. The new reconstruction equation, based on a pulse train with duration τ, can be obtained by modifying Eq. (1.25) to become

$$x(t) = \sum_{n=-\infty}^{\infty} \frac{\sin(n\pi\tau/T_s)}{n\pi} x(nT_s)\,\mathrm{sinc}(2f_m t - n) \tag{1.27}$$

1.7 ADVANCED SAMPLING TECHNIQUES

1.7.1 Sampling with Bandpass Signal

The sampling theory discussed in Section 1.6 generally applies to baseband signals with a maximum frequency component f_m. In the case where a signal is band-limited with $f_l \leq f \leq f_m$ and its center frequency is greater than zero, the Nyquist sampling period $T_s \leq 1/(2f_m)$ must be modified to reduce the sample rate for signal processing.

Figure 1.14 shows the frequency spectra $|G(\omega)|$ of a bandpass signal $g(t)$. Figure 1.14a shows the frequency spectra $|G(\omega)|$ of $g(t)$, while Fig. 1.14b shows the frequency spectra $|G_s(\omega)|$ of a sampled signal $g_s(t)$. From the sampling theory, the spectrum of the sampled signal will consist of the attenuated spectrum from the original signal and its replica located at $\pm n\omega_s$, where $n = 0,1,2,\ldots$. The two-sided spectra, a and b, shown in Fig. 1.14a are repeated around $\pm\omega_s$, $\pm 2\omega_s \ldots$ in Fig. 1.14b. The spectra na and nb are the repeated spectra around $\pm n\omega_s$.

From Fig. 1.14, it can be seen that no spectrum is overlapped as long as the conditions of $f_s = 2(f_m - f_l)$ and $f_l = kf_s$ with k as an integer are satisfied. If the latter condition is not true, then a higher sampling frequency of $f_s > 2(f_m - f_l)$ will be needed.

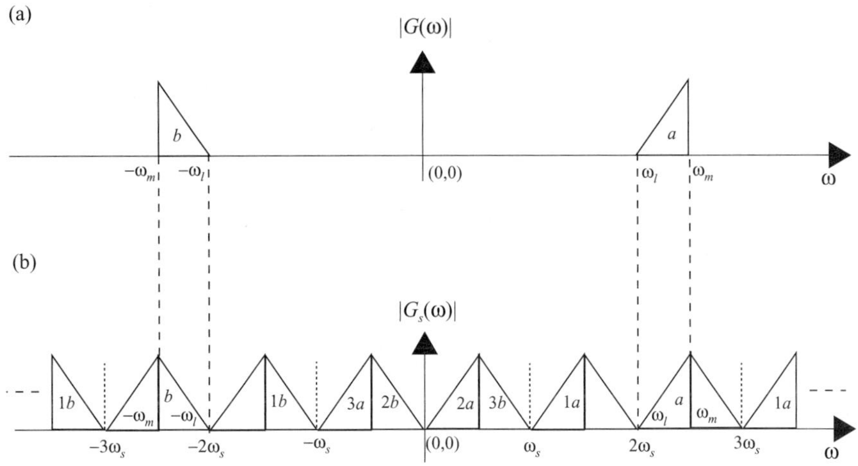

FIGURE 1.14 Frequency spectra of a digitized bandpass signal.

Resampling is necessary for some applications, such as digital communication or audio/video processing. Resampling is used to reduce the computation complexity for signals with a finite bandwidth. Some resampling techniques will now be discussed.

1.7.2 Resampling by Evenly Spaced Decimation

Evenly spaced sample decimation can be used to reduce the sample rate by an integral factor. Consider the case of downsampling on a sequence $x(n)$ by an integer $M = f_{\text{old}}/f_{\text{new}}$, where f_{old} and f_{new} are the old and new sampling rates, respectively. Such a downsampling process can simply retain every Mth sample and discard the others if the frequency aliasing does not occur after downsampling. However, frequency overlap indeed occurs in most downsampling processes. Therefore, a lowpass Finite Impulse Response (FIR) filter is normally used to limit the bandwidth of the signal at a new sampling rate. Both the input sequence $x(n)$ and the FIR filter function $h(n)$ operate at an old sample rate f_{old}, but the output sequence $y(m)$ operates at a new sample rate f_{new}. Figure 1.15a shows the block diagram of downsampling with a FIR filtering. Figures 1.15b and 1.15c show the signal frequency spectra before and after the decimation for the case when $M = 2$. As can be seen from Fig. 1.15c, the signal frequency spectra must be band-limited with $f_{\text{new}} \geq 2f_m$ to avoid the aliasing problem.

1.7.3 Resampling by Evenly Spaced Interpolation

Evenly spaced sample interpolation can be used to increase the sample rate by an integral factor. Consider the case of upsampling on a sequence $x(n)$ by an integer $M = f_{\text{new}}/f_{\text{old}}$, where f_{old} and f_{new} are the old and new sampling rates, respectively. Upsampling by M can be implemented by inserting $M-1$ zeros for every sample of

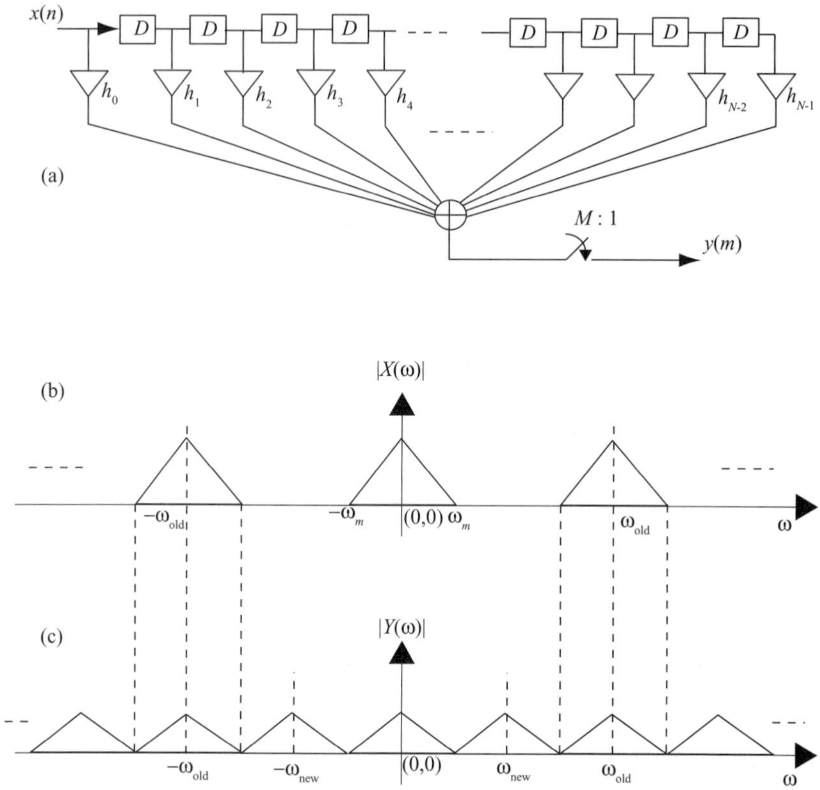

FIGURE 1.15 Downsampling with a FIR filter.

$x(n)$, and then passing it through a lowpass filter operating at the f_{new} sampling rate. The lowpass filter is required to filter out the frequency spectra at sampling rate f_{new}.

Figure 1.16 illustrates the process of interpolation with the sample rate increased by $M = 3$. The overall system diagram is shown in Fig. 1.16a, with $x(n)$ as the input and $y(m)$ as the output. Figures 1.16b and 1.16c show the original signal $x(n)$ and its corresponding spectra $|X(\omega)|$ with sample rate f_{old}. By inserting two zeros for every sample of $x(n)$, one obtains a new signal sequence $x'(m)$. Figures 1.16d and 1.16e show the new sequence $x'(m)$ and its corresponding spectra $|X'(\omega)|$ with sample rate f_{new}. After passing $x'(m)$ through a lowpass filter (LPF), one obtains the interpolated signal $y(m)$ as shown in Fig. 1.16f, and its frequency spectra are shown in Fig. 1.16g.

1.7.4 Resampling by Fractional Rate Interpolation

The interpolation process described in the previous section increases the sample rate with an integral factor. In cases where the sample rate change factor M is not an

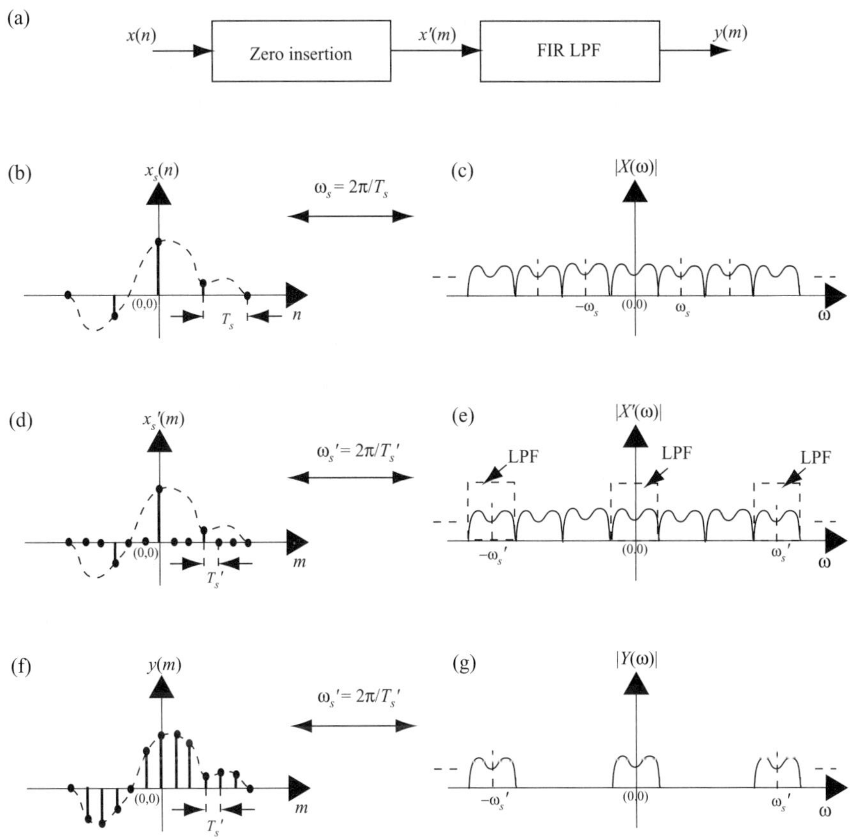

FIGURE 1.16 Interpolation with zero insertion and FIR LPF.

integer, interpolation followed by decimation can be used to obtain any sample rate change. Let the new sample rate change factor be $M = I + F$, where I is an integer and F is a fractional value. Then $M = U/D$, where U and D correspond to the upsampling and downsampling rates, respectively.

Consider the case of $M = 2.5$ with $U = 5$ and $D = 2$. Let the original digitized signal and sampling frequency be $x(n)$ and f_s, respectively. To implement the interpolation at a resampling rate of 2.5, one can first upsample $x(n)$ with sampling frequency equal to $5f_s$ by inserting four zeros between each sample of $x(n)$. The new sequence of samples $x'(n)$ are then lowpass-filtered (operating at $5f_s$) and followed by downsampling with sampling frequency equal to $2f_s$. Figure 1.17a illustrates the overall process of interpolation with $U = 5$ and $D = 2$. Figure 1.17b shows the original signal spectra $|X(\omega)|$ with sample rate f_s. By inserting four zeros for every sample of $x(n)$ followed with a lowpass FIR filter operating at new sampling rate $f_{\text{new}} = 5f_s$, one obtains a new sequence $x'(m)$. Figure 1.17c shows the spectra of $|X'(\omega)|$. The new sequence $x'(m)$ is then decimated by dumping one out of every sample of $x'(m)$.

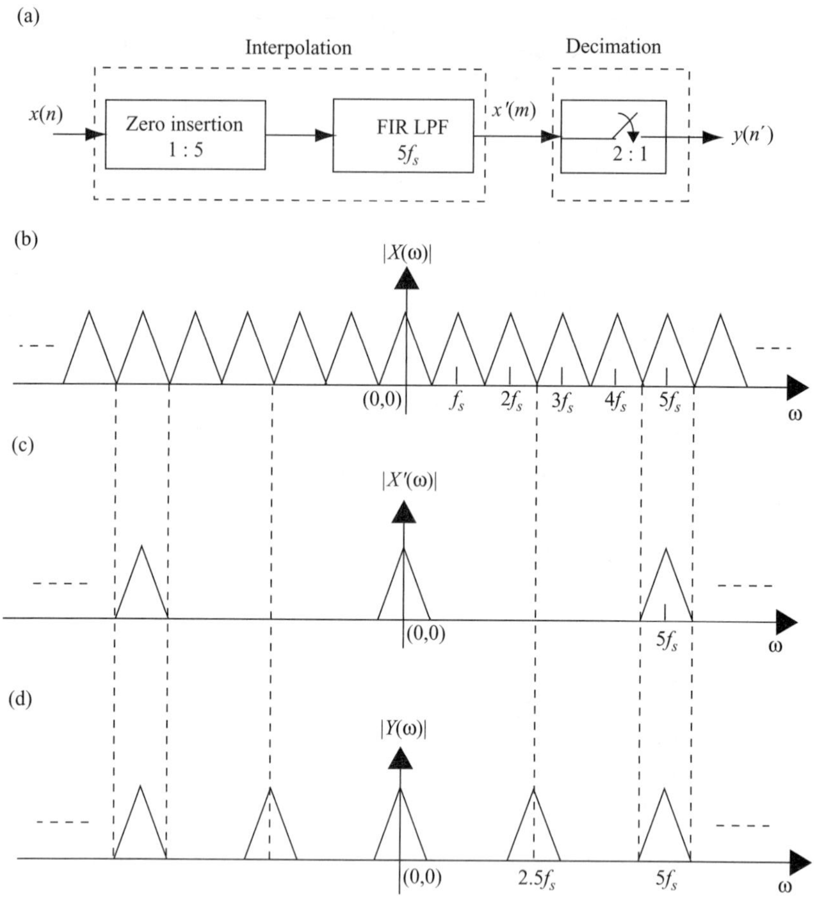

FIGURE 1.17 Fractional rate interpolation with $M = 2.5$.

The spectra of interpolated signal $y(n')$ running at $2.5f_s$ is shown in Fig. 1.17d. Alternatively, one can just compute one FIR filter output for every two samples shifted into the FIR filter.

1.7.5 Resampling from Unevenly Spaced Data

Before discussing the resampling method based on unevenly spaced data, the *Jacobian of transformation* will be explained. This process is used extensively during the transformation between different coordinate systems.

1.7.5.1* Jacobian *of Transformation. A signal can be transformed from one coordinate system into a different coordinate system. Variables inevitably change during a transformation from one system to another. The transformation can be linear or nonlinear. The transformation from one system to another may therefore cause an extra factor to appear. This extra factor is called the *Jacobian of transformation*.

Consider the case where $u = g(v)$, then $(du/dv) = (dg(v)/dv) = g'(v)$ and

$$\int_a^b f(u)du = \int_c^d f(g(v))g'(v)dv.$$

The extra factor $g'(v)$ in this integration is called the *Jacobian of transformation*.

In general, let $x = g(u,v)$ and $y = h(u,v)$ be a transformation from (x,y) domain to (u,v) domain. Then, the *Jacobian* of transformation, or simply Jacobian, is defined as

$$J(u,v) = \frac{\partial(x,y)}{\partial(u,v)} = \begin{vmatrix} \frac{\partial x}{\partial u} & \frac{\partial x}{\partial v} \\ \frac{\partial y}{\partial u} & \frac{\partial y}{\partial v} \end{vmatrix} = \frac{\partial x}{\partial u}\frac{\partial y}{\partial v} - \frac{\partial x}{\partial v}\frac{\partial y}{\partial u}. \tag{1.28}$$

Let $f(x,y)$ be a function in the $R(x,y)$ region, and a transformation of $x = g(u,v)$ and $y = h(u,v)$ transforms $f(x,y)$ from $R(x,y)$ into $S(u,v)$ region. If $g(u,v)$ and $h(u,v)$ have continuous partial derivatives such that the Jacobian is never zero, then

$$\begin{aligned} \iint_R f(x,y)\,dx\,dy &= \iint_S f(g(u,v),h(u,v)) \left|\frac{\partial(x,y)}{\partial(u,v)}\right| du\,dv \\ &= \iint_S f(g(u,v),h(u,v))J(u,v)\,du\,dv. \end{aligned}$$

As an example, let $x = g(r,\theta) = r\cos\theta$ and $y = h(r,\theta) = r\sin\theta$. From Eq. (1.28), the Jacobian of transformation is

$$\begin{aligned} J(r,\theta) &= \frac{\partial x}{\partial r}\frac{\partial y}{\partial \theta} - \frac{\partial x}{\partial \theta}\frac{\partial y}{\partial r} \\ &= \cos\theta \cdot r\cos\theta + r\sin\theta \cdot \sin\theta \\ &= r. \end{aligned}$$

Therefore, the integration shown above becomes

$$\iint_R f(x,y)\,dx\,dy = \iint_S f(g(r,\theta),h(r,\theta))\,r\,dr\,d\theta.$$

This is the well-known rectangular-to-polar coordinate transformation.

The operation of integration in the analog domain is equivalent to summation in the digital domain. The principle of the Jacobian of transformation can therefore be applied to the interpolation or resampling in the digital signal processing field.

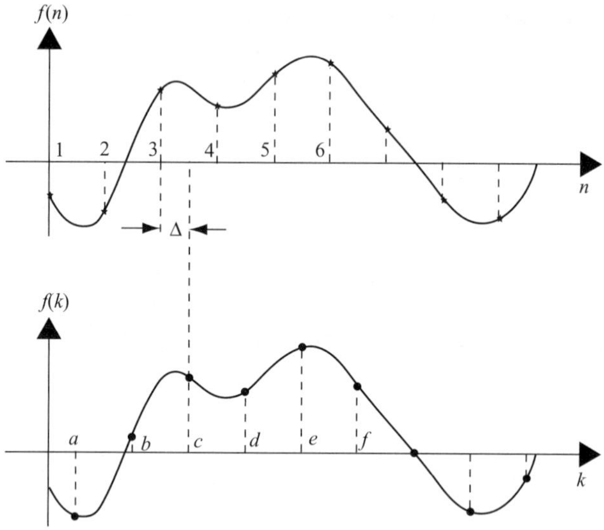

FIGURE 1.18 Interpolation on evenly spaced data.

The sampling theory states that given a set of evenly spaced data $x(n)$ with sampling frequency $f_s = 2f_m$, one can interpolate $x(n)$ with another set of evenly spaced data $x(k)$. Assume that the sampling frequency is normalized to one, that is, $t = kT_s = k$. Then Eq. (1.24) becomes

$$x(k) = \sum_{n=-\infty}^{\infty} x(n)\,\mathrm{sinc}\,(k-n); \quad k = -\infty \text{ to } \infty. \tag{1.29}$$

The interpolated data $x(k)$ and the original data $x(n)$ can be considered as in k and n coordinate systems, respectively, and they are related as

$$\begin{aligned} n &= k - \Delta \\ &= g(k), \end{aligned}$$

where Δ is a fractional number and $0 < \Delta < 1$. The summation of Eq. (1.29) is the discrete convolution of $x(n)$ with sinc (n), and the data sequence $x(n)$ is transformed from n space to k space. Figure 1.18 shows the interpolation of evenly spaced samples from the view point of transformation between coordinate systems.

The Jacobian of transformation is then

$$J(k) = \frac{dn}{dk} = 1.$$

Therefore, Eq. (1.29) holds without any extra factor for transformation from an evenly spaced system to another evenly spaced system.

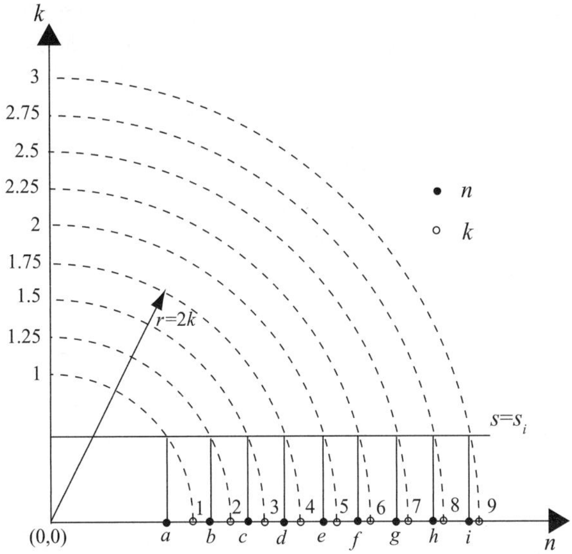

FIGURE 1.19 Generation of unevenly spaced data index.

Now, consider the case where one tries to resample a set of evenly spaced data $f(k)$ from a set of unevenly spaced data $f(n)$ with no aliasing. Before the resampling process, the unevenly spaced signal $f(n)$ must be defined and generated. As an example, the index n of data $f(n)$ is generated nonlinearly through the transformation $g(k)$ as

$$\begin{aligned} n &= g(k) \\ &= \sqrt{(2k)^2 - s^2}, \end{aligned} \tag{1.30}$$

where k is an evenly spaced variable and equal to half of the radius of a circle with origin at (0,0). The symbol s is a constant and $s < 2k$.

Figure 1.19 shows the mapping relation of Eq. (1.30), with the unevenly spaced data index n obtained from Eq. (1.30) when $k = 1, 1.25, 1.5, \ldots, 3$, and $s = s_i < 1$. The evenly spaced data index k is the intersection point between the circle with radius $r = 2k$ and the horizontal axis (abscissa), which is labeled as 1,2,3, ... The unevenly spaced data index n is obtained by mapping the intersection point between the circle of radius $r = 2k$ and the line $s = s_i$ to the horizontal axis, which is labeled as $a, b, c \ldots$

The Jacobian of transformation from the n space to k space is therefore

$$\begin{aligned} J(k) &= \frac{dg(k)}{dk} \\ &= \frac{4k}{\sqrt{4k^2 - s_i^2}} \end{aligned}$$

This Jacobian is then used to transform the data $f(n)$ in n domain to $f(k)$ in k domain. The interpolation equation for evenly spaced data shown in Eq. (1.29) therefore becomes

$$\begin{aligned} f(k) &= \sum_{n=-\infty}^{\infty} J(k) f(n) \operatorname{sinc}(k-n) \\ &= \frac{4k}{\sqrt{4k^2 - s_i^2}} \sum_{n=-\infty}^{\infty} f(n) \operatorname{sinc}(k-n); \quad k = -\infty \text{ to } \infty, \quad 2k > s_i. \end{aligned} \tag{1.31}$$

Notice that in this equation, the variable n shown in summation serves as an index of the unevenly spaced sample. Figure 1.20 shows, as an example, how the unevenly

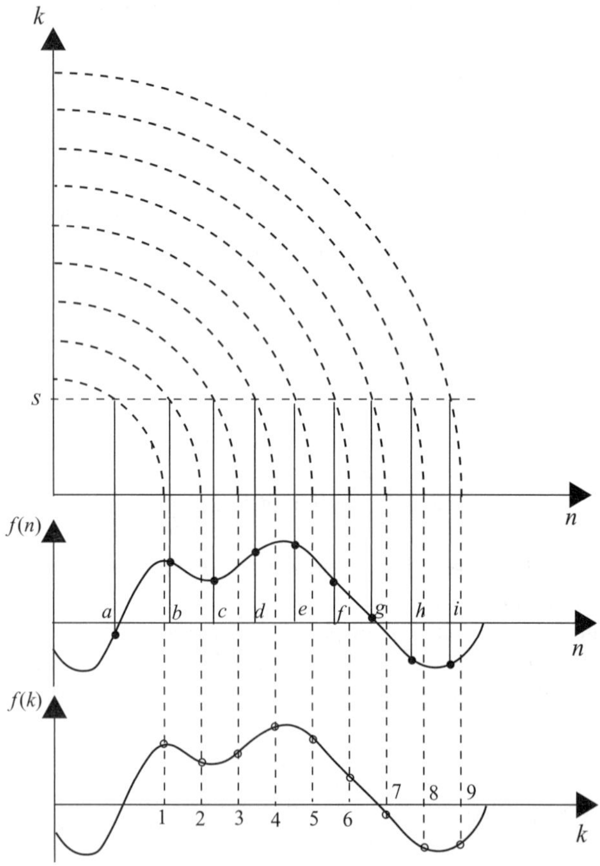

FIGURE 1.20 Relationship between evenly and unevenly spaced data.

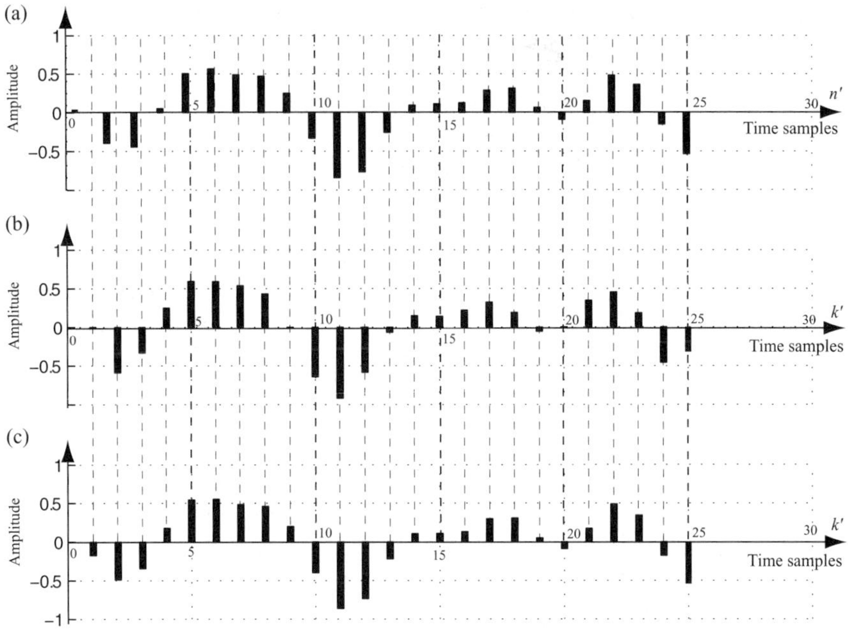

FIGURE 1.21 Interpolation on unevenly spaced data.

spaced data $f(n)$ and the evenly spaced data $f(k)$ were transformed from the data indices n and k. Given the unevenly spaced data $f(n)$, shown as dots (•), the interpolated and evenly spaced data $f(k)$ can therefore be generated from Eq. (1.31), which are denoted by small hollow circles (∘).

The evenly spaced signal $f(k)$ shown in Fig. 1.13 will now be used to illustrate the interpolation on the basis of the unevenly spaced samples. The signal $f(k)$ is expressed as

$$f(k) = 0.35\,\cos(2\pi k f_1) + 0.2\,\sin(2\pi k f_2) - 0.4\,\cos(2\pi k f_3),$$

where $f_1 = 0.25, f_2 = 0.45$, and $f_3 = 0.15$ are all normalized frequencies. We will reverse the process by generating an unevenly spaced data $f(n)$ from $f(k)$ first, then try to interpolate from $f(n)$ to obtain $f(k)$.

The unevenly spaced samples $f(n)$ are generated by computing the unevenly spaced index n for various k from Eq. (1.30). Once the index n is computed, the value of $f(n)$ can be obtained.

Consider the case of k ranging from 0.5 to 6.5 with $\Delta k = 0.25$ and $s = 0.75$; the corresponding Jacobian can be computed as

$$J(k) = \frac{4k}{\sqrt{4k^2 - (0.75)^2}}.$$

The 8-tap filters shown in Table 1.1 will be used in this example, and Eq. (1.31) then becomes

$$f(k) = \frac{4k}{\sqrt{4k^2 - (0.75)^2}} \sum_{n=-3}^{4} f(n)\,\mathrm{sinc}\,(k-n)\,.$$

Figure 1.21a shows the plot of unevenly spaced $f(n)$ with $n = \sqrt{4k^2 - (0.75)^2} \approx 2k - \Delta_k$ and $k = 0.5, 0.75, \ldots, 6.5$, $0 < \Delta_k = (0.75)^2/(4k) < 1$. A total of 25 sinc (x) functions are generated, each corresponding to different Δ_k. Figure 1.21b displays the interpolated plot of evenly spaced samples $f(k)$, $k = 0.5, 0.75, \ldots, 6.5$. For comparison purposes, the original evenly spaced signal is displayed in Fig. 1.21c. Notice that Fig. 1.21a is plotted based on the index $n' = 2n$, Figs. 1.21b and 1.21c are plotted based on the index $k' = 4k - 1$. Notice also that $f(n)$ appears to be unevenly spaced at the left part of the display, and becomes approximately evenly spaced at the right end of the plot as expected. Just like the original digitized samples shown in Fig. 1.21c, the interpolated signal is evenly spaced and matches quite well with the bottom signal. Only a few samples at both ends differ from the original, because of the edge effect of convolution, which is covered in the next chapter.

2

DISCRETE TIME AND FREQUENCY TRANSFORMATION

This chapter addresses discrete signal transformations in both time and frequency domains. The relationship between continuous and discrete Fourier transform is discussed in Section 2.1. Section 2.2 reviews several key properties of discrete Fourier transform. The effect of Window functions on discrete Fourier transform is described in Section 2.3, and fast discrete Fourier transform techniques are covered in Section 2.4. The discrete cosine transform is reviewed in Section 2.5. Finally, the relationship between continuous and discrete signals in both time and frequency domains is illustrated graphically with an example in Section 2.6.

2.1 CONTINUOUS AND DISCRETE FOURIER TRANSFORM

It was shown in Chapter 1 that a periodic signal $g_p(t)$ can be expressed as

$$g_p(t) = \sum_{k=-\infty}^{\infty} g(t - kT_0),$$

where T_0 is the period of $g(t)$, and

$$\begin{aligned} g(t) &= A \quad \text{for} \quad -\frac{\tau}{2} \leq t \leq \frac{\tau}{2}, \\ &= 0 \quad \text{otherwise}. \end{aligned}$$

Digital Signal Processing Techniques and Applications in Radar Image Processing, by Bu-Chin Wang.

It also was shown that $g_p(t)$ can be represented in terms of Fourier series coefficients as

$$g_p(t) = \sum_{n=-\infty}^{\infty} G_n e^{jn\omega_0 t}$$

$$G_n = \frac{1}{T_0}\int_0^{T_0} g_p(t)\, e^{-jn\omega_0 t}\, dt,$$

where $\omega_0 = 2\pi/T_0$.

Now, let $t = n\Delta T$ and $\tau = N_1\Delta T$, where $\Delta T = 1/f_s$, and f_s is the sampling frequency that satisfies the Nyquist requirement. Assume that all frequencies are normalized with respect to f_s, or $f_s = 1$. Then $t = n\Delta T = n$, $T_0 = N\Delta T = N$, $\tau = N_1\Delta T = N_1$, and $\omega_0 = 2\pi/(N\Delta T) = 2\pi/N$. The analog or continuous signal $g_p(t)$ therefore becomes discrete or digital signal $g_p(n)$, which can be expressed as

$$g_p(n) = \sum_{k=-\infty}^{\infty} g(n - kN).$$

and

$$g(n) = A \quad \text{for } -\frac{N_1}{2} \le n \le \frac{N_1}{2},$$
$$= 0 \quad \text{otherwise.}$$

A graphical representation of a periodic signal or sequence $g_p(n)$ is shown in Fig. 2.1.

A digitized and periodic signal $g_p(n)$, with period N, can therefore be expressed as

$$g_p(n) = \sum_{k=-\infty}^{\infty} G_p(k) e^{jk(2\pi/N)n}. \qquad (2.1a)$$

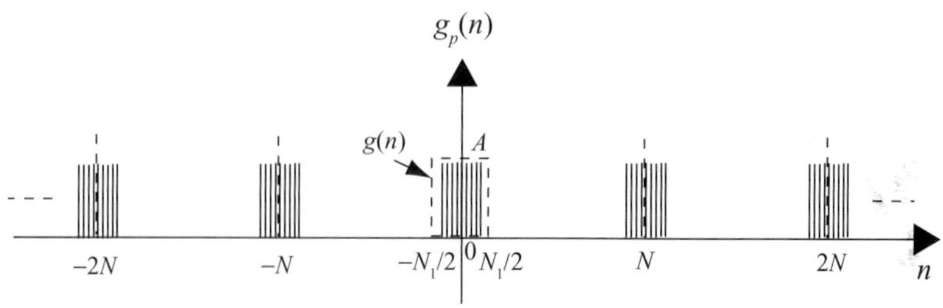

FIGURE 2.1 A periodic sequence.

Let $k = k_0 \pm mN$, where

$$k \in [-\infty, \infty],$$
$$k_0 \in [0, N-1].$$

Then

$$e^{j(2\pi/N)(k_0 \pm mN)n} = e^{j(2\pi/N)k_0 n} \cdot e^{\pm j(2\pi/N)mNn}$$
$$= e^{j(2\pi/N)k_0 n}.$$

Similarly, the variable n can also be represented as $n = n_0 \pm mN$, and the exponential term shown above also holds true for n_0. For simplicity, we will drop the subscript 0 and use k and n directly. Equation (2.1a) then becomes

$$g_p(n) = \sum_{k=0}^{N-1} G(k) e^{jk(2\pi/N)n}. \tag{2.1b}$$

Here $G(k)$ is the N-term sequence of the periodic sequence $G_p(k)$.

Let $g(n)$ be the N-term sequence of $g_p(n)$ in the region for $n = 0$ to $N-1$. By taking summation of $g_p(n) e^{-jr(2\pi/N)n}$ for $n = -\infty$ to ∞, we see that Eq. (2.1b) becomes

$$\sum_{n=-\infty}^{\infty} g_p(n) e^{-jr(2\pi/N)n} = \sum_{n=0}^{N-1} g(n) e^{-jr(2\pi/N)n}$$
$$= \sum_{n=0}^{N-1} \sum_{k=0}^{N-1} G(k) e^{j(k-r)(2\pi/N)n}$$
$$= \sum_{k=0}^{N-1} G(k) \sum_{n=0}^{N-1} e^{j(k-r)(2\pi/N)n}. \tag{2.2a}$$

The second summation of Eq. (2.2a) satisfies

$$\sum_{n=0}^{N-1} e^{j(k-r)(2\pi/N)n} = N \text{ for } \quad k = r,$$
$$= 0 \text{ for } \quad k \neq r.$$

Equation (2.2a) then becomes

$$\sum_{n=0}^{N-1} g(n) e^{-jr(2\pi/N)n} = NG(r),$$

or

$$G(r) = \frac{1}{N}\sum_{n=0}^{N-1} g(n)e^{-jr(2\pi/N)n}. \qquad (2.2b)$$

Equations (2.1b) and (2.2b) constitute a discrete Fourier transform pair. Equation (2.2b) serves as the discrete Fourier transform (DFT) from $\{g(n)\}$ to $G(r)$ and Eq. (2.1b) serves as the inverse discrete Fourier transform (IDFT) from $\{G(r)\}$ to $g(n)$. Notice that $g(n) = g_p(n)$ for $n = 0$ to $N-1$ and $G(r) = G_p(r)$ for $r = 0$ to $N-1$.

To comply with the popular notation of analog domain Fourier transform, Eqs. (2.1b) and (2.2b) are then modified to become

$$\text{IDFT}: \quad g(n) = \frac{1}{N}\sum_{k=0}^{N-1} G(k)e^{jk(2\pi/N)n}, \qquad (2.2c)$$

$$\text{DFT}: \quad G(k) = \sum_{n=0}^{N-1} g(n)e^{-jk(2\pi/N)n}. \qquad (2.2d)$$

Here $n = 0$ to $N-1$ and $k = 0$ to $N-1$. Equations (2.2c) and (2.2d) serve as the discrete Fourier transform pair.

By defining $W_N = e^{-j(2\pi/N)}$, Eqs. (2.2c) and (2.2d) can be modified to become another form of DFT:

$$\text{DFT}: \quad G(k) = \sum_{n=0}^{N-1} g(n)W_N^{nk}, \qquad (2.2e)$$

$$\text{IDFT}: \quad g(n) = \frac{1}{N}\sum_{k=0}^{N-1} G(k)\, W_N^{-nk}. \qquad (2.2f)$$

The definition of W_N satisfies the following two basic properties:

$$W_N^{k+N/2} = -W_N^k,$$

$$W_N^{k+N} = W_N^k.$$

2.2 KEY PROPERTIES OF DISCRETE FOURIER TRANSFORM

The discrete Fourier transform serves as a very important tool in the digital signal processing fields. Some important and useful properties of DFT will now be described.

2.2.1 Shifting and Symmetry

As shown in Eq. (1.18), a delay in the time domain will cause a phase shift in the frequency domain. Similarly, for a periodic sequence, a shift in discrete time will cause a phase shift in the discrete frequency domain. This phase shift characteristic can be derived as shown below.

Let $g(n)$ be a N-sample periodic sequence with the corresponding DFT as $G(k)$. Then the $g(n-m)$ will have its DFT as follows, from Eq. (2.2c):

$$
\begin{aligned}
g\,(n-m) &= \frac{1}{N}\sum_{k=0}^{N-1} G(k)e^{jk(2\pi/N)(n-m)} \\
&= \frac{1}{N}\sum_{k=0}^{N-1} \left[G(k)e^{-j(2\pi/N)km}\right]e^{jk(2\pi/N)n}.
\end{aligned}
$$

This implies that $g(n-m)$ and $G(k)e^{-j(2\pi/N)km}$ form a DFT transform pair, and the exponential term represents a phase shift in frequency domain.

If the values of periodic sequence $g(n)$ are real numbers, then the real part of $G(k)$ for $k = 1$ to $k = (N/2) - 1$ are identical to that of $G(k)$ for $k > (N/2)$. This can be seen from Eq. (2.2d):

$$
\begin{aligned}
G(N-k) &= \sum_{n=0}^{N-1} g(n)e^{-j(N-k)(2\pi/N)n} \\
&= \sum_{n=0}^{N-1} g(n)e^{jk(2\pi/N)n}e^{-j2\pi n} \\
&= \sum_{n=0}^{N-1} g(n)e^{jk(2\pi/N)n}, \\
G^*(N-k) &= \sum_{n=0}^{N-1} g(n)e^{-jk(2\pi/N)n} \\
&= G(k).
\end{aligned}
$$

Furthermore, if $g(n)$ is real and symmetric around its center, that is, if $g(n) = g(N-n)$, then $G(k)$ can easily be proved to be real and $G(N-k) = G(k)$.

2.2.2 Linear and Circular Convolution

Leting $x(n)$ and $h(n)$ be two nonperiodic N-sample sequences, the linear convolution of $x(n)$ and $h(n)$ is defined as

$$
y\,(n) = \sum_{m=0}^{N-1} h\,(m)x\,(n-m), \tag{2.3}
$$

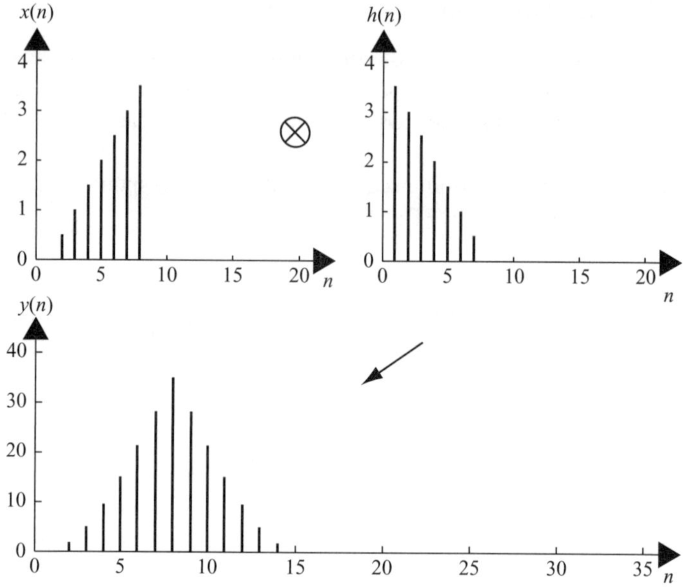

FIGURE 2.2 Linear convolution.

where $h(m)$ and $x(n-m)$ are zero outside the duration of N samples. Figure 2.2 displays the linear convolution of $x(n)$ and $h(n)$ with $y(n)$ as the output and $N = 8$. As can be seen in the figure, the output $y(n)$ has duration of $2N - 1$ samples and is free of spectrum aliasing. The symbol $\otimes$ in Fig. 2.2 represents convolution.

Now, consider the two new sequences $x(n)$ and $h(n)$, which are periodic with period $N = 8$. These two sequences are identical to the above two nonperiodic sequences for $n = 0,1,\ldots,N-1$. The DFTs of these two new sequences are

$$X(k) = \sum_{n=0}^{N-1} x(n)e^{-jk(2\pi/N)n},$$
$$H(k) = \sum_{n=0}^{N-1} h(n)e^{-jk(2\pi/N)n}.$$

The circular convolution of $x(n)$ and $h(n)$ with output $y(n)$ is defined as

$$y(n) = \frac{1}{N}\sum_{k=0}^{N-1} X(k)H(k)e^{jk(2\pi/N)n}. \tag{2.4}$$

where $y(n)$ is the IDFT of $Y(k) = H(k)X(k)$. The function $y(n)$ is also a periodic sequence with period N.

Figure 2.3 displays the circular convolution of two sequences $x(n)$ and $h(n)$ with $n = 0,1,\ldots, 7$. By comparing the values of $y(n)$ with those shown in Fig. 2.2, one

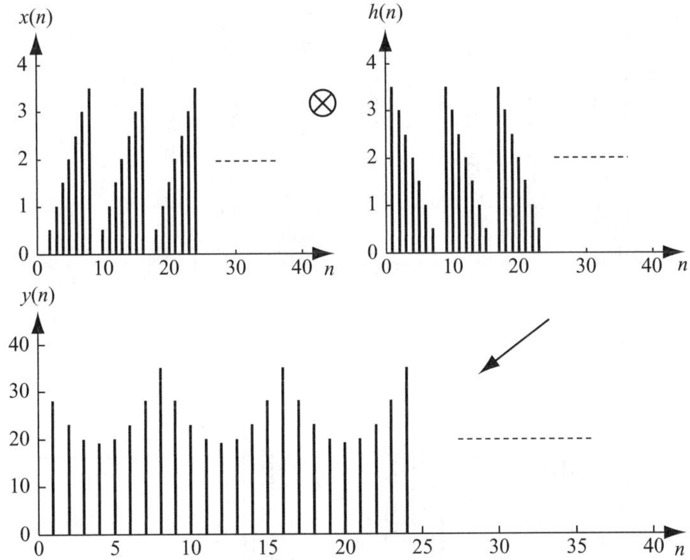

FIGURE 2.3 Circular convolution.

can see that the N-term circular convolution output $y(n)$ is distorted by the overlap with a neighboring sequence when performing multiplication and summation. This type of distortion is called "the edge effect of circulation convolution."

The edge effect of circular convolution is a built-in characteristic of DFT. To ensure that the circular convolution of any two sequences with period N will have nonaliased output with $2N - 1$ samples, one can establish the length of both $x(n)$ and $h(n)$ at $2N$ samples by padding N zeros after the N-term sequence. For practical applications, $L = 2N$ is chosen to be a power of 2. The same N-point data sequences $x(n)$ and $h(n)$ used in Figs. 2.2 and 2.3 are again used in Fig. 2.4 to demonstrate the process of linear convolution based on a circular convolution scheme. As expected, the output $y(n)$ is identical to that of linear convolution as shown in Fig. 2.2.

2.2.3 Sectioned Convolution

For most applications, the two sequences $x(n)$ and $h(n)$ may not have the same length of non-zero-valued samples. For some applications, one could have a much longer sample length than the other has. In that case, the above mentioned zero-padded $2N$-term sequence with $2N = L$ for both $x(n)$ and $h(n)$ may not be practical to compute the convolution. To solve this problem, two techniques are developed. These methods partition the larger sequence into several smaller sequences and compute the partial result first and add them together later to form the complete results. This will be discussed next.

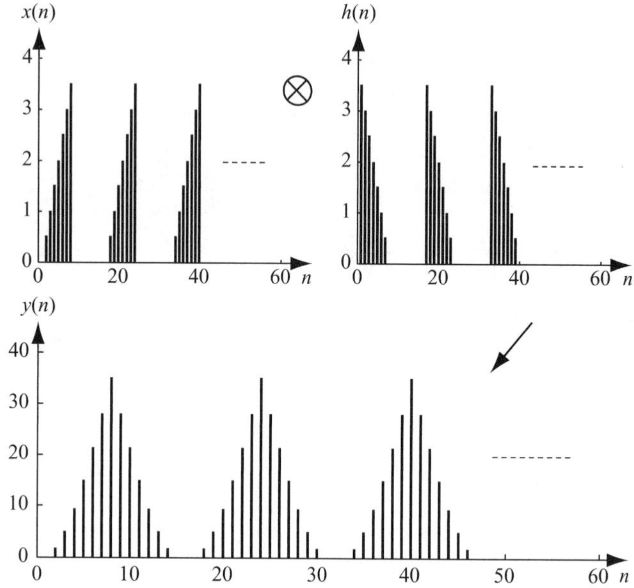

FIGURE 2.4 Linearized circular convolution.

2.2.3.1 *Overlap-and-Add Method.* Let $x(n)$ be the input time sequence with N_1 samples and $h(n)$ be the impulse response with N_2 samples. Assume that $N_1 \gg N_2$ and N_1 can be equally divided into M blocks with each block having N_0 samples, that is, $N_1 = M\ N_0$. The samples of the ith data block, $x_i(n)$, is defined to be equal to $x(n)$ for $n \in [(i-1)N_0, \ldots, iN_0 - 1]$ and zero elsewhere, where $i = 1, 2, \ldots, M$. The output of linear convolution with input data $x_i(n)$ and impulse response $h(n)$ equals $y_i(n)$, which has length $L = N_0 + N_2 - 1$. The linear convolution can be implemented by circular convolution if L is chosen to be a power of 2.

Figure 2.5 demonstrates how the Overlap-and-Add method performs the convolution. As can be seen, the ith block of $x(n)$, or $x_i(n)$ with $i = 1,2,\ldots, M$, convolves with $h(n)$ to generate an ith data block $y_i(n)$. The section of $y_i(n)$ that extends beyond the range $n \in [(i-1)N_0, \ldots, iN_0-1]$ will coincide with part of the next $y_{i+1}(n)$. This extra portion, with N_2-1 samples in length, is called the *overlap part*, which will be added to $y_{i+1}(n)$ to form the convolution sum for the $(i+1)$th section of output $y(n)$.

2.2.3.2 *Overlap-and-Save Method.* Unlike the overlap-and-add method, the overlap-and-save method requires that the input blocks overlap with each other to process the convolution. Consider the input time sequences $x(n)$ and the impulse response $h(n)$, which have N_1 and N_2 samples, respectively. Assume that $N_1 \gg N_2$, and the input data blocks can be divided into M sections of data blocks $x_i(n)$ with size of $N_0 + N_2 - 1$ samples each. The end portion of the input data $x_i(n)$ is overlapped with $N_2 - 1$ samples of the next block of input data $x_{i+1}(n)$. Samples of the ith data block $x_i(n)$ are defined to be equal to $x(n)$ over $n \in [(i-1)N_0, \ldots,$

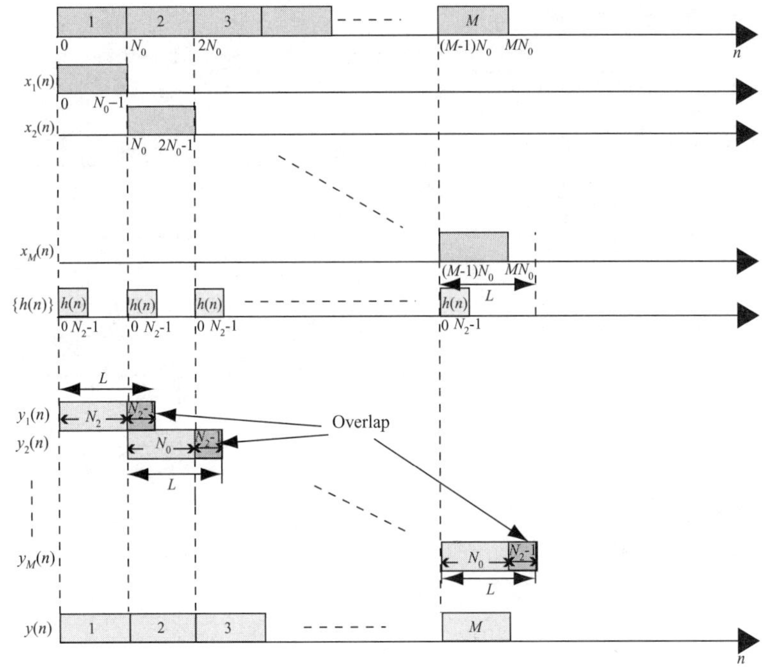

FIGURE 2.5 Convolution using overlap-and-add method.

$iN_0 + N_2 - 1]$, where $i = 1,2,\ldots, M$. The data blocks $x_i(n)$ are then circularly convolved with the impulse response $h(n)$. Because of the overlap redundancy at the input, the circular artifacts in the output (the last $N_2 - 1$ samples) can simply be discarded. Figure 2.6 illustrates the overlap-and-save method.

2.2.4 Zero Stuffing and Discrete Fourier Transform (DFT) Resolution

One of the fundamental characteristics of discrete signals is that zero padding in one domain results in an increased sampling rate in the other domain. The N-sample Hanning window will be used to illustrate the zero-padding technique. Mathematically, the Hanning window is expressed as follows:

$$\begin{aligned} w(n) &= 0.5 - 0.5\cos\left(\frac{2\pi}{N}\right) \text{ for } n = 0,1,2,\ldots,N, \\ &= 0 \quad \text{otherwise.} \end{aligned}$$

The discrete time sequence $w(n)$ with $N = 32$ is shown in Fig. 2.7a. By taking a 32-point DFT on $w(n)$, the power spectrum of Hanning window, $W_{\mathrm{dB}}(m)$, is shown in Fig. 2.7b. For comparison, a 128-point Hanning window $w(n)$ and its corresponding power spectrum $W_{\mathrm{dB}}(m)$ are also displayed in Figs. 2.7c and 2.7d, respectively. Note

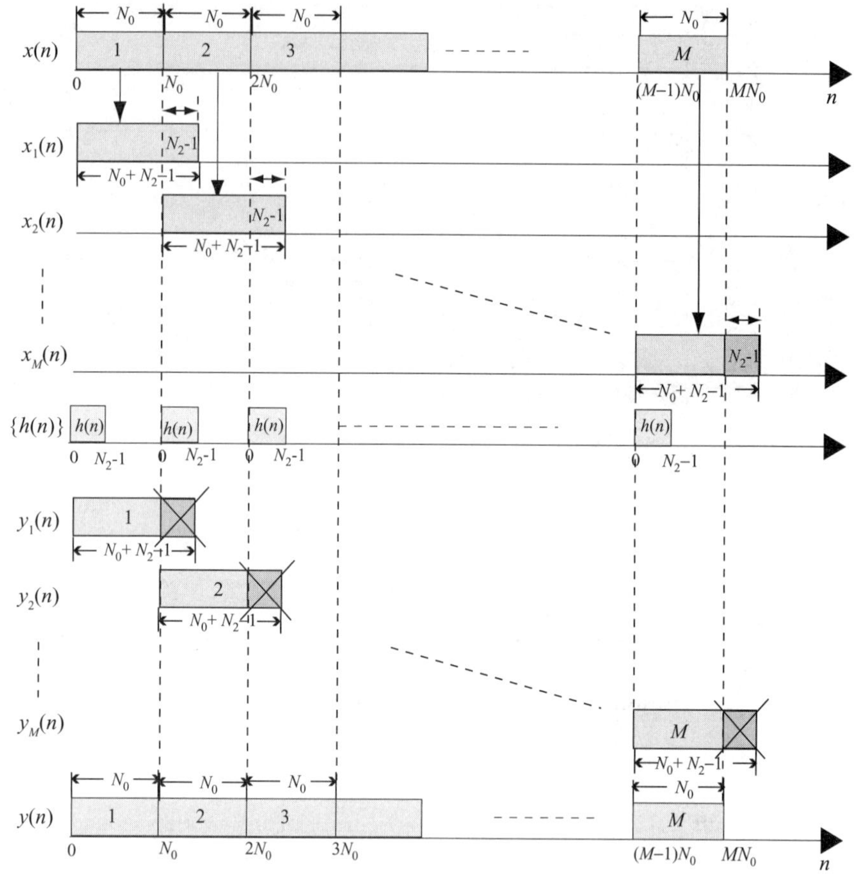

FIGURE 2.6 Convolution using overlap-and-save method.

that, for convenience, the zero-frequency components have been moved to the center of the spectra.

On comparison of the two sets of power spectra, discribed above, it is clear that at a higher sampling rate in the time domain (increasing from 32 samples to 128 samples), the frequency spectrum was hardly improved. The number of frequency samples inside the signal spectrum over –60 dB is about the same in both cases. The signal's frequency bandwidth stays the same in both cases, although it appears wider in the 32-point plot than in the 128-point plot. This is because the sampling frequency for the 128-point is 4 times higher than that of the 32-point.

The zero-padding technique can be used to increase a signal's resolution in either time or frequency domain. The 32-point $w(n)$ is again shown in Fig. 2.8a, together with the real and imaginary parts of its 32-point DFT, $X(m)$, shown in Figs. 2.8b and 2.8c, respectively. For convenience, the zero-frequency components are moved to the center of the spectra in Figs. 2.8b and 2.8c, respectively.

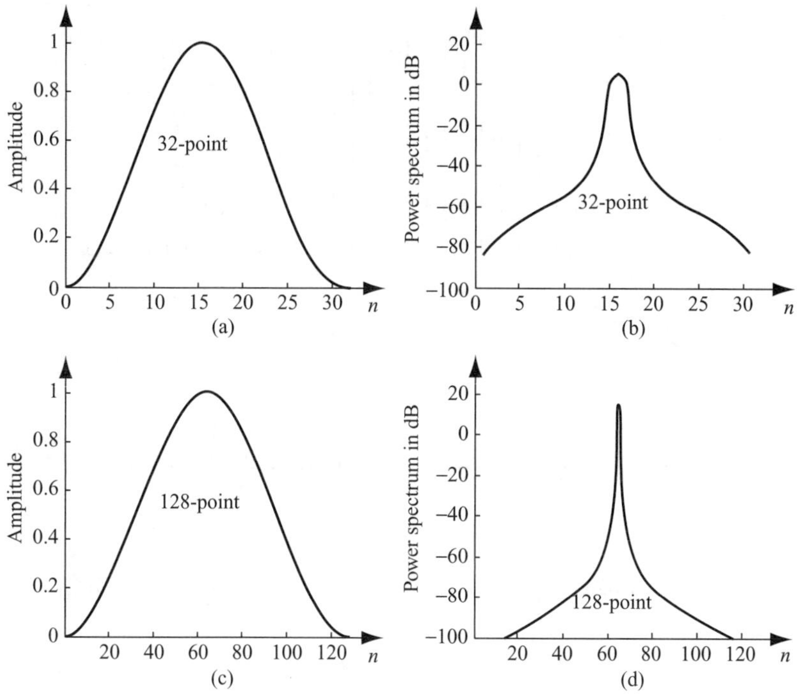

FIGURE 2.7 Hanning window with different sampling frequencies.

Now, by padding (appending) 96 zero-valued samples to the end of $w(n)$, a new 128-sample $w'(n)$ sequence is obtained (Fig. 2.9a). By computing the DFT of $w'(n)$, one obtains a more detailed power spectrum as shown in Fig. 2.9b. This new 128-point spectrum provides a frequency structure that is more detailed than that of Fig. 2.7d. The number of frequency components inside the signal spectrum over –60 dB is about 4 times that of Figs. 2.7b or 2.7d. Figures 2.9c and 2.9d display the real and imaginary parts of the frequency spectrum corresponding to the zero-padded Hanning window.

Again, the power spectrum was displayed with the zero-frequency component moved to the center of the spectrum. The real and imaginary parts of the frequency spectrum are also rearranged accordingly.

The preceding example demonstrates that zero padding in the time domain results in an increased sample rate in the frequency domain. Here, the zero-padding technique increased the frequency-domain sample rate (or resolution) by a factor of 4 (=128/32).

It is interesting to see that doubling the sampling rate in the time domain did not improve the signal's frequency resolution when compared to that of the zero-padding method for an identical new sample rate. This is due to the fact that doubling the sampling rate in the time domain from N samples to $2N$ samples also increases the number of frequency samples in the frequency domain from N to $2N$. The bandwidth

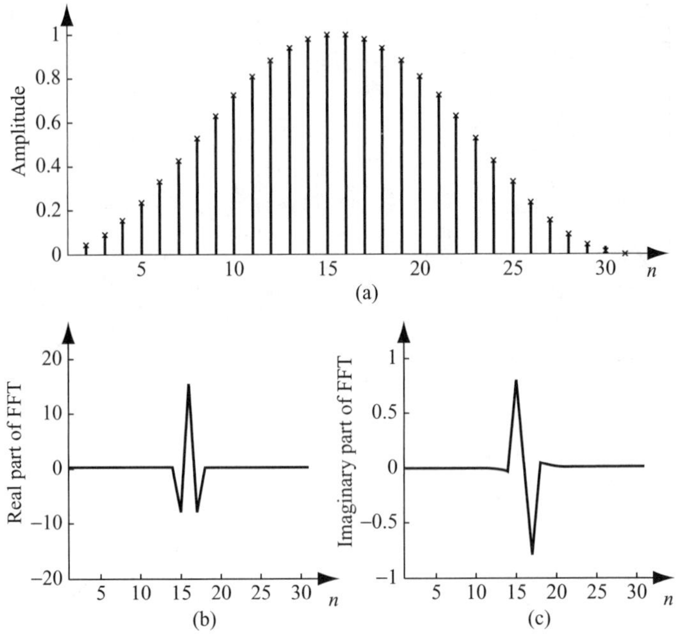

FIGURE 2.8 A 32-point Hanning window.

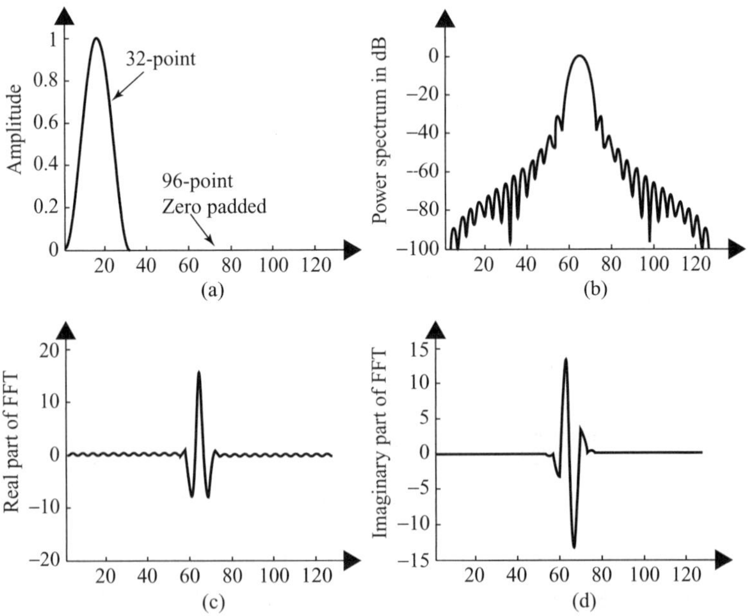

FIGURE 2.9 Hanning window with time-domain zero padding.

of the signal remained the same, but the sampling frequency was increased twofold. In the frequency domain the signal spectrum was still represented by N samples only; the extra N samples were wasted to represent the stretched frequency region.

The zero-padding technique used in the frequency domain is similar to that in the time domain, we will use a 64-point interpolation, not 128-point interpolation as in the time domain for illustration. The same 32-point $w(n)$ is first Fourier-transformed to become $W(m)$. Then, 32 zero-valued complex samples are inserted in the middle of $W(m)$ to form a new 64-point complex spectrum $W'(m)$. The real and imaginary parts of $W'(m)$ are shown in Figs. 2.10a and 2.10b, respectively. Notice that the complex data shown in Figs. 2.8b and 2.8c are used to insert zeros as shown in Figs. 2.10a and 2.10b. However, Figs. 2.8b and 2.8c were arranged with zero frequency in the center, while Figs. 2.10a and 2.10b are zero-padded directly on DFT outputs $W(m)$ with zero frequency at bin 1. In general, the zeros are inserted after the first $N/2$ spectrum samples in order to maintain spectrum symmetry. A 64-point IDFT is then operated on the zero-padded $W'(m)$ to generate the interpolated time sequence $w'(n)$ as shown in Figs. 2.10c and 2.10d. Since $|W(m)|$ is an even function and the phase of $W(m)$ is an odd function, the IDFT outputs are all real numbers with all imaginary parts equal to zeros. The 64 samples shown in Fig. 2.10c represent the new interpolated sequence $w'(n)$. It is obvious that zero padding in the frequency domain results in an increased resolution in the time domain.

The filter-based time-domain interpolation technique, discussed in Chapter 1, inserts zero-valued samples between each of the original time samples; the zero-inserted sequence is then lowpass-filtered to attenuate the sidelobes of spectrum

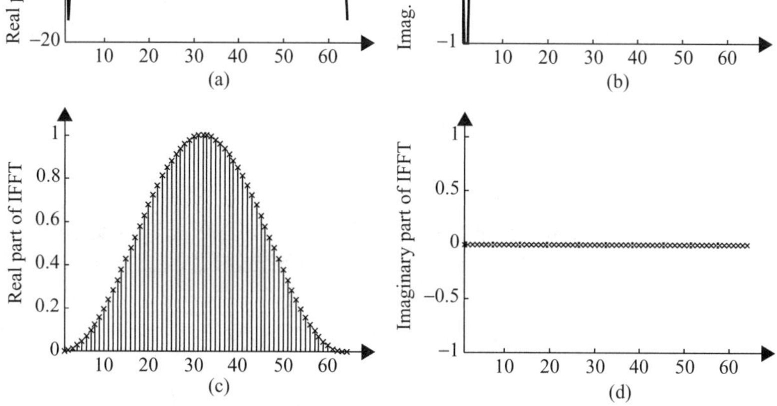

FIGURE 2.10 Hanning window with frequency-domain zero padding.

caused by the inserted zeros. The accuracy of this technique depends on the quality of the filter. The greater the stopband attenuation by lowpass filtering, the more accurate the interpolation becomes. The zero-padding technique does not require lowpass filtering, yet the sidelobe rejection is ideal in the sense that all sidelobes have zero values due to zero padding in the frequency domain. Therefore, this frequency-domain zero-padding technique is called "exact interpolation" for periodic time-domain sequences. This technique fails if the sampling rate of $w(n)$ violates the Nyquist requirement.

2.3 WIDOWS AND DISCRETE FOURIER TRANSFORM

Let the N-sample discrete time sequence $x(n)$ be a sinusoidal function with frequency equal to f_0, and let f_s be the sampling frequency. If the sampling frequency f_s is not an integer multiple of f_0, the frequency spectrum of $x(n)$ will not show up as a sharp pulse at frequency f_0. Instead, the spectrum expands wider over several frequency bins around f_0. One practical solution to reduce this type of energy spreading problem is to use a windowing function. The windowing function involves the imposition of a prescribed profile on the time signal prior to performance of the Fourier transformation. Let $x(n)$ and $w(n)$ be the original time sequence and windowing function, respectively; the new sequence is then given as

$$x'(n) = x(n)w(n). \tag{2.5}$$

When the windowing function is applied to $x(n)$, the resultant frequency spectrum will suffer because of the loss of energy at both ends of time sequence $x(n)$. An averaging scheme that is used to compensate the loss of energy is discussed next.

Assume that an N-sample data block with appropriate window function is used to compute the DFT. To generate the ith block of data DFT, the $(i - \frac{1}{2})$th data DFT, the ith data DFT and $(i + \frac{1}{2})$th data DFT must be added together and divided by 3. That is, using $i = 2$ as an example, one obtains

$$\begin{aligned}\overline{\mathrm{DFT}_2(k)} &= \frac{\mathrm{Re}\{\mathrm{DFT}_{1.5}(k)\} + \mathrm{Re}\{\mathrm{DFT}_2(k)\} + \mathrm{Re}\{\mathrm{DFT}_{2.5}(k)\}}{3} \\ &\quad + j\frac{\mathrm{Im}\{\mathrm{DFT}_{1.5}(k)\} + \mathrm{Im}\{\mathrm{DFT}_2(k)\} + \mathrm{Im}\{\mathrm{DFT}_{2.5}(k)\}}{3}.\end{aligned} \tag{2.6}$$

Figure 2.11 illustrates the generation of DFT based on the averaging scheme.

There are many windowing functions available for different types of applications. The following are four popular ones, and their mathematical expressions are shown below:

1. Rectangular windowing function:

$$\begin{aligned} w(n) &= 1 \quad \text{for} \quad n = 0,1,\ldots, N-1, \\ &= 0 \quad \text{otherwise.} \end{aligned}$$

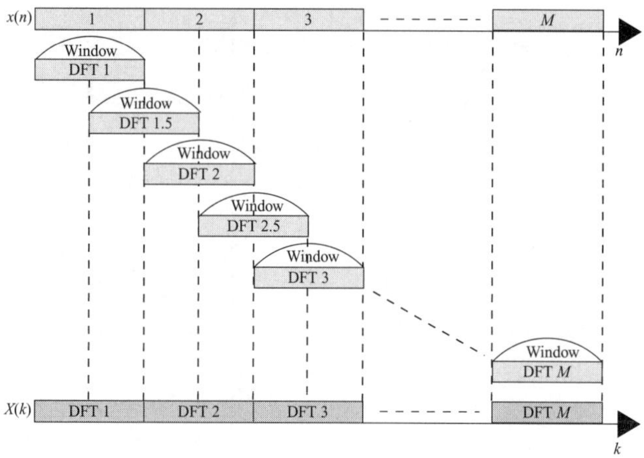

FIGURE 2.11 DFT with sliding (overlapping).

2. Hanning windowing function:

$$w(n) = 0.5 - 0.5\cos\frac{2\pi}{N} \quad \text{for} \quad n = 0, 1, \ldots, N-1,$$
$$- 0 \quad \text{otherwise.}$$

3. Hamming windowing function:

$$w(n) = 0.54 - 0.46\cos\frac{2\pi}{N} \quad \text{for} \quad n = 0, 1, \ldots, N-1,$$
$$= 0 \quad \text{otherwise.}$$

4. Blackman windowing function:

$$w(n) = 0.42 - 0.5\cos\frac{2\pi n}{N} + 0.08\cos\frac{4\pi n}{N} \quad \text{for} \quad n = 0, \ldots, N-1,$$
$$= 0, \quad \text{otherwise.}$$

When a long sequence of a signal is partitioned into smaller sections for processing, this partition process is equivalent to multiplying the long sequence by the rectangular windowing function, which in fact serves no purpose. The Hanning window function has been discussed extensively in previous sections. The two new windowing functions, namely, the Hamming window and the Blackman window, are shown in Fig. 2.12, where the time-domain waveforms and frequency spectra are displayed.

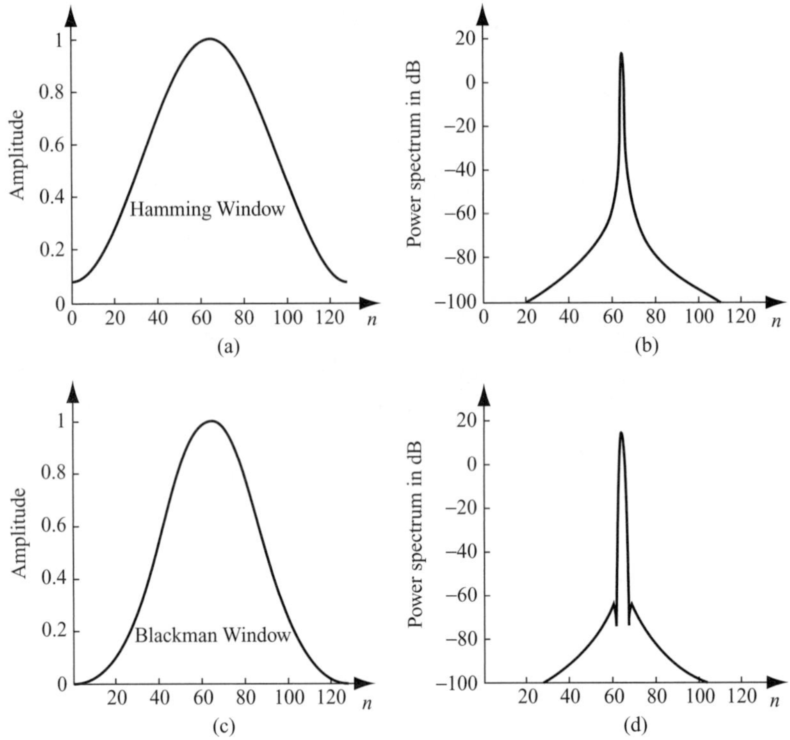

FIGURE 2.12 Hamming and Blackman window functions.

2.4 FAST FOURIER TRANSFORM

The fast Fourier transform (FFT) is a time-saving computation technique of the discrete Fourier transform (DFT). Various schemes are used to realize the FFT. The radix-2 is the most popular one and is discussed next.

2.4.1 Radix-2 Fast Fourier Transform (FFT) Algorithms

Consider the computation of $N = 2^L$ points DFT by the divide-and-conquer approach. Let $f_1(n)$ and $f_2(n)$ be the two $N/2$-point data sequences, corresponding to the even-numbered and odd-numbered samples of N-point data sequence $x(n)$; that is

$$\begin{aligned} f_1(n) &= x(2n), \\ f_2(n) &- x(2n+1), \end{aligned} \tag{2.7}$$

for $n = 0, 1, \ldots, N/2-1$.

Since $f_1(n)$ and $f_2(n)$ are obtained by decimating $x(n)$ by a factor of 2, the resulting FFT algorithm is called the "decimation-in-time" algorithm.

The N-point DFT can be expressed in terms of the decimated sequences as follows:

$$\begin{aligned} X(k) &= \sum_{n=0}^{N-1} x(n) W_N^{kn} \\ &= \sum_{n \text{ even}} x(n) W_N^{kn} + \sum_{n \text{ odd}} x(n) W_N^{kn} \\ &= \sum_{m=0}^{(N/2)-1} x(2m) W_N^{2mk} + \sum_{m=0}^{(N/2)-1} x(2m+1) W_N^{k(2m+1)}, \end{aligned}$$

where $k = 0,1,2,\ldots, N$. With $W_N^2 = W_{N/2}$ and leting $f_1(m) = x(2m)$, $f_2(m) = x(2m+1)$, this equation becomes

$$\begin{aligned} X(k) &= \sum_{m=0}^{(N/2)-1} f_1(m) W_{N/2}^{mk} + W_N^k \sum_{m=0}^{(N/2)-1} f_2(m) W_{N/2}^{km} \\ &= F_1(k) + W_N^k F_2(k), \quad \text{for} \quad k = 0, 1, \ldots, N-1, \end{aligned} \tag{2.8}$$

where $F_1(k)$ and $F_2(k)$ are the $N/2$-point DFTs of the sequences $f_1(m)$ and $f_2(m)$, respectively.

Since $F_1(k)$ and $F_2(k)$ are periodic with period $N/2$, we have $F_1(k + N/2) = F_1(k)$ and $F_2(k + N/2) = F_2(k)$. In addition, $W_N^{k+N/2} = -W_N^k$; hence, Eq. (2.8) can be expressed as

$$\begin{aligned} X(k) &= F_1(k) + W_N^k F_2(k), \qquad k = 0, 1, \ldots, \frac{N}{2} - 1, \\ X\left(k + \frac{N}{2}\right) &= F_1(k) - W_N^k F_2(k), \qquad k = 0, 1, \ldots, \frac{N}{2} - 1. \end{aligned} \tag{2.9}$$

The procedure used in this computation can be repeated through decimation of the $N/2$-point sequences $X(k)$ and $X(k + N/2)$. The entire process involves $L = \log_2 N$ stages of decimation. The computation of N-point DFT via the decimation-in-time FFT requires $(N/2)\log_2 N$ complex multiplications and $N\log_2 N$ complex additions.

Figure 2.13 depicts the computation of $N = 8$ points DFT. The computation is performed in three stages, beginning with the computations of four 2-point DFTs, then two 4-point DFTs, and finally one 8-point DFT.

Figure 2.14 illustrates the implementation of FFT through the decimation-in-time algorithm for the case of $N = 8$. The 8-point DFT is split into two 4-point DFTs and then four 2-point DFTs. The well-known butterfly pattern is clearly visible.

Another important radix-2 FFT scheme, called the "decimation-in-frequency" algorithm, is obtained by also using the divide-and-conquer approach. This method splits the DFT formula into two summations, one involving the sum over the first

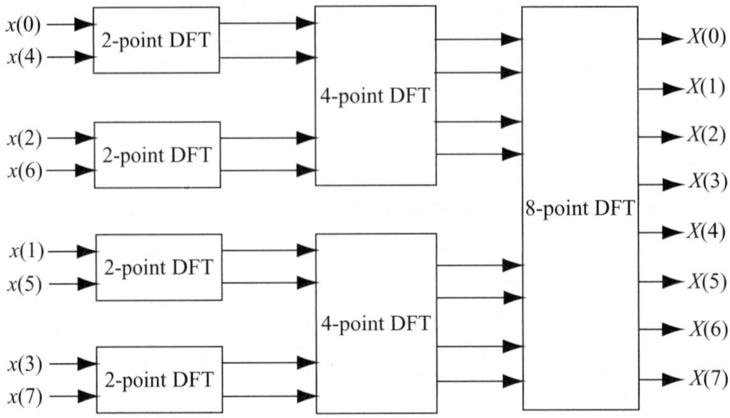

FIGURE 2.13 Three-stage computation of an 8-point DFT.

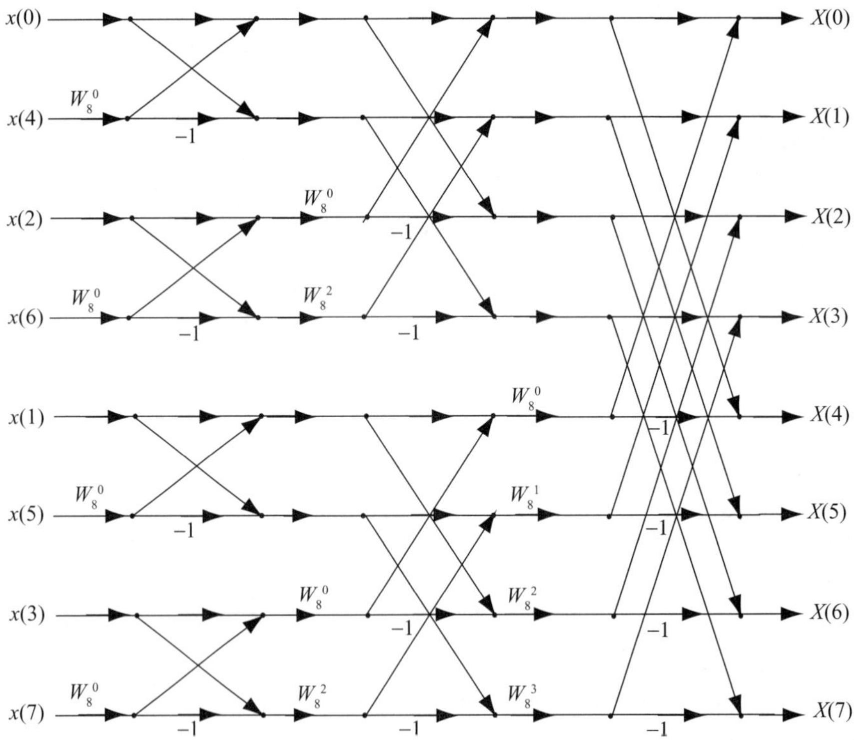

FIGURE 2.14 An 8-point FFT with decimation-in-time algorithm.

$N/2$ data points and the second sum involving the last $N/2$ data points:

$$\begin{aligned} X(k) &= \sum_{n=0}^{(N/2)-1} x(n)W_N^{kn} + \sum_{n=N/2}^{N-1} x(n)W_N^{kn} \\ &= \sum_{n=0}^{(N/2)-1} x(n)\,W_N^{kn} + W_N^{Nk/2} \sum_{n=0}^{(N/2)-1} x\left(n+\frac{N}{2}\right) W_N^{kn}. \end{aligned} \tag{2.10}$$

Since

$$W_N^{kN/2} = e^{-j(2\pi/N)(kN/2)},$$

It follows that

$$e^{-j\pi k} = (-1)^k .$$

Therefore

$$X(k) = \sum_{n=0}^{(N/2)-1} \left[x(n) + (-1)^k x\left(n+\frac{N}{2}\right)\right] W_N^{kn}.$$

By splitting (or decimate) $X(k)$ into even- and odd-numbered samples, one obtains

$$\begin{aligned} X(2k) &= \sum_{n=0}^{(N/2)-1} \left[x(n) + x\left(n+\frac{N}{2}\right)\right] W_{N/2}^{nk}, \quad k = 0,1,\ldots,\frac{N}{2}-1, \\ X(2k+1) &= \sum_{n=0}^{(N/2)-1} \left\{\left[x(n) - x\left(n+\frac{N}{2}\right)\right] W_N^{n}\right\} W_{N/2}^{nk}, \quad k = 0,1,\ldots,\frac{N}{2}-1. \end{aligned} \tag{2.11}$$

Here we have used the fact that $W_N^2 = W_{N/2}$.

The computational procedure shown above can be repeated through decimation of the $N/2$-point DFT data sequences $X(2k)$ and $X(2k+1)$, $k = 0,1,\ldots,(N/2 - 1)$. The entire process involves $L = \log_2 N$ stages of decimation, where each stage involves $N/2$ butterflies. The first stage of an 8-point butterfly is shown in Fig. 2.15.

Consequently, the computation of N-point DFT via the decimation-in-frequency FFT requires $(N/2)\log_2 N$ complex multiplications and $N\log_2 N$ complex additions, the same as in the decimation-in-time algorithm. An 8-point decimation-in-frequency example is shown in Fig. 2.16.

2.5 DISCRETE COSINE TRANSFORM (DCT)

The rapid growth of digital imaging applications, including desktop publishing, multimedia, teleconferencing, high-definition television (HDTV), and radar imaging, has increased the need for effective and standardized image compression techniques. Among the emerging standards are JPEG (Joint Photographics Expert Group)

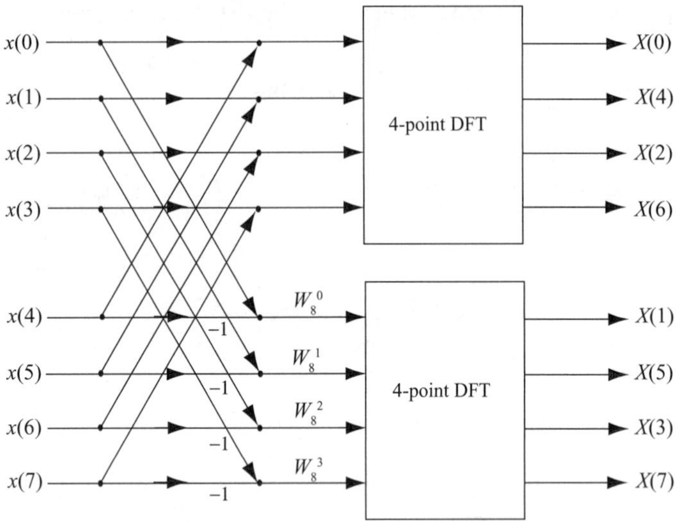

FIGURE 2.15 First stage of the decimation-in-frequency FFT algorithm.

for compression of still images, MPEG (Moving Picture Experts Group) for compression of motion video, and CCITT [International Consultative Committee on Telecommunications and Telegraphy (now ITU-T)] H.261 (also known as Px64) for compression of videotelephony and videoteleconferencing. All three standards employ a basic technique known as the discrete cosine transform DCT. We will start our discussion with the regular DCT, followed by the two-dimensional (2D) DCT later.

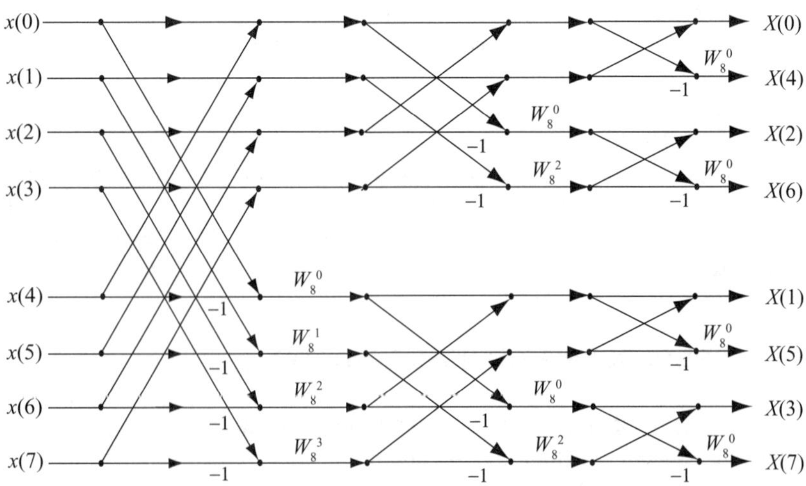

FIGURE 2.16 The 8-piont decimation-in-frequency FFT algorithm.

The DCT of a periodic and finite-length sequence $x(n)$ is defined as

$$\begin{aligned} y(k) &= \sqrt{\frac{2}{N}}\alpha(k)\sum_{n=0}^{N-1} x(n)\cos\frac{(2n+1)k\pi}{2N}, \quad k = 0,1,\ldots,N-1, \\ \alpha(k) &= 1 \quad \text{for} \quad k \neq 0, \\ &= \frac{1}{\sqrt{2}}, \quad \text{for} \quad k = 0. \end{aligned} \tag{2.12}$$

The DCT is closely related to the discrete Fourier transform (DFT), but has better energy compaction properties. This implies that the majority of the signal energy in time sequence can be represented by just a few of the DCT-transformed coefficients. The energy compaction properties of the DCT also make it useful in applications of data communication.

The inverse discrete cosine transform (IDCT) computes the inverse DCT of a finite length sequence. It reconstructs a signal from a complete or partial set of DCT coefficients. The IDCT is defined as

$$\begin{aligned} x(n) &= \sqrt{\frac{2}{N}}\sum_{k=0}^{N-1} \alpha(k)y(k)\cos\frac{(2n+1)k\pi}{2N}, \quad n = 0,1,\ldots,N-1, \\ \alpha(k) &= 1 \quad \text{for} \quad k \neq 0, \\ &= \frac{1}{\sqrt{2}}, \quad \text{for} \quad k = 0. \end{aligned} \tag{2.13}$$

Comparing Eqs. (2.12) and (2.13), one can see that the forward and backward (inverse) DCT have identical transformation kernels. These basis vectors are sampled cosines and can be expressed as

$$b_k = \left[\sqrt{\frac{2}{N}}\alpha(k)\cos\frac{(2n+1)k\pi}{2N}\right]_{n=0,1,\ldots,N-1}, \quad \text{for} \quad k = 0,1,\ldots,N-1. \tag{2.14}$$

The attractiveness of the DCT is twofold: (1) it is nearly optimal with high positive values of adjacent-sample correlation, and (2) it can be computed via the DFT using a FFT algorithm. The computation of DCT through FFT consists of extending the N samples of DCT to form a $2N$-block data with even symmetry and then taking a $2N$-point DFT and saving N terms in it. The following procedure is used to compute DCT through FFT:

Step 1. Define a $2N$ data block samples $\{x'(n)\}$ as

$$\begin{aligned} x'(n) &= x(n) \quad \text{for} \quad n = 0,1,\ldots,N-1, \\ &= x(2N-1-n) \quad \text{for} \quad n = N, N+1,\ldots,2N-1, \end{aligned} \tag{2.15}$$

Step 2. Compute the 2N DFT of $x'(n)$ as

$$
\begin{aligned}
y'(k) &= \frac{1}{\sqrt{2N}} \sum_{n=0}^{2N-1} x'(n) e^{-j(2\pi kn/2N)} \\
&= \frac{1}{\sqrt{2N}} \left[\sum_{n=0}^{N-1} x(n) e^{-j(2\pi kn/2N)} + \sum_{n=N}^{2N-1} x(2N-1-n) e^{-j(2\pi kn/2N)} \right] \\
&= \sqrt{\frac{2}{N}} e^{jk\pi/2N} \sum_{n=0}^{N-1} x(n) \cos\left[\frac{(2n+1)k\pi}{2N}\right]. \qquad (2.16)
\end{aligned}
$$

By comparing $y'(k)$ with $y(k)$ of Eq. (2.12), it is clear that

$$y(k) = \alpha(k) e^{-j(k\pi/2N)} y'(k). \qquad (2.17)$$

Figure 2.17 illustrates the end effects of DFT and DCT. Figure 2.17a shows the original N-sample data block, while Fig. 2.17b shows the end effect of the N-sample-based DFT. Figure 2.17c shows how 2N data sequence $x'(n)$ is formed from an

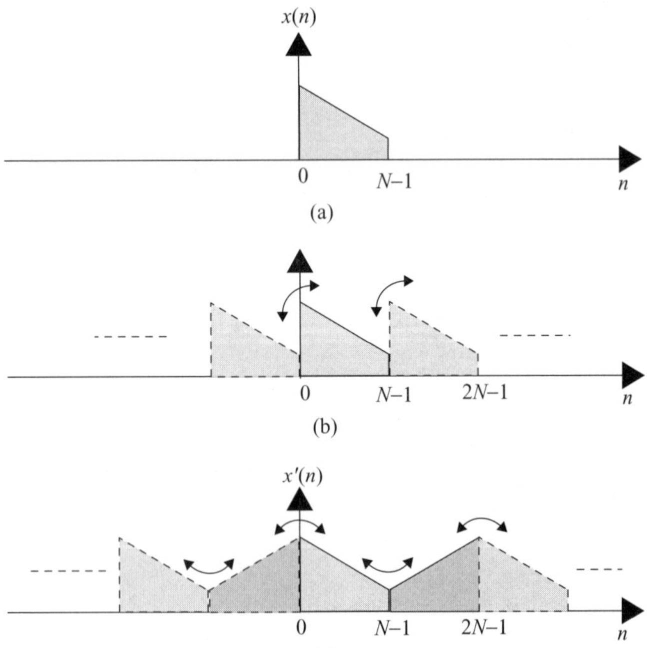

FIGURE 2.17 Input block (a) and end effects in DFT (b) and DCT (c).

N data sequence $x(n)$. It also shows the end effect of DCT. One can see that the $2N$-extension DCT has a smaller end effect (caused by a discontinuity at the border of one block and its repetition) compared to DFT operations based on the N-sample data block.

Since the N-sample DCT can be computed through the FFT by extending the input samples to $2N$, the computation requirement for N-sample DCT is then $2N \log_2 N$ multiply–add operations. Notice that the computation requirements for the regular N-sample FFT is $(N/2) \log_2 N$ complex multiplications and $N \log_2 N$ complex additions.

2.5.1 Two-Dimensional DCT

The one-dimensional DCT is useful in processing one-dimensional signals such as speech waveforms. For two-dimensional (2D) signals such as images, a 2D version of the DCT is needed. The 2D DCT can be expressed as follows:

Forward 2D DCT:

$$y(k, l) = \frac{2}{N}\alpha(k)\alpha(l)\sum_{m=0}^{N-1}\sum_{n=0}^{N-1} x(m, n)\cos\frac{\pi k(2m+1)}{2N}\cos\frac{\pi l(2n+1)}{2N}. \quad (2.18)$$

Inverse 2D DCT:

$$x(m, n) = \frac{2}{N}\sum_{k=0}^{N-1}\sum_{l=0}^{N-1} \alpha(k)\alpha(l)\, y(k, l)\cos\frac{\pi k(2m+1)}{2N}\cos\frac{\pi l(2n+1)}{2N}, \quad (2.19)$$

where

$$\alpha(0) = \frac{1}{\sqrt{2}}, \ \alpha(j) = 1 \quad \text{for} \quad j \neq 0.$$

Here $x(m, n)$ is an $N \times N$ data array. The parameters k, l, m, and n are all integers and range from 0 to $N-1$. Since the transformation kernels are separable, the 2D DCT can be performed in two steps, each involving a 1D DCT. Similar to the 1D case, the 2D DCT image coding with an even symmetry has fewer-edge effect problems than that of 2D DFT image coding.

2.6 CONTINUOUS AND DISCRETE SIGNALS IN TIME AND FREQUENCY DOMAINS

2.6.1 Graphical Representation of DFT

In the following discussion, a signal is assumed to be of finite length in the time domain and band-limited in the frequency domain. In general, the frequency spectrum of a finite-length time signal will extend to infinity. The assumption here is that if the frequency components of a finite-length time sequence exceed a certain range, they are considered small enough to be neglected. The relationship between a continuous signal and a discrete signal in both time and frequency domains, based on Fourier transform, leads to the definition of the discrete Fourier transform, described next.

Let an analog sinusoidal function $s(t)$ with single frequency f_1, namely, $s(t) = \cos(2\pi f_1 t)$, be the band-limited time-domain signal. The time-limited, or gated, signal of $s(t)$ can then be represented as $s(t)g(t)$, where $g(t)$ is a gate function with time duration T_p and is defined as

$$\begin{aligned} g(t) &= 1 \quad \text{for} \quad |t| \leq \frac{T_p}{2}, \\ &= 0 \quad \text{for} \quad |t| > \frac{T_p}{2}. \end{aligned}$$

The band-limited signal $s(t)$ is shown on the left side of Fig. 2.18a, and its frequency spectrum $S(f)$ is shown on the right side of Fig. 2.18a. For simplicity, the time duration T_p of $g(t)$ is chosen to be $T_p = 4/f_1$. The gate function $g(t)$ and the gated signal $s(t)g(t)$, together with their corresponding frequency spectra, are shown in Figs. 2.18b and 2.18c, respectively. An impulse train $i_1(t)$ with period $\Delta t = 1/(2f_1)$ is shown on the left side of Fig. 2.18d, together with its frequency spectrum, shown on the right side of Fig. 2.18d. The gated signal $s(t)g(t)$ is then multiplied by an impulse train $i_1(t)$ to become a finite-length discrete signal as shown on the left side of Fig. 2.18e, together with its spectrum shown on the right side of Fig. 2.18e. An impulse train $I_2(f)$ with period $\Delta f = 1/T_p$ is shown on the right side of Fig. 2.18f, together with its time-domain signal shown on the left side of Fig. 2.18f. The signal spectrum shown on the right side of Fig. 2.18e becomes discrete by multiplying it by $I_2(f)$ and is shown on the right side of Fig. 2.18g, together with its time-domain signal shown on the left side of Fig. 2.18g.

Leting $T_p = N\Delta t$, with $N = 8$ in this example, one can express the gated and digitized signal as

$$\begin{aligned} s(t) \cdot g(t) \cdot i_1(t) &= s(t)g(t) \sum_{n=-\infty}^{\infty} \delta(t - n\Delta t) \\ &= \sum_{n=-4}^{3} s(n\Delta t)\,\delta(t - n\Delta t). \end{aligned}$$

Notice that this discrete and finite-time signal has a continuous frequency spectrum and is repeated at the sampling frequency $f_s = 1/\Delta t$, where $\Delta t = 1/(2f_1)$. The pulse train in the frequency domain, displayed on the right side of Fig. 2.18f, has $\Delta f = 1/T_p = f_1/4$. The digitized and periodic spectrum is obtained by multiplying the analog (or continuous) and periodic spectrum by the pulse train in the frequency domain.

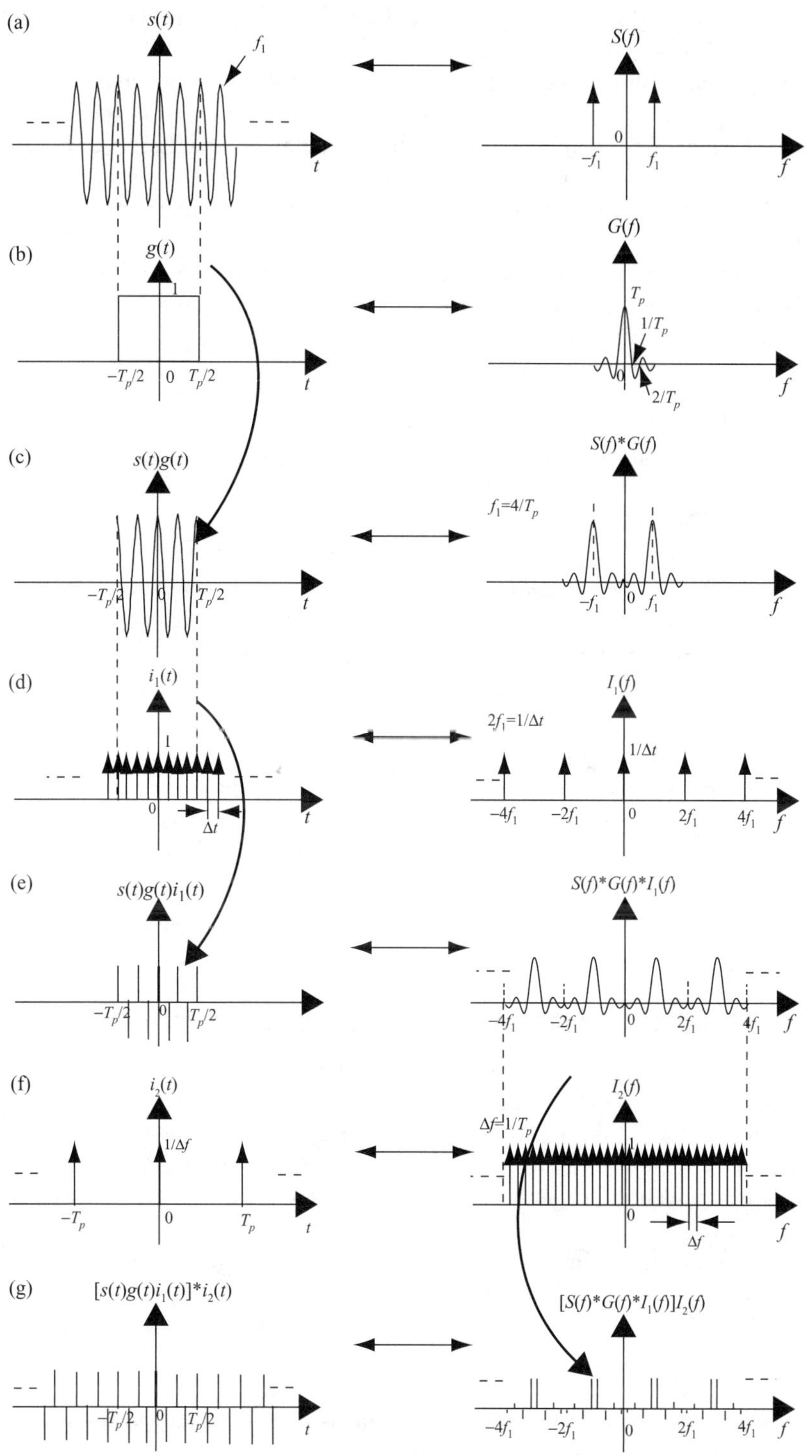

FIGURE 2.18 Graphical representations of DFT.

The corresponding digitized and periodic time-domain signal is shown on the left side of Fig. 2.18g.

The preceding discussion, which starts with an analog or continuous time signal and goes through various gating and multiplications with impulse trains in either time or frequency domain, finally comes to the results of two discrete and periodic signals in both time and frequency domains. These two discrete and periodic signals, shown on left and right sides of Fig. 2.18g, form the discrete Fourier transform pair!

From the preceding discussion and graphical representation on time–frequency and analog–discrete relations for a time-gated and frequency band-limited signal, the following observations can be obtained: (Assume that the signal $x(t)$ is bandlimited with $f \le 2f_m$, and the sampling frequency $f_s \ge 2f_m$)

1. By digitizing $x(t)$ in the time domain to render a discrete sequence $x(n)$ with each time sample spaced by Δt, the corresponding Fourier spectrum will be a periodic and continuous (or analog) waveform of $X_p(f)$ with period equals to $f_s = 1/\Delta t$.
2. By digitizing the analog (or continuous) frequency spectrum $X(f)$ to render a discrete sequence $X(m)$ with each frequency component spaced by Δf, the time-domain signal $x(t)$ becomes a periodic signal $x_p(t)$ with period $T = 1/\Delta f$.
3. If one tried to digitize the periodic waveform of frequency spectra $X_p(f)$ in observation 1, with the frequency bin spaced by Δf, then the time sequence $x(n)$ would become a periodic signal $x_p(n)$ with period $T = 1/\Delta f$.
4. Similarly, if one tried to digitize the periodic signal $x(t)$ in observation 2, with the time sample spaced by Δt, then the digitized frequency spectrum $X(m)$ would become a periodic signal $X_p(m)$ with period $f_s = 1/\Delta t$.
5. If $f_s = N\,\Delta f$ and $T = N\,\Delta t$, then observations 3 and 4 would both lead to the same results contained in the definition of the discrete Fourier transform. Here N is the number of samples in both time and frequency domains.

2.6.2 Resampling with Fractional Interpolation Based on DFT

Hardware implementation of real-time resampling and interpolation on a discrete time sequence based on FIR filtering was addressed in Section 1.6 (of Chapter 1). A second method of zeropadding in the frequency domain followed by IDFT was described in Section 2.2. This section describes another method for fractional interpolation, which is based on DFT and provides some advantages over other methods.

The interpolation on an N-sample data sequence $x(n)$ to generate a new N-sample sequence $x'(n) = x(n + \Delta)$, where $n = 0,1, \ldots, N - 1$ and $0 < \Delta < 1$, is shown in Fig. 2.19. The continuous signal $x(t)$ appears as an envelope of the discrete signal $x(n)$, and is shown as a dashed curve line. The discrete signal $x(n)$ is shown as a vertical solid line with a dot on top. The new interpolated signal $x'(n) = x(n + \Delta)$ is the signal with Δ away from the signal $x(n)$ and is shown as a dashed vertical line.

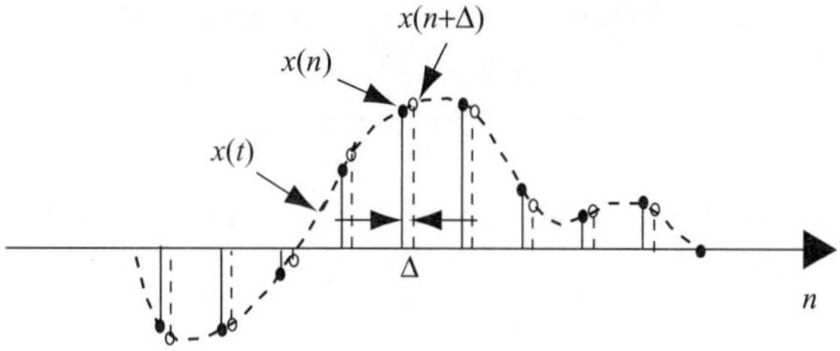

FIGURE 2.19 Example of resampling.

To compute $x'(n)$ from $x(n)$, we will use Eqs. (2.2c) and (2.2d),which are repeated here:

$$\text{DFT}: \quad X(k) = \sum_{n=0}^{N-1} x(n) W_N^{nk},$$

$$\text{IDFT}: \quad x(n) = \frac{1}{N} \sum_{k=0}^{N-1} X(k)\, W_N^{-nk}.$$

Leting $x'(n) = x(n + \Delta)$, Eq. (2.2d) becomes

$$x'(n) = \frac{1}{N} \sum_{k=0}^{N\ 1} X(k) W_N^{-(n+\Delta)k}$$

$$= \frac{1}{N} \sum_{k=0}^{N-1} X(k)\, W_N^{-\Delta k} W_N^{-nk}.$$

Defining $X'(k) = X(k)W_N^{-\Delta k}$, then

$$x'(n) = \frac{1}{N} \sum_{k=0}^{N-1} X'(k)\, W_N^{-nk}.$$

Therefore

$$X'(k) = \sum_{n=0}^{N-1} x'(n) W_N^{nk}.$$

This implies that the interpolated signal $x(n+\Delta)$ can be obtained by

1. Computing $X(k)$, $k = 0,1,\ldots, N-1$, by applying FFT on $x(n)$ for $n = 0,1,\ldots, N-1$.

2. Computing $X'(k) = X(k)W_N^{-\Delta k}$ for every k where $k = 0,1,\ldots,N-1$.
3. Applying IFFT on $X'(k) = X(k)W_N^{-\Delta k}$ for $k = 0,1,\ldots, N-1$, to obtain the interpolated signal $x(n+\Delta) = x'(n)$ for $n = 0,1,\ldots,N-1$.

This DFT–based resampling scheme can be applied to any fractional interpolation based on evenly spaced data. It fails on unevenly spaced data.

3

BASICS OF ANTENNA THEORY

The antenna is a device that radiates or receives electromagnetic waves. In general, antennas can be classified as either single-element radiators or multiple (usually identical) radiating elements. Dipoles, horns, printed patch radiators, and reflectors are considered single-element radiators. Linear, circular, and area arrays consist of multiple (usually identical) radiating elements. Antennas play a very important role in radar-related applications. In this chapter, antenna theory is reviewed, starting with the Maxwell equation in Section 3.1. The infinitesimal dipole is described in Section 3.2, and the half-wavelength dipole is covered in Section 3.3. The 1D linear phase array is briefly reviewed in Section 3.4, and the 2D linear array is discussed in Section 3.5. The key antenna-related parameters are covered in Section 3.6. Some commonly used antennas, including the microstrip antenna, are reviewed in Section 3.7.

3.1 MAXWELL AND WAVE EQUATIONS

The classical Maxwell equations serve as the fundamental basis for all electromagnetic phenomena and are shown below:

$$
\begin{aligned}
\nabla \times \vec{E} &= -\frac{\partial \vec{B}}{\partial t}, \\
\nabla \times \vec{H} &= \vec{J} + \frac{\partial \vec{D}}{\partial t}, \\
\nabla \cdot \vec{D} &= \rho, \\
\nabla \cdot \vec{B} &= 0.
\end{aligned}
\tag{3.1}
$$

Digital Signal Processing Techniques and Applications in Radar Image Processing, by Bu-Chin Wang.

Since $\nabla \cdot \nabla \times \vec{A} = 0$ for any vector $\vec{A}$, the equation $\nabla \cdot \vec{B} = 0$ implies the existence of $\vec{A}$, such that $\vec{B} = \nabla \times \vec{A}$; that is

$$\nabla \times \vec{E} = -\frac{\partial \vec{B}}{\partial t} = -\nabla \times \frac{\partial \vec{A}}{\partial t}$$

or

$$\nabla \times \left(\vec{E} + \frac{\partial \vec{A}}{\partial t} \right) = 0.$$

The quantity $\vec{E} + (\partial \vec{A}/\partial t)$ is curlless and, mathematically speaking, $\nabla \times \nabla\varphi = 0$ for any scalar φ. Thus $\vec{E} + (\partial \vec{A}/\partial t)$ can be represented as the gradient of a scalar potential φ, namely, $\vec{E} + (\partial \vec{A}/\partial t) = -\nabla\varphi$. Therefore, the two Maxwell equations

$$\nabla \cdot \vec{B} = 0$$
$$\nabla \times \vec{E} = -\frac{\partial \vec{B}}{\partial t}$$

imply the existence of the magnetic and electric potentials $\vec{A}(\vec{r}, t)$ and $\varphi(\vec{r}, t)$, such that

$$\begin{aligned} \vec{E} &= -\nabla\varphi - \frac{\partial \vec{A}}{\partial t} \\ \vec{B} &= \nabla \times \vec{A}. \end{aligned} \tag{3.2}$$

The potentials $\vec{A}$ and φ are not uniquely defined and may be changed by adding constants to them. This freedom in selecting the potentials allows us to impose some convenient constraints between them. In discussing radiation problems, it is customary to impose the Lorenz condition:

$$\nabla \cdot \vec{A} + \frac{1}{c^2}\frac{\partial \varphi}{\partial t} = 0 \qquad \text{(Lorenz condition)}. \tag{3.3}$$

The remaining two Maxwell equations become, with $\vec{D} = \varepsilon\vec{E}$, $\vec{B} = \mu\vec{H}$ and $c^2 = 1/\mu\varepsilon$,

$$\begin{aligned} \nabla \cdot \vec{E} &= \frac{1}{\varepsilon}\rho \\ \nabla \times \vec{B} &= \mu\vec{J} + \frac{1}{c^2}\frac{\partial \vec{E}}{\partial t}. \end{aligned} \tag{3.4}$$

Substituting Eqs. (3.2) and (3.3) into Eq. (3.4), together with the equation $\nabla \cdot (\nabla \vec{A}) = \nabla^2 \vec{A}$ and $\nabla \times \nabla \times \vec{A} = \nabla\nabla \cdot \vec{A} - \nabla^2 \vec{A}$, one obtains the Maxwell equations as the equivalent wave equation forms, in terms of electric and magnetic potentials φ and $\vec{A}$:

$$\begin{aligned} \frac{1}{c^2}\frac{\partial^2 \varphi}{\partial t^2} - \nabla^2 \varphi &= \frac{1}{\varepsilon}\rho \\ \frac{1}{c^2}\frac{\partial^2 \vec{A}}{\partial t^2} - \nabla^2 \vec{A} &= \mu \vec{J} \end{aligned} \tag{3.5a}$$

It is assumed that sinusoidal time dependence exists for all quantities φ, ρ, $\vec{A}$, and $\vec{J}$:

$$\varphi(r,t) = \varphi(r)e^{j\omega t}, \quad \rho(r,t) = \rho(r)e^{j\omega t}, \quad \ldots .$$

With the sinusoidal time dependence characteristics, Eq. (3.5a) can be further simplified as

$$\begin{aligned} \nabla^2 \varphi + k^2 \varphi &= -\frac{1}{\varepsilon}\rho, \\ \nabla^2 \vec{A} + k^2 \vec{A} &= -\mu \vec{J}, \end{aligned} \tag{3.5b}$$

where $k = \omega/c$. The Maxwell equations will now be explored further.

3.1.1 Harmonic Time Dependence

Consider the following equation, where $\delta^{(3)}(\vec{r})$ is the delta function and serves as the excitation source:

$$\nabla^2 G + k^2 G = -\delta^{(3)}(\vec{r}). \tag{3.6}$$

The solution $G(\vec{r})$ of Eq. (3.6) can be considered as an impulse response of the system to have the form

$$G(\vec{r}) = \frac{e^{-jkr}}{4\pi r}. \tag{3.7}$$

Equation (3.6) states that $\delta^{(3)}(\vec{r})$ can be considered as the excitation source, while $G(\vec{r})$, from Eq. (3.7), is the impulse response of Eq. (3.6). With ρ/ε and $\mu\vec{J}$ as the excitation sources, from linear system theory, the solutions of φ and $\vec{A}$ in Eq. (3.5b)

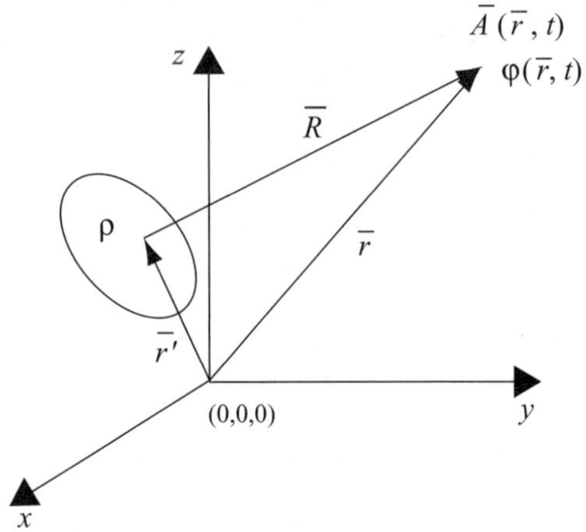

FIGURE 3.1 Potentials generated by current/charge distribution.

can be obtained as the convolution of excitation function with the impulse response function:

$$\varphi(\vec{r}) = \int_V \frac{1}{\varepsilon}\rho(\vec{r}\,')G(\vec{r} - \vec{r}\,')\,d^3\vec{r}\,'$$

$$\vec{A}(\vec{r}) = \int_V \mu J(\vec{r}\,')G(\vec{r} - \vec{r}\,')\,d^3\vec{r}\,'.$$

Figure 3.1 shows the geometry of magnetic and electric potentials $\vec{A}(\vec{r}, t)$ and $\varphi(\vec{r}, t)$, located at distance $\vec{r}$ and caused by the excitation source ρ located at distance $\vec{r}\,'$.

With $R = |\vec{r} - \vec{r}\,'|$, then

$$\begin{aligned} \varphi(\vec{r}) &= \int_V \frac{\rho(\vec{r}\,')e^{-jkR}}{4\pi\varepsilon R} d^3\vec{r}\,' \\ \vec{A}(\vec{r}) &= \int_V \frac{\mu\vec{J}(\vec{r}\,')e^{-jkR}}{4\pi R} d^3\vec{r}\,' \end{aligned} \tag{3.8}$$

Following replacement of $\partial/\partial t$ with $j\omega$, the Lorenz condition takes the form

$$\nabla \cdot \vec{A} + j\omega\mu\varepsilon\varphi = 0. \tag{3.9}$$

The electric and magnetic fields of Eq. (3.2) then become

$$\begin{aligned} \vec{E} &= -\nabla\varphi - j\omega\vec{A} \\ \vec{H} &= \frac{1}{\mu}\nabla \times \vec{A}. \end{aligned} \tag{3.10a}$$

From Eqs. (3.9) and (3.10a), with $\omega^2\mu\varepsilon = \omega^2/c^2 = k^2$, one obtains the $\vec{E}$ and $\vec{H}$ fields as

$$\vec{E} = \frac{1}{j\omega\mu\varepsilon}[\nabla(\nabla \cdot \vec{A}) + k^2\vec{A}],$$
$$\vec{H} = \frac{1}{\mu}\nabla \times \vec{A}. \tag{3.10b}$$

In the following sections, we will investigate the electric and magnetic fields due to various current sourcing dipoles. We will start from the point source, followed by the line source, and then the planar source. Fields as a function of distance will also be discussed.

3.2 RADIATION FROM AN INFINITESIMAL CURRENT DIPOLE

The infinitesimal dipole is a dipole whose length dl is much smaller than the wavelength λ of the excited wave, namely, $dl \ll \lambda$. An infinitesimal dipole is equivalent to a current element Idl with I as the current. The infinitesimal dipole, with $dl \approx 0$, serves as a point radiator or excitation source of the electric and magnetic fields. A point radiator is also called an *isotrope*. Physically, it does not exist. Yet, theoretically it serves as a reference for many types of antennas whose performance is expressed in terms of a basic radiator. It is used as a building block to solve many problems related to complicated antenna design. It radiates energy equally in all directions, and the radiation pattern in any plane is a circle.

Figure 3.2 is a graphical representation of radiation from an infinitesimal dipole (point radiator), where the point radiator is located at the origin of the coordinate.

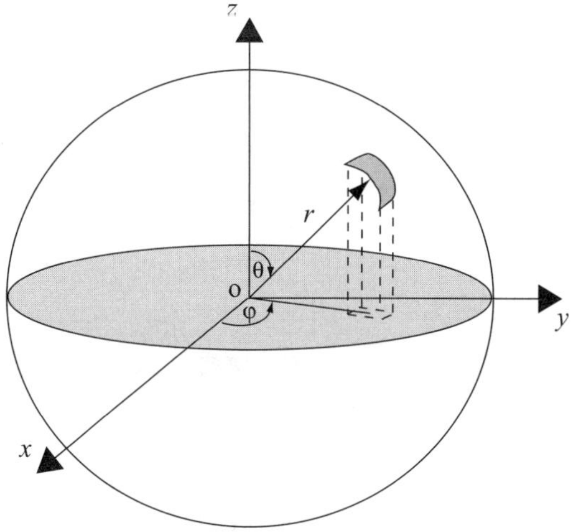

FIGURE 3.2 Radiation from a point radiator.

3.2.1 Magnetic Vector Potential Due to a Small but Finite Current Element

The magnetic vector potential $\vec{A}$ due to a current source and observed at point P, with distance R from the dipole, can be obtained from Eq. (3.8) as

$$\vec{A}(\vec{r}) = \int_V \frac{\mu \vec{J}(\vec{r}')e^{-jkR}}{4\pi R} d^3\vec{r}'.$$

For a linear current source, $\vec{J}(\vec{r}')d^3\vec{r}' = I\, d\vec{l}$, it follows that

$$\vec{A}(P) = \int_L \frac{\mu I e^{-jkR}}{4\pi R} d\vec{l}. \tag{3.11a}$$

Since current I is a constant along Δl, and the dipole is very small, the integration shown above can be approximated to its integrand. In addition, the distance $R \approx r$ will be replaced by r in both the exponential term and in the denominator. Therefore, with $d\vec{l} = \Delta l \hat{z}$,

$$\vec{A}(P) = \mu I \Delta l \frac{e^{-jkr}}{4\pi r}\hat{z},$$

which can be represented in spherical coordinates as

$$\begin{aligned} A_r &= A_z \cos\theta \\ &= \mu I\ \Delta l \frac{e^{-jkr}}{4\pi r} \cos\theta, \\ A_\theta &= -A_z \sin\theta \\ &= -\mu I\ \Delta l \frac{e^{-jkr}}{4\pi r} \sin\theta, \\ A_\varphi &= 0, \end{aligned} \tag{3.11b}$$

where θ is the angle between $\vec{A}(\vec{r})$ and the z axis.

3.2.2 Field Vectors Due to Small but Finite Current Radiation

From Eqs. (3.10b) and (3.11b), the fields $\vec{E}$, $\vec{H}$ can then be obtained as

$$\vec{H} = \frac{1}{\mu}\nabla \times \vec{A}$$

$$= \frac{1}{\mu r^2 \sin\theta}\begin{vmatrix} \hat{r} & r\hat{\theta} & r\sin\theta\hat{\varphi} \\ \dfrac{\partial}{\partial r} & \dfrac{\partial}{\partial \theta} & \dfrac{\partial}{\partial \varphi} \\ A_r & rA_\theta & r\sin\theta A_\varphi \end{vmatrix}$$

$$= \frac{1}{\mu r}\left[\frac{\partial}{\partial r}(r\cdot A_\theta) - \frac{\partial A_r}{\partial \theta}\right]\hat{\varphi}.$$

Therefore

$$H_\varphi = jk\cdot(I\Delta l)\cdot\sin\theta\cdot\left[1 + \frac{1}{jkr}\right]\frac{e^{-jkr}}{4\pi r} \tag{3.12a}$$

$$H_\theta = H_r = 0.$$

Since

$$\vec{E} = -j\omega\vec{A} - \frac{j}{\omega\mu\varepsilon}\nabla\nabla\cdot\vec{A},$$

it follows that

$$E_r = 2\eta(I\ \Delta l)\cos\theta\left[\frac{1}{r} + \frac{1}{jkr^2}\right]\frac{e^{-jkr}}{4\pi r},$$

$$E_\theta = j\eta k(I\ \Delta l)\sin\theta\left[1 + \frac{1}{jkr} - \frac{1}{(kr)^2}\right]\frac{e^{-jkr}}{4\pi r}, \tag{3.12b}$$

$$E_\varphi = 0,$$

where $\eta = \sqrt{\mu/\varepsilon}$ is the impedance of free space.

The radiation field vectors can then be tabulated as

$$H_\varphi = \frac{I\Delta l}{2\lambda}\cdot\sin\theta\cdot\left[j\frac{e^{-jkr}}{r} + \frac{e^{-jkr}}{kr^2}\right],$$

$$E_r = \frac{-\eta I\Delta l}{\lambda}\cdot\cos\theta\cdot\left[-\frac{e^{-jkr}}{kr^2} + j\frac{e^{-jkr}}{k^2r^3}\right],$$

$$E_\theta = \frac{-\eta I\Delta l}{2\lambda}\cdot\sin\theta\cdot\left[j\frac{e^{-jkr}}{r} + \frac{e^{-jkr}}{kr^2} - j\frac{e^{-jkr}}{k^2r^3}\right],$$

$$H_\theta = H_r = 0,$$

$$E_\varphi = 0.$$

3.2.3 Far-Field Region

The far-field region is defined as a region with $kr \gg 1$. By neglecting the second order term $1/r^2$ and keeping only the first order term $1/r$, one can obtain the far field of the infinitesimal dipole from Eqs. (3.12a) and (3.12b) as

$$\begin{aligned} H_\varphi &= jk\cdot(I\,\Delta l)\cdot\sin\theta\cdot\frac{e^{-jkr}}{4\pi r}, \\ E_\theta &= j\eta k(I\,\Delta l)\sin\theta\cdot\frac{e^{-jkr}}{4\pi r}, \\ E_r &= E_\varphi = 0, \\ H_\theta &= H_\varphi = 0. \end{aligned} \tag{3.13}$$

From these electric and magnetic fields, one can see that no radial component exists at the far field and that the E field is perpendicular to the H field.

As an example, consider a case where a time-varying standing wave with the current distribution along the dipole's length is the sinusoidal one:

$$\begin{aligned} I(z') &= I_0 \sin\left[k\left(\frac{l}{2}-z'\right)\right], & 0 \le z' \le l/2 \\ &= I_0 \sin\left[k\left(\frac{l}{2}+z'\right)\right], & -l/2 \le z' \le 0 \end{aligned} \tag{3.14}$$

The finite-length dipole can be considered as formed by an infinite number of infinitesimal dipoles with length dz'. Each infinitesimal dipole produces the elementary far field and can be described as follows, from Eq. (3.13):

$$\begin{aligned} dE_\theta &\approx j\eta k I_e(z')\frac{e^{-jkR}}{4\pi R}\sin\theta\cdot dz', \\ dH_\varphi &\approx jk I_e(z')\frac{e^{-jkR}}{4\pi R}\sin\theta\cdot dz', \\ dE_r &= dE_\varphi = 0, \\ dH_r &= dH_\theta = 0. \end{aligned}$$

Here, $I_e(z')$ is the current value of the current element at dz'. Using the far-field approximations, as shown in Fig. 3.3, then

$R \approx r$, for the amplitude factor.
$R \approx r - z'\cos\theta$, for the phase factor.

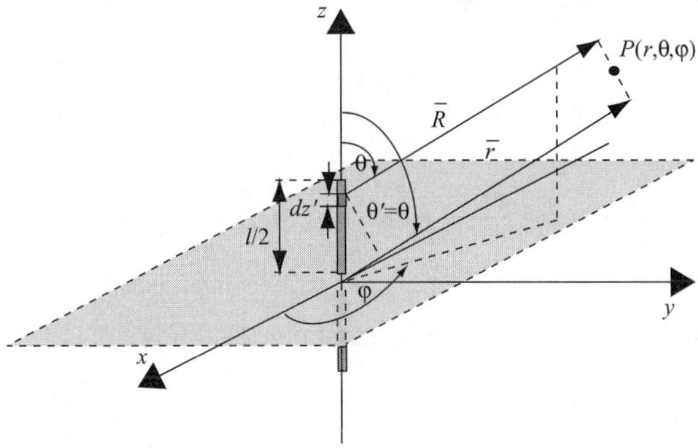

FIGURE 3.3 Far-field approximation of z-oriented dipole.

Therefore

$$
\begin{aligned}
dE_\theta &\approx j\eta k I_e(z')\frac{e^{-jk(r-z'\cos\theta)}}{4\pi r}\sin\theta\cdot dz' \\
&= j\eta k I_e(z')\frac{e^{-jkr}}{4\pi r}e^{jkz'\cos\theta}\cdot\sin\theta\cdot dz',
\end{aligned}
$$

and

$$
\begin{aligned}
E_\theta &= \int_{-l/2}^{l/2} dE_\theta \\
&\approx j\eta k\frac{e^{-jkr}}{4\pi r}\cdot\sin\theta\cdot\int_{-l/2}^{l/2} I_e(z')e^{jkz'\cos\theta}dz' \\
H_\varphi &= \int_{-l/2}^{l/2} dH_\varphi \\
&\approx jk\frac{e^{-jkr}}{4\pi r}\cdot\sin\theta\cdot\int_{-l/2}^{l/2} I_e(z')e^{jkz'\cos\theta}dz'.
\end{aligned}
$$

For current distribution as defined in Eq. (3.14), the integration factor expressed above becomes

$$
\begin{aligned}
&\int_{-l/2}^{l/2} I_e(z')e^{jkz'\cos\theta}dz' \\
&= I_0\left\{\int_{-l/2}^{0}\sin\left[k\left(\frac{l}{2}+z'\right)\right]e^{jkz'\cos\theta}dz' + \int_{0}^{l/2}\sin\left[k\left(\frac{l}{2}-z'\right)\right]e^{jkz'\cos\theta}dz'\right\}
\end{aligned}
$$

By utilizing the following two integrals,

$$\int e^{ax} \sin bx \, dx = \frac{e^{ax}}{a^2 + b^2} (a \sin bx - b \cos bx)$$

$$\int e^{ax} \cos bx \, dx = \frac{e^{ax}}{a^2 + b^2} (a \cos bx + b \sin bx),$$

one obtains the far field of the finite-length dipole as follows

$$\begin{aligned}
E_\theta &= j\eta I_0 \frac{e^{-jkr}}{2\pi r} \left[\frac{\cos\left(\frac{kl}{2}\cos\theta\right) - \cos\left(\frac{kl}{2}\right)}{\sin\theta} \right], \\
H_\varphi &= j I_0 \frac{e^{-jkr}}{2\pi r} \left[\frac{\cos\left(\frac{kl}{2}\cos\theta\right) - \cos\left(\frac{kl}{2}\right)}{\sin\theta} \right], \\
E_r &= E_\varphi = 0, \\
H_\theta &= H_\varphi = 0.
\end{aligned} \tag{3.15}$$

3.2.4 Summary of Radiation Fields

On the basis of the preceding discussion, we can divide the antenna radiation field into three regions, which are functions of wavelength λ and dimension of antenna D. The definition of these three regions are listed below.

Reactive Near-Field Region. This is the region immediately surrounding the antenna where the reactive field (stored energy – standing waves) is dominant. In this region, $r \ll \lambda$, fields with $1/r^3$ term will prevail and remain. This region is defined by a sphere with radius $R_1 = 0.62\sqrt{D^3/\lambda}$.

Near-Field (Fresnel) Region. This is the region between the reactive near field and the far field where the radiation fields predominate and the field distribution is dependent on the distance from the antenna. In this region, $r < \lambda$, fields with $1/r^2$ term will prevail and remain. This region is defined by a sphere with radius $R_2 = 2D^2/\lambda$.

Far-Field (Fraunhofer) Region. In this region, the field distribution is essentially independent of the distance from the antenna. In this region, $r \gg \lambda$, fields with the $1/r$ term will prevail and remain. This region is defined by a sphere with radius $R_3 > R_2$.

3.3 RADIATION FROM A HALF-WAVELENGTH DIPOLE

For current distributed along a dipole with length $l = \lambda/2$, the electric and magnetic fields can be obtained from Eq. (3.15) as follows:

$$\begin{aligned} E_\theta &= j\eta I_0 \frac{e^{-jkr}}{2\pi r} \frac{\cos\left(\frac{\pi}{2}\cos\theta\right)}{\sin\theta}, \\ H_\varphi &= \frac{E_\theta}{\eta}, \\ E_r &= E_\varphi = 0, \\ H_\theta &= H_\varphi = 0. \end{aligned} \tag{3.16}$$

The received power density, or far-field power density, is defined as

$$W(\theta, \varphi) = \frac{\left|\vec{E}(\theta, \varphi)\right|^2}{\eta}, \tag{3.17}$$

where η is the impedance of the free space, or $\eta = |E/H|$. For a half-λ dipole, this is

$$W(\theta, \varphi) = \frac{\eta I_0^2}{(2\pi r)^2} \left[\frac{\cos\left(\frac{\pi}{2}\cos\theta\right)}{\sin\theta}\right]^2.$$

The normalized power pattern is the ratio of received power density to the maximum power density:

$$\begin{aligned} W_n(\theta, \varphi) &= \left[\frac{\cos\left(\frac{\pi}{2}\cos\theta\right)}{\sin\theta}\right]^2 \\ &\approx \sin^3\theta. \end{aligned} \tag{3.18}$$

The radiation pattern can be either a normalized field $|\vec{E}(\theta, \phi)|$ or a normalized power density $|\vec{E}(\theta, \varphi)|^2/\eta$. The radiation pattern can be plotted as either a 2D or 3D pattern. Figure 3.4 displays the radiation pattern of a half-wavelength dipole in both 3D and 2D views. Figure 3.4a is a 3D cross-sectional view. The left side of Fig. 3.4b shows the azimuth pattern, looking down from the z axis into the 3D pattern. The right side of Fig. 3.4b shows the elevation pattern, looking from the x axis into the 3D pattern.

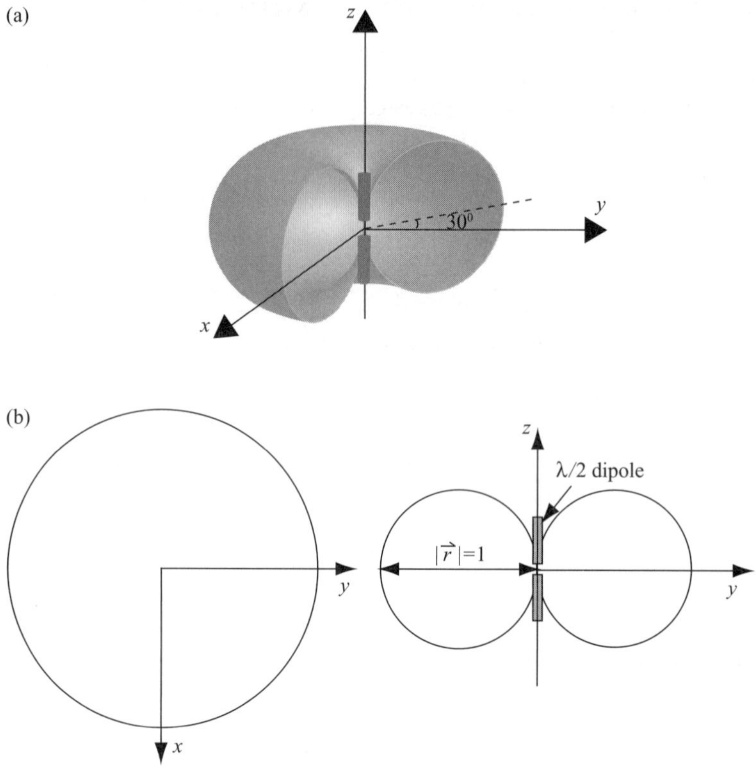

FIGURE 3.4 Radiation pattern of a half-wavelength dipole.

3.4 RADIATION FROM A LINEAR ARRAY

A single antenna with a given radiation pattern may be grouped and arranged to yield a different radiation pattern. There are several design variables that can be modified to achieve the desired antenna radiation pattern. These variables consist of the array shape (linear, planar, and circular), element spacing, element-exciting amplitude and phase, and element radiation pattern.

Linear array is one of the patterns that utilize identical radiating elements arranged along a straight line. Figure 3.5 is an example of a 10-element linear array, with the infinitesimal dipole serving as the radiating element. Each dipole is equally spaced with distance d along the x axis. Assuming that the current amplitudes and phases are all equal for the array elements, the far field at observation point $P(r,\theta,0)$ due to the infinitesimal dipole located at the origin is, from Eq. (3.13), expressed as follows:

$$\begin{aligned} E_{\theta 0} &= j\eta k(I\,\Delta l)\sin\theta \cdot \frac{e^{-jkr}}{4\pi r} \\ &= E_0. \end{aligned}$$

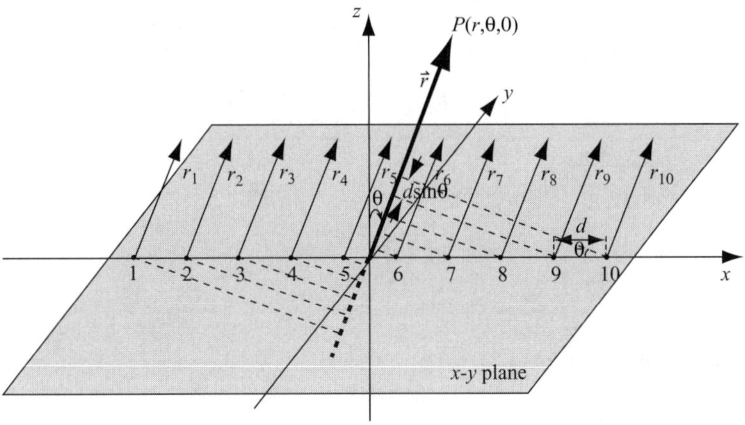

FIGURE 3.5 A 10-element linear array.

Notice that $E_{\theta 0} = E_0$ is a fictitious field with no radiating element at the origin and serves as a reference field. Notice also that θ is the observation angle between the receiver (point *P*) to origin vector $\vec{r}$ and the *z* axis. The far field due to the individual element *n* of the 10-element array is

$$E_{\theta n} = j\eta k(I\,\Delta l)\sin\theta \cdot \frac{e^{-jkr_n}}{4\pi r_n},$$

where the parameter r_n can be approximated as $r_n \approx r$ for the magnitude term. For the phase term, r_n can be replaced with $r_n \approx r - (n - 5.5)\, d \sin\theta$. In terms of reference signal E_0, the immediately preceding equation becomes

$$\begin{aligned} E_{\theta n} &\approx j\eta k\,(I\,\Delta l)\sin\theta \cdot \frac{e^{-jk[r-(n-5.5)d\cdot\sin\theta]}}{4\pi r} \\ &= E_0 e^{j(n-5.5)kd\cdot\sin\theta}. \end{aligned}$$

The array's far field can then be obtained as

$$\begin{aligned} E_\theta &= E_{\theta 1} + E_{\theta 2} + \ldots + E_{\theta 10} \\ &= E_0\left(e^{-j4.5kd\cdot\sin\theta} + e^{-j3.5kd\cdot\sin\theta} + \ldots + e^{j3.5kd\cdot\sin\theta} + e^{j4.5kd\cdot\sin\theta}\right). \end{aligned}$$

Let

$$\psi = kd\ \sin\theta;$$

then

$$
\begin{aligned}
E_\theta &= E_0 e^{-j4.5\psi} \sum_{n=1}^{10} e^{j(n-1)\psi} \\
&= E_0 e^{-j4.5\psi} \frac{1 - e^{j10\psi}}{1 - e^{j\psi}} \\
&= E_0 \frac{\sin\,(5\psi)}{\sin\,(\psi/2)}.
\end{aligned}
$$

In general, if the number of radiating elements in the array is N, then the total field observed at far distance can be derived as

$$
\begin{aligned}
E_\theta &= E_0 \frac{\sin(N\psi/2)}{\sin(\psi/2)} \\
&= F_\theta(\theta) F_A(\psi).
\end{aligned}
\tag{3.19}
$$

Here $F_\theta(\theta) = E_0 = j\eta k(I\,\Delta l)\sin\theta \cdot (e^{-jkr}/4\pi r)$ is the element factor, and

$$
F_A\,(\psi) = \frac{\sin\,(N\psi/2)}{\sin\,(\psi/2)}
$$

is the array factor.

The array factor $F_A(\psi)$ has its maximum value equal to N, which occurs at the peak of the mainlobe. Therefore, the normalized array factor is

$$
F_A(\psi) = \frac{1}{N} \frac{\sin(N\psi/2)}{\sin(\psi/2)}.
\tag{3.20a}
$$

Figure 3.6 displays the magnitude of normalized array factor $F_A(\psi)$ in terms of degrees for $N = 10$.

As can be seen in Fig. 3.6, there are one mainlobe and eight sidelobes for the angle ψ ranging from 0° to 360°. The nulls of the array function can be computed by determining the zeros of the numerator term where the denominator is not simultaneously zero:

$$
\sin(N\psi/2) = 0 \quad \text{or} \quad N\psi/2 = \pm n\pi.
$$

Therefore

$$
\psi = kd \cdot \sin\theta_n = \pm\frac{2n\pi}{N},
$$

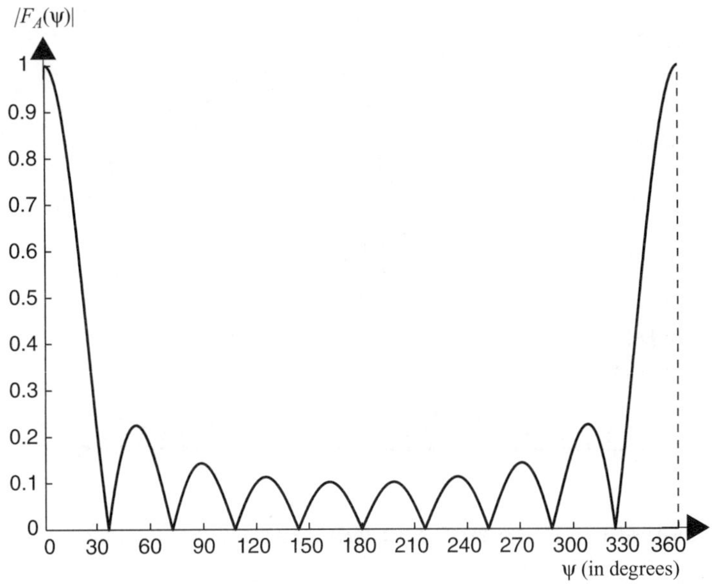

FIGURE 3.6 Normalized linear antenna array factor for $N = 10$.

and

$$\theta_n = \sin^{-1}\left[\pm\frac{2\pi n}{N}\cdot\frac{\lambda}{2\pi d}\right], \qquad n = 1,2,\ldots \quad \text{but} \quad n \neq 0, N, 2N, \ldots. \tag{3.20b}$$

The peaks of the array factor can be found by determining the zeros of the numerator term where the denominator is simultaneously zero:

$$\begin{aligned}\sin(\psi/2) = 0 \quad \text{or} \quad \psi/2 &= \pm m\pi,\\ \psi &= kd\cdot\sin\theta_m\\ &= \pm 2m\pi.\end{aligned}$$

Therefore

$$\theta_m = \sin^{-1}\left[\pm\frac{m\lambda}{d}\right], \qquad m = 0,1,2,\ldots. \tag{3.20c}$$

The phase of the linear array elements, based on the infinitesimal dipole, may be chosen such that the mainlobe of the array pattern lies along the array axis (x axis in Fig. 3.5) or normal to the array axis (z axis in Fig. 3.5). The former is known as the *end-fire array*, while the latter is called the *broadside array*. For finite-length dipoles with the same configuration as shown in Fig. 3.5, each dipole is equally spaced along the x axis but is oriented along the y axis direction in the end-fire array

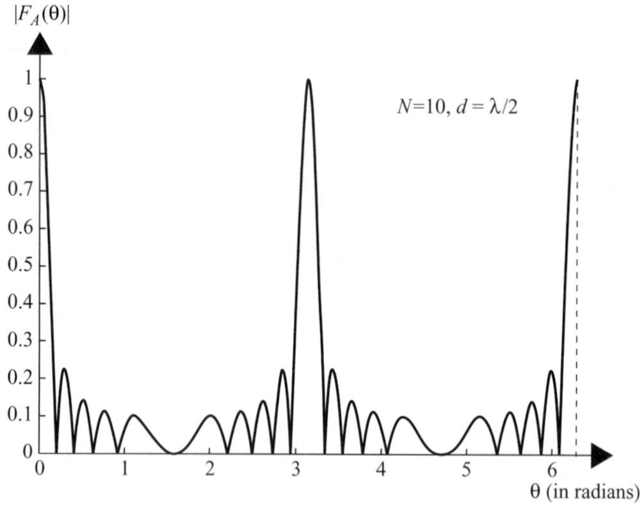

FIGURE 3.7 Normalized linear antenna array factor for $N = 10$, $d = \lambda/2$.

for the mainlobe along the x axis, and along the z axis direction in the broadside array for mainlobe along the y axis. The dipoles in the end-fire array are usually closer together than they are in the broadside array, and the currents are usually 180° out of phase with each other. The dipoles in the broadside array usually are driven by currents with the same phase.

From Eq. (3.20), one can observe that by varying the angle θ, the received far field will change. Figure 3.7 displays the plot of normalized $F_A(\theta)$ versus θ for a broadside array with radiating elements separated by $d = \lambda/2$ and $N = 10$. As can be seen, the total-field array factor has the maximum value at $\theta = 0$ and π, which is the angle when the receiver is perpendicular to the antenna array.

3.4.1 Power Radiation Pattern from a Linear Array

The normalized power radiation pattern of a linear array is defined as the square of the normalized array factor. According to Eq. (3.20a), the power radiation pattern,

$$G_a(\theta) = |F_A(\theta)|^2$$

$$= \frac{1}{N^2} \frac{\sin^2\left[\dfrac{N\pi d}{\lambda}\sin\theta\right]}{\sin^2\left[\dfrac{\pi d}{\lambda}\sin\theta\right]}. \tag{3.21a}$$

Since $(N\pi d/\lambda)\sin\theta \gg (\pi d/\lambda)\sin\theta$ for large N, with $Nd = L$ (the length of the antenna) and $\sin\theta \approx \theta$ for small angle θ, the radiation pattern becomes

$$G_a(\theta) \approx \frac{\sin^2\left[\dfrac{\pi L}{\lambda}\sin\theta\right]}{\left[\dfrac{\pi L}{\lambda}\sin\theta\right]^2}. \tag{3.21b}$$

Let $[(L\sin\theta)/\lambda] = x$; then

$$\begin{aligned} G_a(\theta) &\approx \frac{\sin^2[\pi x]}{[\pi x]^2} \\ &= \operatorname{sinc}^2 x. \end{aligned} \tag{3.21c}$$

The half-power beamwidth $\theta_{3\text{dB}}$ is computed by setting $G_a(\theta = \theta_{3\text{dB}}) = 0.5$:

$$\sin^2\left[\frac{\pi L}{\lambda}\sin\theta_{3\text{dB}}\right] = 0.5\left[\frac{\pi L}{\lambda}\sin\theta_{3\text{dB}}\right]^2. \tag{3.21d}$$

This equation can be solved graphically, resulting in

$$\frac{\pi L}{\lambda}\sin\theta_{3\text{dB}} = \pm 1.39 \quad \text{in radians}$$

or

$$\sin\theta_{3\text{dB}} = \pm\frac{0.44\,\lambda}{L}.$$

If $\theta_{3\text{dB}}$ is small, then $\sin\theta_{3\text{dB}} \approx \theta_{3\text{dB}}$. Therefore, the one-way half-power beamwidth can be computed as follows:

$$\begin{aligned} \theta_{3\text{dB}} &= 0.88\,\frac{\lambda}{L} \\ &\approx \frac{\lambda}{L} \quad \text{in radians} \\ &= 50.5\,\frac{\lambda}{L} \quad \text{in degrees.} \end{aligned} \tag{3.22}$$

Equation (3.22) serves as a very important parameter used in radar detection and radar image processing.

For a N-element linear array with $d = \lambda/2$, $Nd = L$ implies $\lambda/L = 2/N$, and

$$\sin\theta_{3\text{dB}} = \pm\frac{0.88}{N}$$

With small-angle approximation, one obtains

$$\theta_{3\text{dB}} \approx \frac{1.76}{N} \quad \text{in radians.}$$

$$\theta_{3\text{dB}} \approx \frac{102}{N} \quad \text{in degrees.}$$

If N is sufficiently large, the antenna can be considered as a uniformly illuminated aperture.

The peak of the first sidelobe of $G_a(\theta)$, from Eq. (3.21c), can be computed by finding the first sidelobe peak of sinc(x). Since sinc(x) equals zeros at $x =$ 1,2,3, ..., the peak of the first sidelobe of $G_a(\theta)$ appears at $x = 1.5$; therefore

$$G_a(\theta) = \frac{[\sin(3\pi/2)]^2}{[(3\pi/2)]^2} = \frac{4}{9\pi^2}$$

or

$$G_a(\theta)\,|_{\text{dB}} = 10\log\left(\frac{4}{9\pi^2}\right) = -13.2\text{ dB}.$$

3.5 POWER RADIATION PATTERN FROM A 2D RECTANGULAR ARRAY

The 2D power radiation pattern may be approximated as the product of the patterns of the two planes that contain the principal axes of the 1D antenna, namely

$$G(\theta,\varphi) = G(\theta)G(\varphi), = \frac{\sin^2\left[\dfrac{\pi L}{\lambda}\sin\theta\right]\sin^2\left[\dfrac{\pi W}{\lambda}\sin\varphi\right]}{\left[\dfrac{\pi L}{\lambda}\sin\theta\right]^2\left[\dfrac{\pi W}{\lambda}\sin\varphi\right]^2}, \tag{3.23}$$

where L and W are the lengths of the two arrays in the θ and φ directions, respectively.

The half-power beamwidth of $G(\theta,\varphi)$ in the θ and φ directions are, from Eq. (3.22), $\theta_{3\text{dB}} \approx \lambda/L$ and $\varphi_{3\text{dB}} \approx \lambda/W$, respectively.

3.6 FUNDAMENTALS OF ANTENNA PARAMETERS

The antenna parameters describe the antenna performance with respect to space distribution of the radiated energy. Key antenna parameters are radiation beamwidth, directivity, gain, antenna efficiency, antenna impedance, and radiation efficiency. A brief description of these parameters now follows.

3.6.1 Radiation Beamwidth

The radiation beamwidth can be defined in two ways, namely, the half-power beamwidth and the first-null beamwidth. *The half-power beamwidth* (HPBW) is the angle between two vectors where the electric field intensity equals 0.707 of the maximum mainbeam field intensity. *The first-null beamwidth* (FNBW) is the angular difference between the first null on each side of the mainbeam. In general, FNBW is about twice the size of the HPBW. Radiation beamwidth can be defined in terms of either power radiation pattern or field pattern. The field pattern (the absolute value of array factor) will be used for illustration.

A graphical representation of the radiation beamwidth, together with the relative power level against the angle θ, is shown in both rectangular and polar formats in Figs. 3.8 and 3.9, respectively. A linear array with 6 dipoles, located at half wavelength from each other, is used here for illustration. Figure 3.8 displays the field pattern in rectangular format. Here the mainlobe with normalized field = 1 is

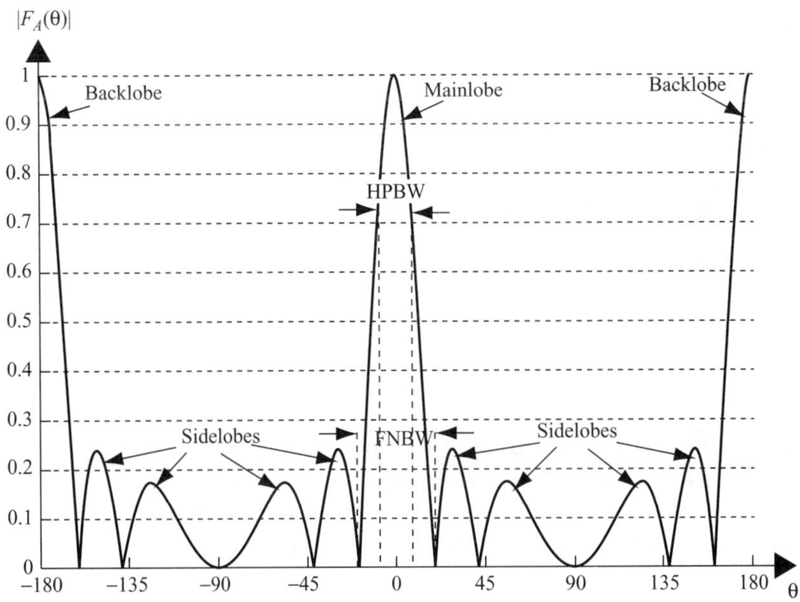

FIGURE 3.8 Field pattern in rectangular format for $N = 6$.

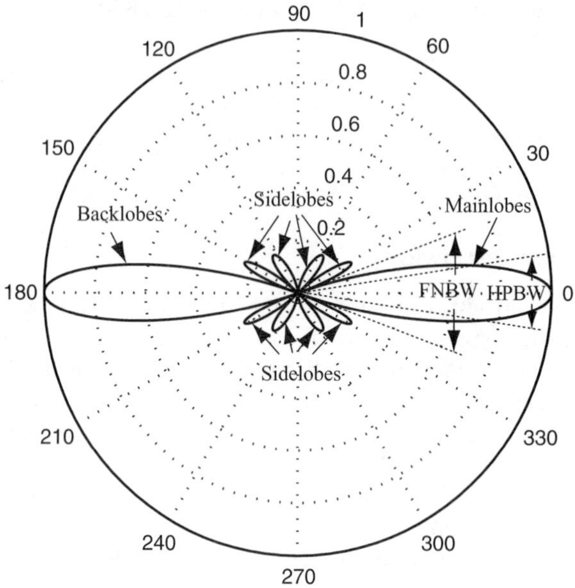

FIGURE 3.9 Field pattern in polar format for $N = 6$.

located at angle $\theta = 0°$. A backlobe with normalized field $= 1$ is located at angle $\theta = 180°$. In between, there are 4 sidelobes symmetrically located at angles $\theta = 0°$ to $\theta = 180°$, and $\theta = 180°$ to $\theta = 360°$, respectively. The HPBW and FNBW are also shown around the mainlobe at $\theta = 0°$.

Figure 3.9 shows the same field pattern of Fig. 3.8 with relative field level plotted against the angle θ in polar format. The outside circle, shown as a solid line, corresponds to the field level of 1. The inner circles, shown as dashed lines, correspond to various levels of the field.

3.6.2 Solid Angle, Power Density, and Radiation Intensity

Given a sphere of radius r (see Fig. 3.10), the solid angle Ω is defined to be a spherical surface S_Ω divided by r^2:

$$\Omega = \frac{S_\Omega}{r^2}. \tag{3.24}$$

The definition of one steradian (1 sr) is the solid angle with its vertex at the center of a sphere with radius r, which is subtended by a spherical surface area equal to that of a square with each side of length r. There are 4π steradians in a sphere.

The infinitesimal area ds on a surface of a sphere, with radius r, in spherical coordinates is

$$ds = r^2 \sin\theta \, d\theta \, d\varphi.$$

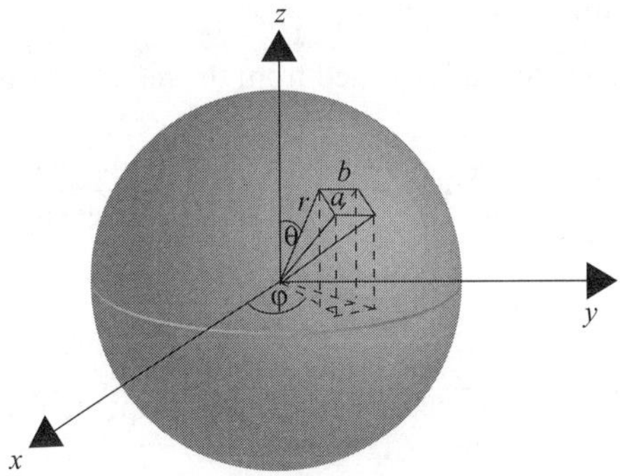

FIGURE 3.10 Graphical representation of a solid angle.

Therefore

$$d\Omega = \frac{ds}{r^2} = \sin\theta \; d\theta \; d\varphi.$$

Letting P_T be the radiator transmitting power, the power density P_d is defined as

$$P_d = \frac{dP_T}{ds}. \tag{3.25}$$

The radiation intensity U is defined as the power per solid angle:

$$U = \frac{dP_T}{d\Omega}. \tag{3.26}$$

For an ideal isotropic point radiator, the power densities P_d at all points on the surface of the sphere of radius r are equal, and can be expressed as

$$P_d = \frac{P_T}{4\pi r^2}. \tag{3.27}$$

The radiation intensity on any surface surrounding the sphere is given by

$$U = \frac{P_T}{4\pi}. \tag{3.28}$$

The relationship between power density and radiation intensity is then

$$U = r^2 P_d. \tag{3.29}$$

3.6.3 Directivity and Gain

The *directivity* D of an antenna is defined from the radiation pattern as

$$D = \frac{U_{\max}}{U_0} = \frac{\text{maximum radiation intensity}}{\text{average radiation intensity}}$$

or

$$\begin{aligned} D &= \frac{\text{maximum power density}}{\text{average density from an isotropic radiator}} \\ &= \frac{P_{d,\max}}{P_T/4\pi r^2}. \end{aligned} \tag{3.30}$$

The *antenna gain* is defined as the ratio of the maximum radiation intensity $U_{\max}$ to the maximum radiation intensity of a reference antenna U_{ref}, which has the same power input to the measured antenna:

$$G = \frac{U_{\max}}{U_{\text{ref}}}. \tag{3.31}$$

Quite often, the terms *directivity* and the *gain* are used interchangeably. The difference is that the directivity ignores antenna losses, which are quite small for most classes of antennas.

The radiation intensity averaged over all directions is equal to the total power radiated by the antenna divided by $4\pi r^2$. The radiation intensity averaged in a given direction is equal to $r^2\theta_{3\text{dB}}\varphi_{3\text{dB}}$. The antenna gain is then equal to the area of the isotropic sphere divided by the sector (cross section) area:

$$\begin{aligned} G &= \frac{\text{area of sphere}}{\text{area of antenna pattern}} \\ &= \frac{4\pi}{\varphi_{3\text{dB}}\theta_{3\text{dB}}}, \end{aligned} \tag{3.32}$$

where $\varphi_{3\text{dB}}$ and $\theta_{3\text{dB}}$ are the azimuth and elevation beamwidths in radians, respectively.

3.6.4 Antenna Impedance

An antenna is seen from the generator as a load with impedance Z_A represented as

$$Z_A = (R_L + R_{\text{rad}}) + jX_A,$$

were R_L represents the ohmic resistance where energy is lost (or transformed into heat) in the antenna structure and R_{rad} represents the radiation resistance where energy radiates into space. X_A is frequency dependent and represents energy stored in electric and magnetic fields.

The maximum power is delivered to or from the antenna when the impedance of an antenna is matched to the impedance of an equivalent generator or load. Impedance-matching devices, such as a balum or matching transformer, are used to serve this purpose.

3.6.5 Antenna Efficiency

The antenna power radiated into air is represented as

$$P_{\text{rad}} = \frac{I^2 R_{\text{rad}}}{2}.$$

The antenna power loss is represented as

$$P_{\text{loss}} = \frac{I^2 R_L}{2}.$$

The antenna efficiency η_a is a measure of the power dissipated in the ohmic losses of the antenna. It is the ratio of the total power radiated from the antenna to the power delivered to the antenna from the transmission line. It can be expressed as follows:

$$\eta_a = \frac{P_{\text{rad}}}{P_{\text{loss}} + P_{\text{rad}}}. \tag{3.33}$$

3.6.6 Effective Area and Antenna Gain

The antenna has an effective area A_e by which the power density is multiplied to obtain power delivered to the load. The effective area is also referred to as the *capture area* of the antenna. The effective area is related to the physical size and shape of the antenna.

From Eqs. (3.32) and (3.22), the antenna gain is expressed as

$$G = \frac{4\pi}{\varphi_{3\text{dB}}\theta_{3\text{dB}}},$$

and

$$\theta_{3\text{dB}} = \frac{\lambda}{L},$$

$$\theta_{3\text{dB}} = \frac{\lambda}{W}.$$

Therefore

$$G = \frac{4\pi}{(\lambda/L)\cdot(\lambda/W)}$$
$$= \frac{4\pi A_e}{\lambda^2},$$

where $A_e = LW$.

The effective area and antenna gain are therefore related as

$$A_e = \frac{\lambda^2}{4\pi} G. \tag{3.34}$$

The following examples are used to illustrate the above mentioned antenna parameters.

Example 3.1 By approximating the antenna pattern as a rectangular area, with $\theta_{3\text{dB}} = \varphi_{3\text{dB}} = 1^\circ$ and an average efficiency of 70%, the antenna gain G can be computed as follows.

From Eq. (3.32), one obtains

$$G = \frac{\text{area of sphere}}{\text{area of antenna pattern}}$$

The area of the antenna radiation pattern can be computed as follows, from Fig. 3.11:

$$a = r\ \sin\theta_{3\text{dB}},$$
$$b = r\ \sin\varphi_{3\text{dB}},$$
$$\text{Area} = ab,$$
$$= r^2 \sin\theta_{3\text{dB}} \sin\varphi_{3\text{dB}}.$$

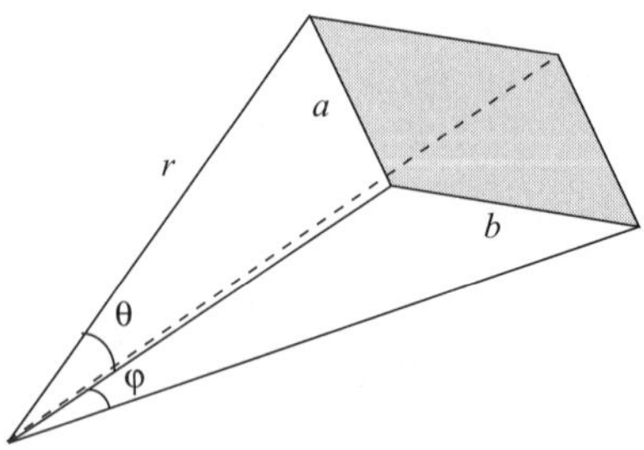

FIGURE 3.11 Antenna radiation pattern approximated as a rectangular area.

Therefore

$$G = \frac{4\pi r^2}{r^2 \sin\theta_{3dB} \sin\varphi_{3dB}}$$
$$= \frac{4\pi}{\sin\theta_{3dB} \sin\varphi_{3dB}}.$$

For a small angle θ, $\sin\theta \approx \theta$ and $\sin\varphi \approx \varphi$. Therefore

$$G \approx \frac{4\pi}{\varphi_{3dB}\theta_{3dB}} \quad \text{in radians}$$
$$= \frac{41{,}253}{\varphi_{3dB}\theta_{3dB}} \quad \text{in degrees}$$
$$= 10\log_{10}\frac{41{,}253}{\varphi_{3dB}\theta_{3dB}} \quad \text{in decibels.}$$

Given $\theta_{3dB} = \varphi_{3dB} = 1^\circ$, one obtains

$$G = 41{,}253 \text{ in degrees,}$$
$$G_{dB} = 10\log_{10} 41{,}253$$
$$= 46.2 \text{ dB.}$$

For antenna efficiency equal to 70%, one obtains

$$G = 0.7(41{,}253)$$
$$= 28{,}877 \text{ in degrees,}$$
$$G_{dB} = 44.6 \text{ dB.}$$

Example 3.2 By approximating the antenna radiation pattern as an elliptical area with $\theta_{3dB} = \varphi_{3dB} = 1^\circ$, and an average efficiency of 55%, the antenna gain G can be computed using Eq. (3.32):

$$G = \frac{\text{area of aphere}}{\text{area of antenna pattern}}.$$

From Fig. 3.12, the area of ellipse can be computed as

$$\text{Area of elliptical area} = \pi ab$$
$$= \pi\left(\frac{r\sin\theta_{3dB}}{2}\frac{r\sin\varphi_{3dB}}{2}\right)$$
$$= (\pi r^2 \sin\theta_{3dB}\sin\varphi_{3dB})/4.$$

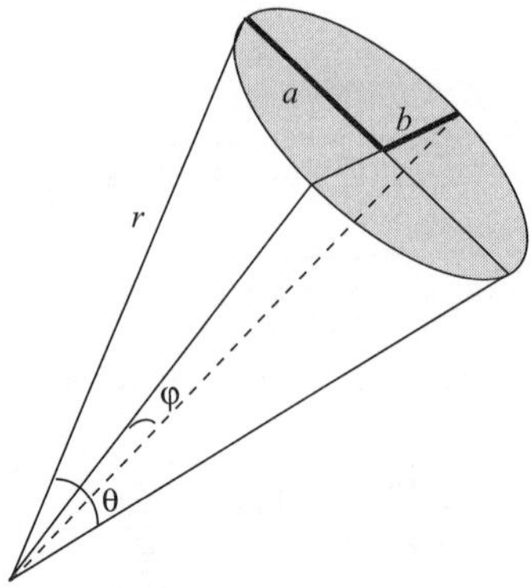

FIGURE 3.12 Antenna radiation pattern approximated as an elliptical area.

The antenna gain is then computed as

$$G = \frac{4\pi r^2}{(\pi r^2 \sin\theta_{3\text{dB}} \sin\varphi_{3\text{dB}})/4}$$
$$= \frac{16}{\sin\theta_{3\text{dB}} \sin\varphi_{3\text{dB}}}.$$

For small values of $\theta_{3\text{dB}}$, one can approximate $\sin\theta_{3\text{dB}} \approx \theta_{3\text{dB}}$ and $\sin\varphi_{3\text{dB}} \approx \varphi_{3\text{dB}}$. Therefore

$$G \approx \frac{16}{\theta_{3\text{dB}}\phi_{3\text{dB}}} \quad \text{in radians}$$
$$= \frac{52{,}525}{\theta_{3\text{dB}}\phi_{3\text{dB}}} \quad \text{in degrees}$$
$$= 10\log_{10}\frac{52{,}525}{\theta_{3\text{dB}}\phi_{3\text{dB}}} \quad \text{in decibels.}$$

Given the values $\theta_{3\text{dB}} = \varphi_{3\text{dB}} = 1^\circ$, one obtains

$$G = 52{,}525 \text{ in degrees}$$
$$G_{\text{dB}} = 10\log_{10} 52{,}525$$
$$= 47.2 \text{ dB}.$$

With efficiency equal to 55%, the antenna gain becomes

$$G = 0.55(52{,}525)$$
$$= 28{,}888 \text{ in degrees.}$$
$$G_{\text{dB}} = 44.6 \text{ dB}.$$

3.6.7 Polarization

The orientation of the fields in the wave is called *polarization*. For a dipole antenna, the original electric field is oriented along the axis of the antenna, and therefore the induced magnetic field will be perpendicular to both the electric field and the direction of travel. As the wave propagates outward, the electric and magnetic fields will remain perpendicular to each other; they will also be perpendicular to the direction of propagation (Fig. 3.13).

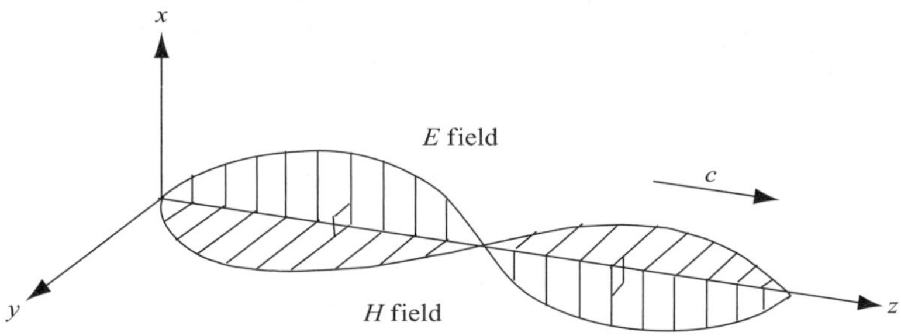

FIGURE 3.13 Polarized fields.

When the wave propagates outward and the field remains in a particular direction, as in the case of waves from the dipole antenna, the wave is considered to be linearly polarized. The field direction will be aligned to the antenna. A vertical antenna will create a vertically, linearly polarized electromagnetic wave. A receiving antenna that is aligned with the polarization will have the greatest sensitivity.

In the case of circular polarization, the electric field rotates as it travels along. If the rotation is clockwise as seen when looking in the direction of propagation, it is called *right-hand circular polarization* (RHCP). The other possibility is *left-hand circular polarization* (LHCP). Circular polarization is often used in satellite communications because it eliminates the need to match the receiving antenna to the orientation of the satellite's antenna.

3.7 COMMONLY USED ANTENNA GEOMETRIES

3.7.1 Single-Element Radiators

Many antennas are designed with a single-element radiator. A half-wavelength dipole antenna is an example of a single-element radiator. The circular loop antenna is another example of a wired single-element radiator. Although this type of antenna is easy to implement, its dimension is wavelength-dependent. Therefore, it has some restrictions on high-frequency applications—typically a few GHz.

An open-ended waveguide is inefficient in radiating energy because of a mismatch of impedance at its mouth. By flaring out the sides into a horn to match the intrinsic impedance of free space, the open-ended waveguide becomes a horn antenna. Horns flared in the E or H plane are called *sectoral horns* and if flared in both E and H plane, *pyramidal horns*.

The parabolic antenna normally consists of a parabolic reflector and a small feed antenna. The reflector is a metallic surface having a focal point. The small feed antenna is typically a half-wavelength dipole or a horn-based antenna.

Figure 3.14 displays the typical shapes of a circular loop antenna, linear polarized horn antenna, and the parabolic antenna. There are many other well-known antennas and their radiation patterns can easily be found in various books or journal publications on antennas, and will not be repeated here.

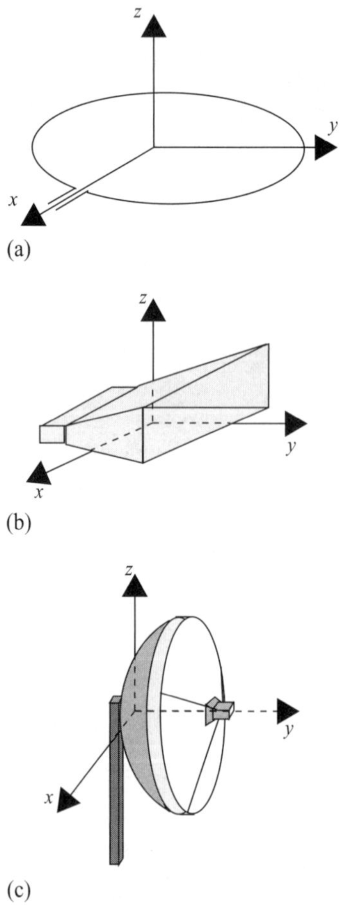

FIGURE 3.14 Popular antennas: (a) circular loop antenna; (b) linear polarized horn antenna; (c) parabolic antenna.

3.7.2 Microstrip Antennas and Antenna Array

There are many types of microstrip antennas; and the most common one is the printed patch antenna, consisting of thin metallic patches etched on a dielectric substrate, with a ground on the other side of the substrate. The dimensions of the patch are usually a fraction of the wavelength, while the thickness of dielectric substrate is normally less than a few percent of the wavelength. Because of its size, the microstrip antenna is usually used for high-frequency applications in the range from a few gigahertz (GHz) to 100 GHz.

The feeding signal to the microstrip patch antenna is an important issue in designing a high-performance antenna. There are four popular techniques for feeding the signal to a microstrip patch antenna: (1) the microstrip transmission line, (2) coaxial probe, (3) aperture coupling, and (4) proximity coupling. Methods 1 and 2 utilize a direct-contact approach, while methods 3 and 4 are noncontacting with the patch antenna. A printed patch antenna with a coaxial probe feed is shown in Fig. 3.15. Figure 3.15a is a 3D view of the circuit; Fig. 3.15b is a side view of the microstrip antenna. The size of the patch, the method of current feed, the thickness and characteristics of the substrate, and the geometry of the circuit will decide both the operating frequency and the radiation pattern of the microstrip antenna.

The radiation pattern of a microstrip antenna is normally shown in both the *E* and *H* planes. The *E* plane is the plane in parallel with the *E* field of the antenna, while the *H* plane is the plane perpendicular to the *E* field.

An antenna array consists of multiple (usually identical) radiating elements. Arranging the radiating elements in an array achieves unique radiation characteristics, which cannot be obtained through a single element. Phased array is the antenna array with multiple radiating elements. It changes the radiation pattern electronically

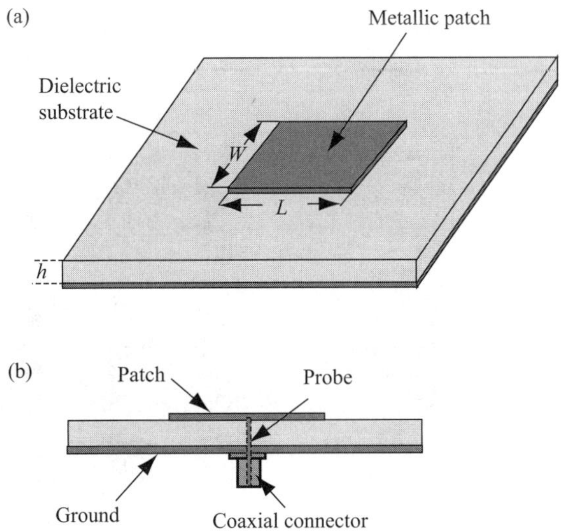

FIGURE 3.15 Printed patch antenna.

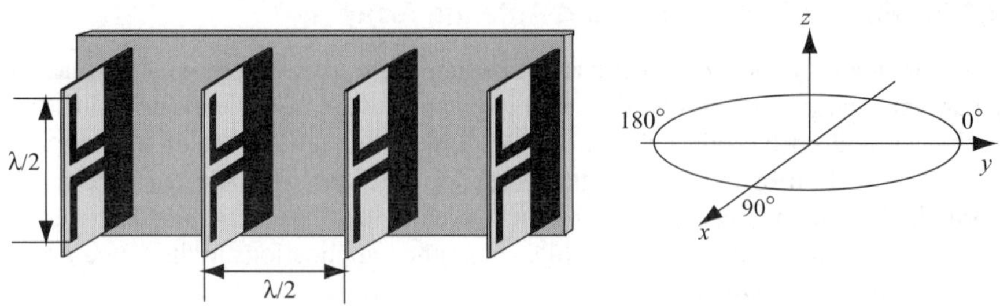

FIGURE 3.16 Configuration of a 4-dipole linear array.

by carefully choosing and controlling the phase shift and the magnitude of the signal fed to each radiating element.

A planar circuit-based dipole can be used to implement the phased antenna array. Figure 3.16 shows a simple configuration of a linear phased array based on four units of planar half-wavelength dipole, each spaced $\lambda/2$ apart from other units and fed with in-phase signal to each of the 4 dipoles. Sean Hum and Rob Randall designed such an array in 2003. [ENEL 619.50 Project: Self-Phased Array System; available at `http://www.enel.ucalgary.ca/People/Okoniewski/ENEL619-50/Projects2003/SPAS`; accessed on 1/10/07.] The metal (copper) strip–based dipole has one layer of substrate and an integrated planar balun. For $\lambda/2$ dipole array with $N = 4$, the antenna length $L = 4 \times \lambda\ /2 = 2\lambda$; therefore the 3-dB beamwidth can be calculated as $\theta = \lambda\ /L = 0.5$ in radians or 28.64°. A measured beamwidth of 28.6° along the x axis matches the theoretical result quite closely.

To design a 2D planar circuit-based antenna array, the signal line must be aligned with equal distance to provide an in-phase signal to each radiating element. One possible implementation of a 4 × 4 antenna with half-wavelength dipole is shown in Fig. 3.17. Figure 3.17a is a front–back view of the planar circuit-based dipole array; Fig. 3.17b shows the configuration of a 2D antenna array.

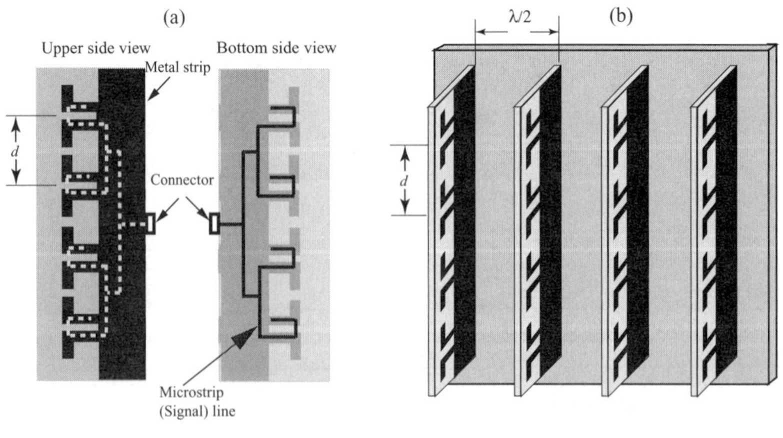

FIGURE 3.17 Half-wavelength dipole-based 2D antenna array.

4

FUNDAMENTALS OF RADAR

This chapter reviews basic theories of radar signal processing. Section 4.1 describes the principles of radar operation. Section 4.2 covers the basic configuration of a radar system. Section 4.3 and 4.4 discuss radar range equations, radar cross section, and clutter. Section 4.5 explains the Doppler frequency shifts from the wave propagation viewpoint. Radar resolution and ambiguity function are covered in Section 4.6.

4.1 PRINCIPLES OF RADAR OPERATION

By transmitting and receiving electromagnetic waves at regular time intervals, the radar is capable of identifying the distance from the target to the radar and/or the moving speed of the target. The detection and ranging functions can be accomplished either by timing the delay between transmission of a radio pulse and its subsequent return, or by calculating the difference in frequency between the transmitted signal and the echo signal. The pulse-based timing delay normally employs the amplitude modulation (AM) technique. Conversely, the frequency-difference-based ranging and detection method usually employs the linear frequency modulation scheme.

Consider the simple case of a radar system where a sequence of an AM pulse is transmitted. In this case, the information signal is a single pulse repeated at regular time intervals. The commonly used radar carrier-modulated pulse train is shown in Fig. 4.1.

In Fig. 4.1, the pulse repetition interval (PRI) is represented as T, which is the time interval between the start of one pulse and the start of the next pulse. The pulse

Digital Signal Processing Techniques and Applications in Radar Image Processing, by Bu-Chin Wang.

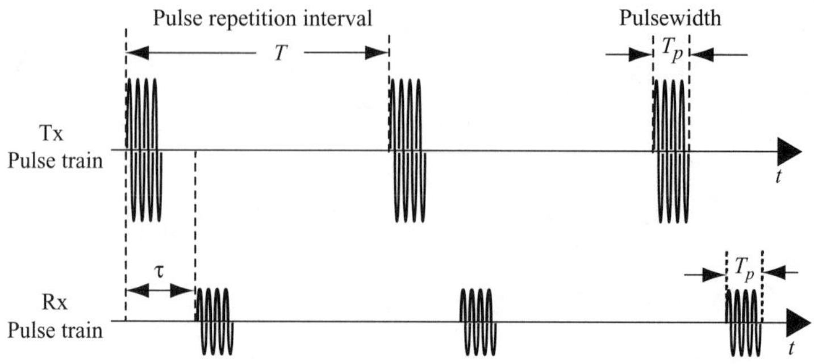

FIGURE 4.1 Transmitter and receiver pulse trains.

repetition frequency f_{PRF} is the inverse of PRI; that is, $f_{\mathrm{PRF}} = 1/\mathrm{PRI}$. The pulse repetition interval T is equal to the sum of the pulsewidth time T_p and the receiver listening time. If the target delay time is τ, then its range can be computed by the following equation with c as the speed of light or EM wave:

$$R = \frac{c\tau}{2}. \tag{4.1a}$$

The minimum and the maximum unambiguous detection range can be computed as follows:

$$\begin{aligned} R_{\min} &= \frac{c\,(\text{pulsewidth} + \text{radar recovery time})}{2} \\ &= \frac{c(T_p + t_r)}{2}, \\ R_{\max} &= \frac{cT}{2}. \end{aligned} \tag{4.1b}$$

Range ambiguity occurs when continuous pulse streams are sent and continuous echoes from the targets are received by the radar. Since all the pulses are identical, the radar cannot identify which echo originated from which pulse, and this gives rise to an ambiguity in the range measurement.

Figure 4.2 shows the problem of range ambiguity due to short pulse repetition period T, where a series of transmitted pulses and received echoes are displayed. In this case, target 2 exceeds the maximum unambiguous range. Echoes from target 1 show up after every transmitted pulse, but echoes from target 2 appear only after the second pulse was transmitted. For this type of pulsed radar waveform, the radar does not have sufficient data to decide which range corresponds to the correct target.

The most common method for resolving range ambiguity involves using multiple f_{PRF} values. Since the amount of range ambiguity depends on the f_{PRF}, it will change when the f_{PRF} is changed, as stated in Eq. (4.1b). Switching f_{PRF} gives the additional advantage of uncovering targets that are at the blind range for one or some of the f_{PRF} values. Such targets would be invisible at one or some f_{PRF} levels but clearly visible at the others.

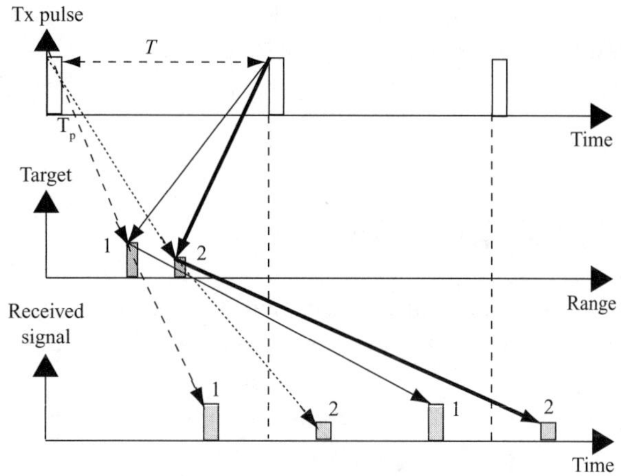

FIGURE 4.2 Pulse repetition period and range ambiguity.

Range resolution is the ability of a radar system to distinguish between two or more targets at different ranges. Two pulses with duration T_p and separated by a time interval Δt are resolvable if $\Delta t \geq T_p$. As shown in Fig. 4.3, the range resolution is then equal to one-half of the pulsewidth time, or

$$\text{Range resolution} = \frac{cT_p}{2}, \tag{4.1c}$$

where the $\frac{1}{2}$ factor comes from the round-trip traveling time of the radar pulse.

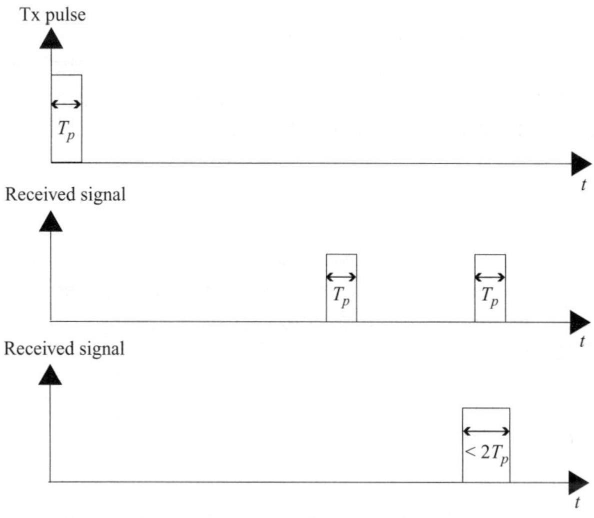

FIGURE 4.3 Range resolution.

The duty cycle of a radar pulse stream is defined as the ratio of the time ON to the time OFF of the transmitter. The equation for computing a duty cycle is

$$\text{Duty cycle} = \frac{T_p}{T}.$$

4.2 BASIC CONFIGURATION OF RADAR

A typical radar system can be functionally divided into six essential blocks as shown in Fig. 4.4. These blocks and their interrelationships are depicted in Fig. 4.4.

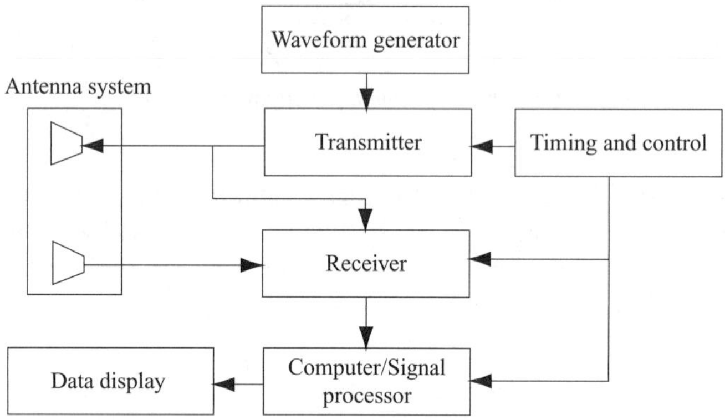

FIGURE 4.4 Block diagram of a radar system.

4.2.1 Waveform Generator

The waveform generator produces and controls the waveform to be modulated and transmitted by the transmitter. Depending on the application, amplitude modulation, linear frequency modulation, and stepped frequency modulation are three examples of signal modulation techniques provided by waveform generators.

4.2.2 Transmitter

The transmitter generates powerful pulses and/or waveforms of electromagnetic energy at precise time intervals and sends them to the antenna block. Radar transmission normally utilizes either resonating microwave tubes or solid-state devices for high-powered transmission.

4.2.3 Antenna System

The antenna system routes the energy from the transmitter and radiates it in a directional beam. It also picks up the returning echo energy and passes it to the receiver with minimal loss of energy. The antenna system includes a transmitting function and a receiving function. The transmitting function has an antenna, as well as a transmission line or waveguide from the transmitter to the antenna. The receiving function consists of the transmission line or waveguide from the antenna to the receiver. When the transmitter and receiver operate on separate antennas for signal transmission and receiving, the system is called a *bistatic radar system*. When the same antenna is used for both transmitter and receiver, it is called a *monostatic radar system*. In this situation, a *duplexer*, which is essentially a microwave circuit or electronic switch, is required. The duplexer connects the antenna to the transmitter and disconnects the antenna from the receiver for the duration of the transmitted pulse. After the transmitting pulse has ended, the duplexer immediately disconnects the transmitter and connects the receiver to the antenna.

4.2.4 Receiver

The receiver receives the target reflected energy from the antenna block. It then performs amplification, filtering, and demodulation on the received signal. The preprocessed signal is then sent to a computer or signal processor for further processing.

4.2.5 Computer/Signal Processor

The computer/signal processor performs complex mathematical computations on the demodulated signal to extract target velocity and/or range information.

4.2.6 Timing and Control

The timing-and-control block provides timing information to synchronize various signals and to control the operation of other radar components.

After the echo signal is processed, the targets' range and/or speed information are/is sent out for display on cathode ray tube–liquid crystal display (CRT/LCD)-based devices.

4.3 THE RADAR RANGE EQUATION

The power density at range R from the transmitter is defined as the transmitted power divided by the surface area $4\pi R^2$ of an imaginary sphere of radius R. Let P_T be the power transmitted by the antenna through an isotropic radiator (one that radiates uniformly in all directions); then the power density can be expressed as

$$\text{Power density from an isotropic antenna} = \frac{P_T}{4\pi R^2} \quad \text{in watts per square meter (W/m}^2\text{)}.$$

Let G_T be the radar transmitter antenna gain; the power density S_I received by the target from a transmitter with antenna gain G_T then becomes

$$S_I = \frac{P_T G_T}{4\pi R^2} \text{ in W/m}^2$$

Let σ be the target's scattering coefficient or radar cross section (which is in units of m^2 and will be discussed later); then the power density back at the input of the receiver antenna will be

$$S_R = \frac{P_T G_T}{4\pi R^2} \cdot \frac{\sigma}{4\pi R^2} \text{ in W/m}^2$$

In Chapter 3 we defined the effective area of receiving antenna A_e as the ratio of the received power at the antenna terminals to the power density of the incident wave. We also mentioned that A_e affects the received power. Therefore, let S be the power received by the receiving antenna; then

$$S = \frac{P_T G_T \sigma}{(4\pi R^2)^2} A_e \text{ in watts}$$

Let G_R be the gain of receiving antenna, from Eq. (3.34), the gain G_R is then related to its effective area by the expression

$$G_R = \frac{4\pi A_e}{\lambda^2}.$$

The total received power S scattered by a target is then

$$S = \frac{P_T G_T \sigma}{\left(4\pi R^2\right)^2} \cdot \frac{G_R \lambda^2}{4\pi} = \frac{P_T G_T G_R \lambda^2 \sigma}{(4\pi)^3 R^4}.$$

For a monostatic radar where the antenna performs both transmitting and receiving functions, then $G_T = G_R = G$. Therefore

$$S = \frac{P_T G^2 \lambda^2 \sigma}{(4\pi)^3 R^4}. \tag{4.2a}$$

There is always some energy loss due to circuit or equipment power dispassion. This energy loss will reduce the total amount of power received. They are lumped together and are denoted by L, where $L < 1$. The total received power then becomes

$$S = \frac{P_T G^2 \lambda^2 \sigma}{(4\pi)^3 R^4} L. \tag{4.2b}$$

This equation can also be expressed in dB values as

$$S_{\mathrm{dB}} = P_{T\mathrm{dB}} + 2G_{\mathrm{dB}} + 10\log_{10}\left(\frac{\lambda^2}{(4\pi)^3}\right) + \sigma_{\mathrm{dB}} - 40\log_{10} R - L_{\mathrm{dB}}. \tag{4.2c}$$

Figure 4.5 illustrates the power density relationships between the transmitter, the target, and the receiver.

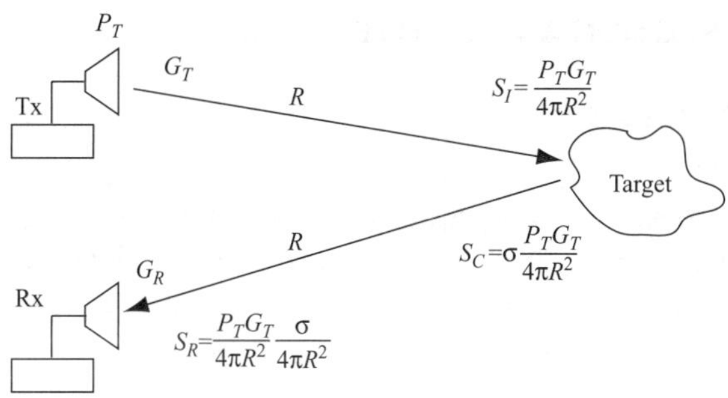

FIGURE 4.5 Key elements of radar range equation.

Equation (4.2b) or (4.2c) is called the *Radar range equation*, which can be used to compute the range of the target. The maximum detection range $R_{\max}$ is the range at which the received power just equals the minimum detectable signal $S_{\min}$. By equating $R_{\max}$ for $S = S_{\min}$, the maximum range can be computed as follows, from Eq. (4.2b):

$$R_{\max} = \left[\frac{P_T G^2 \lambda^2 \sigma}{(4\pi)^3 S_{\min}} L\right]^{1/4}. \tag{4.3}$$

Example 4.1 Calculation of Maximum Radar Range Given the following parameters of a range detection radar,

- Antenna gain: 34 dB
- Transmit peak power: 2 kW
- Radar carrier frequency f_c: 10 GHz
- Overall system losses: 5 dB
- Target RCS: 5 m^2
- Minimum detectable signal power: 1.6×10^{-13} W

The maximum detection range of the radar can be calculated as follows:

For a 10-GHz carrier frequency, the wavelength $\lambda = c/f = 3 \times 10^{-2}$ m. By applying the radar range equation (4.2c), with $S_{\mathrm{dB}} = -128$, one obtains

$$\begin{aligned} -128 &= 33 + 2 \cdot 34 + 10 \log_{10} \left[\frac{9 \cdot 10^{-4}}{(4\pi)^3} \right] + 7 - 40 \log_{10} R - 5 \\ &\approx 40 - 40 \log_{10} R. \end{aligned}$$

Therefore

$$R_{\max} \approx 15{,}848 \text{ m}.$$

4.4 CROSS SECTION AND CLUTTER

4.4.1 Target Cross Section

The "cross section" of a target relates the amount of transmitting power that strikes the target to the amount of power that is reflected back to the receiver. Let the power density of a plane wave incident on the target be S_i W/m^2, and the amount of power scattered isotropically be S_c; then the cross section σ is defined in terms of S_i and S_c as follows:

$$S_c = \sigma S_i .$$

Notice that the incident power density S_i is referred back to the transmitting antenna. For a large value of range R, the power density received at the target can be roughly considered as the amount of power received at the target. The reflected power S_c is therefore a fraction of the incident power S_i.

The power density S_r of the scattered wave at the receiving antenna is

$$S_r = \frac{S_c}{4\pi R^2}.$$

Here S_r is referred back to the target.

The cross section can be rearranged in terms of the power densities S_r and S_i as

$$\sigma = \frac{4\pi R^2 S_r}{S_i}.$$

To ensure that the receiving antenna is in the far field and that the waves are planar, the cross section is expressed as

$$\sigma = \lim_{R \to \infty} 4\pi R^2 \left| \frac{S_r}{S_i} \right| .$$

In terms of the various fields that make up EM waves, this equation can be expanded as follows:

$$\begin{aligned}\sigma &= \lim_{R\to\infty} 4\pi R^2 \left|\frac{E_r^2}{E_i^2}\right| \\ &= \lim_{R\to\infty} 4\pi R^2 \left|\frac{H_r^2}{H_i^2}\right|.\end{aligned} \tag{4.4}$$

Here E_r and H_r are the respective electric and magnetic fields at the radar receiver, while E_i and H_i are the respective electric and magnetic fields at the target.

4.4.2 Cross Section and the Equivalent Sphere

If a target scatters power uniformly over all angles, its cross section would be equal to the area from which power was extracted from the incident wave. Since the sphere has the ability to scatter isotropically, it is convenient to interpret its cross section in terms of the projected area of an equivalent sphere.

A sphere with radius $a \gg \lambda$ will intercept power contained in πa^2 of the incident wave. The cross section of a sphere is therefore equal to the area πa^2 even though the actual area that returns power to the receiver is a very small area where the surface lies parallel to the incident wavefront. The cross section of a sphere is therefore one-quarter of the surface area, and it can be shown that the average cross section of any large object that consists of continuous curved surfaces will be one quarter of its total surface area.

4.4.3 Cross Section of Real Targets

Target cross section is a complicated function of the viewing aspect, incident wave frequency, and polarization. It is often related to target physical size, and under certain circumstances, it may be significantly larger. The effective surface roughness of a target (as a function of λ) also plays an important role in determining its cross section. There are two popular reflection mechanisms: diffuse reflection and specular reflection. The diffuse reflector, such as a cornfield, scatters the EM waves in all directions and returns only a fraction of the energy back to the radar antenna. The specular reflector, such as a smooth paved surface, acts as a mirror and reflects the energy from the EM wave in one well-defined direction. The energy returned to the radar antenna could be either nonexistent or very strong, depending on whether the reflecting surface is at a right (90°) angle to the radar beam. These two reflection mechanisms determine individually, or in combination, the target reflection characteristics.

4.4.4 Radar Cross Section (RCS)

The *radar cross section* (RCS) of an object is a measure of its size as seen at a particular radar wavelength and polarization. RCS has units of m^2 and is often expressed in decibels relative to a square meter (dBsm),

$$\sigma(\text{dBsm}) = 10\log_{10}[\sigma(\text{m}^2)].$$

The characteristics of RCS, or scattering coefficient, depend on the field polarity of the transmitter and the receiver. Since a radar receiver detects electric fields, the RCS is a function of polarity on E^r and E^t. Here E^r and E^t refer to the receiver and transmitter E field, respectively. The relationship between RCS and polarity of the transmitted and received electric fields can be expressed as

$$\begin{bmatrix} E_H^r \\ E_V^r \end{bmatrix} = \begin{bmatrix} \sigma_{HH} & \sigma_{VH} \\ \sigma_{HV} & \sigma_{VV} \end{bmatrix} \begin{bmatrix} E_H^t \\ E_V^t \end{bmatrix}, \tag{4.5}$$

where the subscripts V and H refer to the respective vertical polarized and horizontal polarized E fields. The first subscript of σ_{HH} refers to the scattering coefficient σ due to a transmitted horizontal polarized E field, or E_H^t. The second subscript of σ_{HH} relates the scattering coefficient σ to the received horizontal polarized E field, or E_H^r. Similar relations apply to σ_{VH}, σ_{HV}, and σ_{VV}. For most applications, the radar transmitter and receiver are arranged to possess the same polarity. The target RCS is therefore simplified as a single value that corresponds to either σ_{HH} or σ_{VV} depending on whether the transmitted polarization is horizontal (H) or vertical (V).

The following list displays the radar cross section of some well-known objects for comparison purpose:

Unit	Insects	Birds	Humans	Aircraft	Ships
m^2	0.001	0.01	0.1	100	10,000
dBsm	−30	−20	−10	20	40

4.4.5 Clutter

The EM waves reflected from the objects around the target are referred to as "clutter." These returns may be from the surface surrounding the target (ground), or from the volume of space around it (rainfall) as shown in Fig. 4.6.

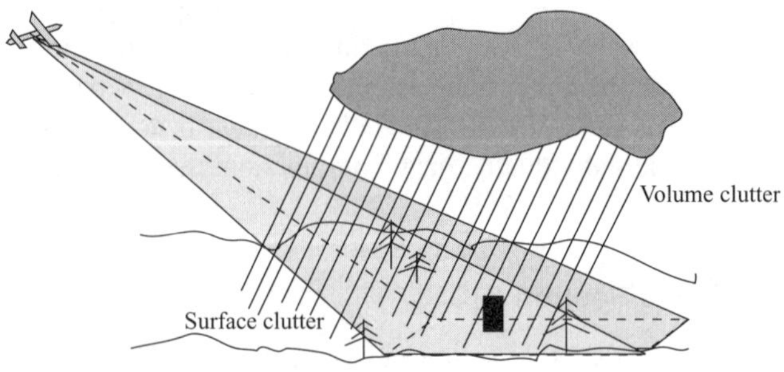

FIGURE 4.6 Surface clutter and volume clutter.

Clutter is thus defined as the undesired return from a physical object or a group of objects. Clutter may be divided into sources distributed over a surface (land or sea), within a volume (weather or chaff), or concentrated at discrete points (structures, birds, or vehicles). The magnitude of the signal reflected from the surface is a function of the material, roughness, and angle. Two main scattering types of clutter are diffuse and specular, as described before. Rain and dust are the two main contributors of volume backscatters.

In addition to detecting the target range, radar is also used extensively to detect the speed of a target. The detection of target speed is primarily based on the detection of Doppler frequency, which is discussed next.

4.5 DOPPLER EFFECT AND FREQUENCY SHIFT

For stationary and separate wave receivers and wave sources, such as sound or light, the frequency or wavelength generated by the wave source will be the same at the receiver site; that is, $f_r = f_s$ or $\lambda_r = \lambda_s$, where the subscripts r and s refer to wave receiver and wave source, respectively. Only electromagnetic waves are considered in this book. Figure 4.7 shows a wave source S and a wave receiver R at some distance away. The concentric circles are the wavefronts emitted from the wave source S located at the center of the circles. The symbol r_i, for i = 1,2,3, denotes the radius corresponding to the wavefront generated at time $T_i = (i - 1)T$ with $T = 1/f_s$ as the wave period. The symbol S_i is the origin of the circle r_i. For a stationary wave source, $S_i = S$ for every T_i, and the distance between any two circles has the same wavelength $\lambda_s = \lambda_r$.

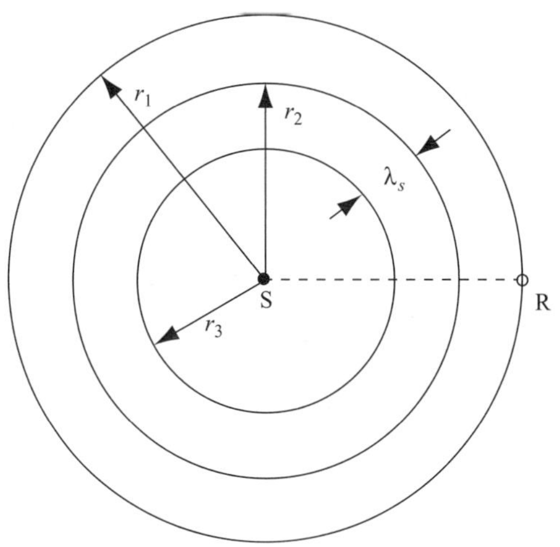

FIGURE 4.7 Wave propagation for stationary source and stationary receiver.

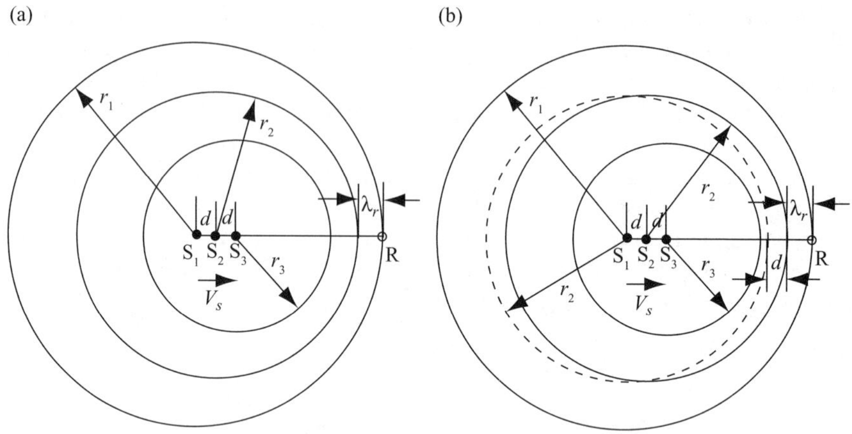

FIGURE 4.8 Wave propagation for moving source and stationary receiver.

4.5.1 Doppler Frequency

When the wave source moves, either toward or away from the receiver, the center of the concentric circles will move as well. As a consequence, a frequency difference occurs between the wave source and the wave receiver. This frequency difference or shift is called the *Doppler frequency.*

Figure 4.8a shows a situation where the wave source S is moving at the speed V_s toward the receiver R. For a wave period of T, the origin of the wave source S has moved to two locations, from S_1 to S_2 at time $T_2 = T$, and to S_3 at time $T_3 = 2T$. The distances between S_1, S_2, and S_3 are $S_2 - S_1 = d$ and $S_3 - S_2 = d$.

Figure 4.8b shows the effects of the waveforms generated at different locations by the moving source. Consider the circle of r_1, which has origin S_1 at time $t = 0$. At time $t = T$, the source S moves to a new position S_2 with a radius of r_2 and having $d = V_sT$ away from S_1. A dotted circle corresponding to the stationary case generated at time T with the same radius r_2 is shown for comparison. The difference between radii r_1 and r_2 along the direction between S and R is the wavelength as seen by the receiver. In other words, the received wavelength λ_r is related to the source wavelength λ_s by

$$\begin{aligned}\lambda_r &= \lambda_s - d \\ &= \lambda_s - V_sT \\ &= \lambda_s - \frac{V_s\lambda_s}{c} \\ &= \frac{\lambda_s(c - V_s)}{c},\end{aligned}$$

where c is the speed of the wave, in this case the speed of light.

Since $\lambda = c/f$, the received frequency can be computed from the source frequency by

$$f_r = \frac{f_s c}{c - V_s}.$$

The frequency difference between receiver and source, or Doppler frequency f_D, is

$$\begin{aligned} f_D &= f_r - f_s \\ &= \frac{f_s V_s}{c - V_s}. \end{aligned} \tag{4.6}$$

The positive value of f_D means that the received frequency is higher than the frequency emitted by the approaching source. If the receiver is stationary, then, after the source passes the receiver, the "speed of approach" V_s becomes negative, and the frequency recorded by the receiver becomes lower than the frequency emitted by the now-receding source. The Doppler frequency, caused by relative movement between source and receiver, is also called the Doppler frequency shift.

The same principle applies when the source is stationary but the receiver is approaching it at a speed V_r. For one receiver wave period T', which equals to $1/f_r$, the receiver will move a distance of $d = V_r T'$. This means that the receiver will receive the source wave by less distance d than when the receiver is stationary. Figure 4.9 displays an example of a receiver moving toward the wave source at a speed V_r.

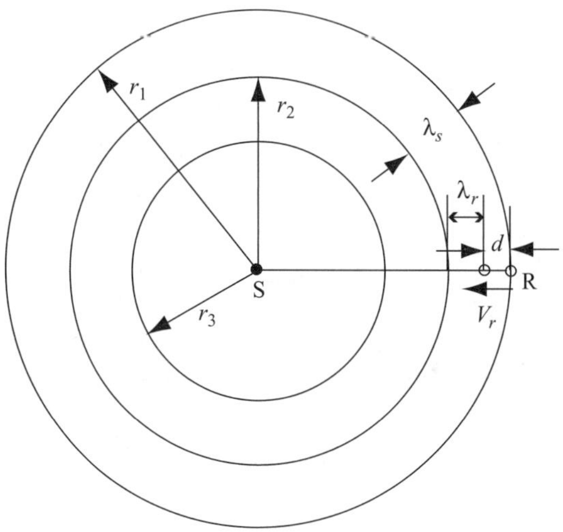

FIGURE 4.9 Wave propagation for stationary source and moving receiver.

The new received wavelength is related to the source wavelength by

$$\begin{aligned}\lambda_r &= \lambda_s - d \\ &= \lambda_s - V_r T' \\ &= \lambda_s - \frac{V_r \lambda_r}{c} \\ \lambda_r &= \frac{\lambda_s c}{c + V_r}.\end{aligned}$$

Since the moving receiver now determines the period of the wave, the received frequency is related to the source frequency by

$$f_r = \frac{f_s(c + V_r)}{c}.$$

The frequency difference between the receiver and source or Doppler frequency is

$$\begin{aligned}f_D &= f_r - f_s \\ &= \frac{f_s V_r}{c}.\end{aligned} \tag{4.7}$$

Consider the case when both the wave source and the wave receiver are moving and V_s and V_r are the speeds with which they are approaching each other. Figure 4.10 displays a scenario in which the source moves toward the receiver at speed V_s for a time interval $t = T = 1/f_s$, with a distance $d_1 = V_s T$. It also shows that the receiver moves toward the source at speed V_r for the same time interval T with a distance $d_2 = V_r T$.

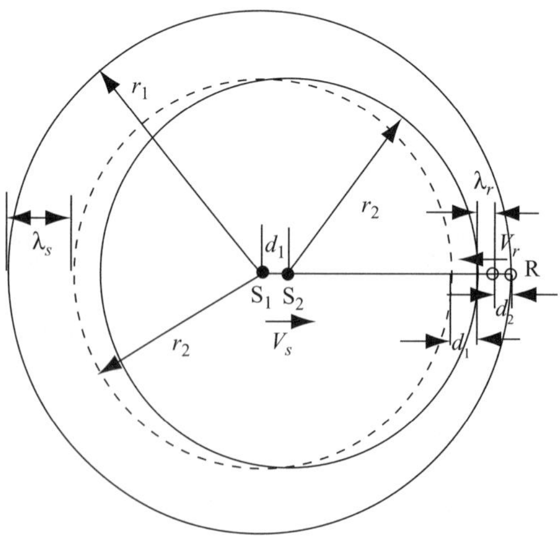

FIGURE 4.10 Wave propagation for moving source and moving receiver.

The received wavelength λ_r is related to the source wavelength λ_s by

$$\begin{aligned}\lambda_r &= \lambda_s - d_1 - d_2 \\ &= \lambda_s - V_s T - V_r T \\ &= \lambda_s - \frac{V_s \lambda_s}{c} - \frac{V_r \lambda_s}{c} \\ &= \lambda_s \left(1 - \frac{V_s}{c} - \frac{V_r}{c}\right).\end{aligned}$$

The received frequency is related to the source frequency by

$$\begin{aligned}f_r &= \frac{f_s}{1 - (V_s + V_r)/c} \\ &= f_s \left(1 - \frac{V_{rs}}{c}\right)^{-1},\end{aligned}$$

where $V_{rs} = V_r + V_s$. For $V_{rs} \ll$ c,

$$f_r \approx f_s \left(1 + \frac{V_{rs}}{c}\right).$$

The frequency difference between receiver and source, or Doppler frequency, is

$$\begin{aligned}f_D &= f_r - f_s \\ &= \frac{f_s V_{rs}}{c} \\ &= \frac{V_{rs}}{\lambda_s}.\end{aligned} \tag{4.8}$$

Notice that the value of V_{rs}, and therefore f_D, is positive if the wave source and wave receiver are moving toward each other; otherwise, it is negative.

For most Doppler frequency detectors, both the transmitter and the receiver are stationary. Figure 4.11 illustrates one possible application where the transmitter (source) and the receiver are stationary, while the target is moving at speed V.

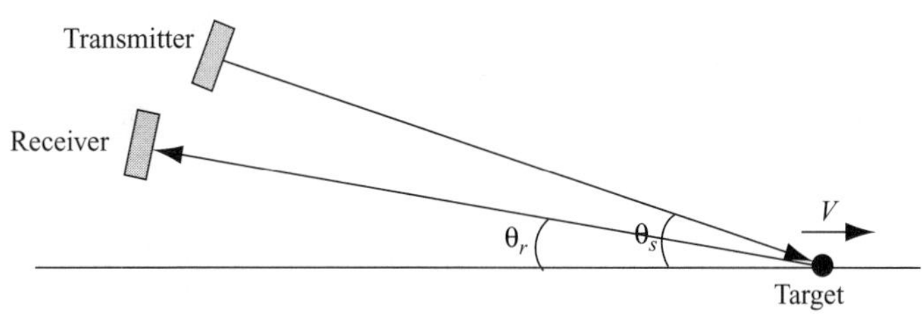

FIGURE 4.11 Doppler radar with separate source and receiver.

As shown in the figure, the radial velocity of the target relative to the transmitter (source) is $-V \cos \theta_s$, and the radial velocity of the target relative to the receiver is $-V \cos \theta_r$. The Doppler frequency arising under these circumstances can be calculated by assuming that

- The target is moving away from the transmitter with a velocity $-V \cos \theta_s$.
- The receiver is moving away from the target with a velocity $-V \cos \theta_r$.

These two assumptions are equivalent to the receiver moving away from the transmitter (source) with velocity $-V \cos \theta_s - V \cos \theta_r$, even though both are stationary.

The Doppler frequency of the separated transducers described above can be computed as follows, from Eq. (4.7):

$$
\begin{aligned}
f_D &= -\frac{f_s V}{c}(\cos\theta_s + \cos\theta_r) \\
&= -\frac{2 f_s V}{c}\cos\left(\frac{\theta_r + \theta_s}{2}\right)\cos\left(\frac{\theta_r - \theta_s}{2}\right).
\end{aligned}
$$

If the transmitter (source) and the receiver are collocated, then $\theta_r = \theta_s = \theta$ and the formula for the Doppler frequency becomes

$$
\begin{aligned}
f_D &= -\frac{2 f_s V}{c}\cos\theta \\
&= -\frac{2V}{\lambda_s}\cos\theta. \qquad (4.9)
\end{aligned}
$$

Equation (4.9) is a very important formula for computing the Doppler frequency, which serves as the basis for detecting the speed of a target.

An alternative way to derive the Doppler frequency and the corresponding relative velocity between the radar and target is as follows. Assume that the distance between radar and target is R, and that the wavenumber over the transmitted and reflected paths is then $2R/\lambda_s$, where λ_s is source wavelength. Since one wavelength λ_s corresponds to an angular phase of 2π, the total phase ϕ is $4\pi R/\lambda_s$. The rate of change in ϕ with respect to time t is the angular Doppler frequency ω_D, which is then

$$
\begin{aligned}
\omega_D &= 2\pi f_D \\
&= \frac{d\phi}{dt} \\
&= \frac{4\pi}{\lambda_s}\frac{dR}{dt} \\
&= \frac{4\pi V_r}{\lambda_s}.
\end{aligned}
$$

Therefore

$$f_D = \frac{2V_r}{\lambda_s}. \tag{4.10}$$

Here f_D is the Doppler frequency and V_r is the relative velocity of the target with respect to the radar. The value of V_r, and therefore f_D, is positive if the radar and target are moving toward each other; otherwise, it is negative.

Example 4.2 Computation of Doppler Frequency Figure 4.12 shows a radar-carrying vehicle flying along the y axis at (velocity) $V = 800$ kilometers per hour (km/h), with a ground target located at $(r_0, 0)$. Let $r_0 = 20$ km and the carrier frequency of radar be 3 GHz. The distance from the three radar positions A, B, and C to the target are r_1, r_0, and r_3, respectively. Position B is at (0,0), and the angles between the x axis and positions A and C are α and β, respectively. Letting $\alpha = \beta = 30°$, the Doppler frequency at positions A, B, and C can be computed as follows. The wavelength of a radar signal is computed as

$$\begin{aligned}\lambda &= \frac{c}{f_c} \\ &= \frac{3 \times 10^8}{3 \times 10^9} \\ &= 0.1\,\text{m}.\end{aligned}$$

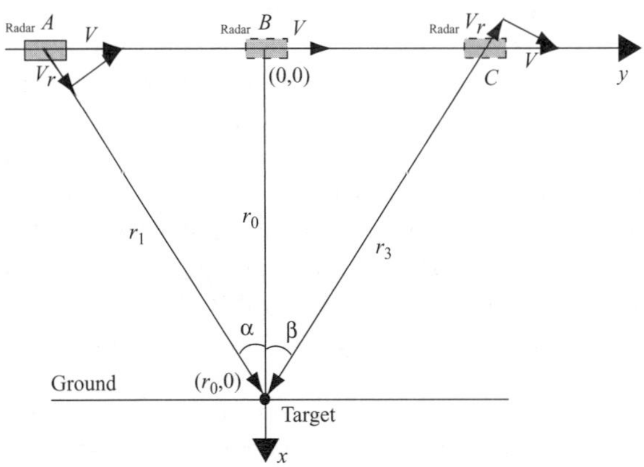

FIGURE 4.12 Example of Doppler frequency.

The radial velocity V_r at position A is

$$\begin{aligned}V_r &= V \sin \alpha \\ &= 800 \sin(30°) \\ &= 111.11\,\text{m/s}.\end{aligned}$$

Given V_r and λ, from Eq. (4.10), the Doppler frequency can be obtained as

$$\begin{aligned} f_D &= \frac{2V_r}{\lambda} \\ &= 2.222\,\text{kHz}. \end{aligned}$$

The radial velocity at position B equals zero; therefore the Doppler frequency $f_D = 0$. The radial velocity at position C is

$$\begin{aligned} V_r &= -V \times \sin\beta \\ &= -800\,\text{km/h} \times \sin(30^\circ) \\ &= -111.11\,\text{m/s}. \end{aligned}$$

therefore,

$$\begin{aligned} f_D &= \frac{2V_r}{\lambda} \\ &= -2.222\,\text{kHz}. \end{aligned}$$

4.6 RADAR RESOLUTION AND AMBIGUITY FUNCTION

The primary function of a modern radar system is to detect targets of interest while estimating each target's position and velocity. Range and velocity determinations depend on the accuracy of measuring the time delay and the Doppler frequency, which in turn are related to the fundamental properties of radar waveform. The continuous-wave (CW) waveform and the pulse waveform are two of the most frequently used techniques in the radar applications. The advantage of CW radar is unambiguous Doppler measurement; that is, each target velocity produces a single unique Doppler frequency of the CW carrier. However, the target range measurements using CW radar are entirely ambiguous. This is because that all the returned waveforms are continuous; the radar cannot identify which echo came from which waveform, and thus an ambiguity in the range measurement occurs.

Most modern radars employ a pulse waveform technology, where a single antenna serves for both transmitting and receiving functions. A pulse-based radar waveform will provide both the range and the velocity information for a target. The waveform of a pulse-based radar can be single-carrier-frequency-based, frequency-modulated, or phase-modulated, as described in more detail in next chapter.

The target range of a pulse-based radar, with echo delay time τ, is computed as $R = c\tau/2$.

Letting T be the pulse repetition period, the maximum unambiguous target range R occurs at $\tau = T$, namely, $R_{\max} = cT/2$. The range ambiguity arises if the target is located at a range exceeding $R_{\max}$.

The Doppler frequency f_D, which causes carrier frequency shift, is induced by a target having velocity V relative to the radar, specifically, $f_D = 2V/\lambda$.

For moving targets, the differential phase change between successive pulses, denoted as $\Delta\phi$, is given by $\Delta\phi = 2\pi(2\Delta R/\lambda)$, where ΔR is the change in target range between successive radar pulses. When the phase change between radar pulses exceeds 2π, the Doppler measurement becomes ambiguous.

The most common method for resolving range and Doppler ambiguities involves using multiple f_{PRF} values. This has the effect of changing the apparent target range estimated by each pulse or pulse burst, and allows for some ambiguity resolution in either Doppler frequency or range, depending on the particular application.

The ambiguity function (AF) is widely used in radar waveform analysis which shows the distortion of a matched filter output due to the Doppler shift of the echo signal from a moving target. It reveals the range–Doppler position of an ambiguous response and defines the range and Doppler resolution. The ambiguity function $\chi(\tau, f_D)$ is defined as the cross-correlation between a waveform $s(t)$ and $s^*(t-\tau)\exp(j2\pi f_D t)$, which is a time-delayed and complex-conjugated, frequency-shifted replica of $s(t)$:

$$\chi(\tau, f_D) = \int_{-\infty}^{\infty} s(t)\, s^*(t-\tau)\exp(j2\pi f_D t)\, dt. \tag{4.11}$$

Along the delay τ axis, $\chi(\tau, f_D)$ represents the autocorrelation function of $s(t)$; along the f_D axis, $\chi(\tau, f_D)$ is proportional to the spectrum of $s^2(t)$:

$$\left|\chi(\tau, f_D = 0)\right|^2 = \left|\int_{-\infty}^{\infty} s(t)s^*(t-\tau)dt\right|^2 \tag{4.12a}$$

$$\left|\chi(\tau = 0, f_D)\right|^2 = \left|\int_{-\infty}^{\infty} |s(t)|^2 \exp(j2\pi f_D t)dt\right|^2. \tag{4.12b}$$

It is common to refer to the absolute value of $\chi(\tau, f_D)$ as the ambiguity surface of the waveform. A normalized expression of the ambiguity surface requires that

$$\int_{-\infty}^{\infty} |s(t)|^2 dt = 1. \tag{4.13}$$

A short monotone pulse will now be used as an example to show the ambiguity surface $|\chi(\tau, f_D)|$. Let the envelope of the rectangular pulse be

$$s(t) = \sqrt{\frac{1}{T_p}}\mathrm{Rect}\left(\frac{t}{T_p}\right), \tag{4.14}$$

where

$$\operatorname{Rect}\left(\frac{t}{T_p}\right) = 1, \qquad |t| \le \frac{T_p}{2}$$
$$= 0, \qquad |t| > \frac{T_p}{2} \tag{4.15}$$

The rectangular pulse and its frequency spectrum are shown in Fig. 4.13. The frequency spectrum is a sinc function with its maximum occurring at the origin, while the first zero is located at $1/T_p$ and $-1/T_p$ away from the origin. It is clear that the narrower the rectangular pulse is, the wider the bandwidth of the spectrum becomes.

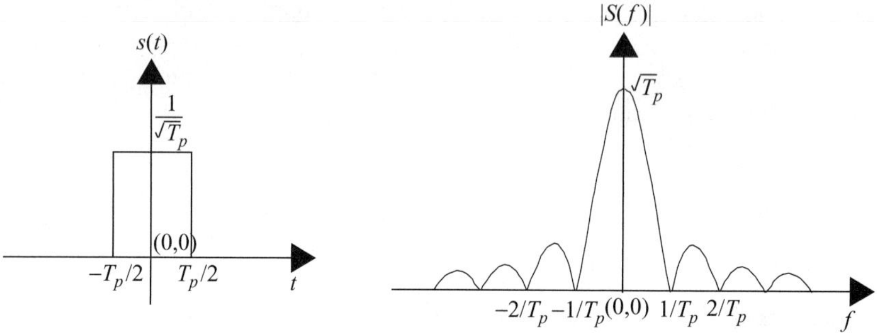

FIGURE 4.13 Rectangular pulse and its frequency spectrum.

The ambiguity surface $|\chi(\tau, f_D)|$ of the rectangular pulse can be expressed as

$$\begin{aligned}
|\chi(\tau, f_D)| &= \left| \int_{-\infty}^{\infty} s(t) s^*(t-\tau) \exp(j2\pi f_D t) dt \right| \\
&= \frac{1}{T_p} \left| \int_{\tau}^{T_p} \exp(j2\pi f_D t)\, dt \right| \\
&= \left| \left(1 - \frac{|\tau|}{T_p}\right) \frac{\sin\left[\pi f_D T_p \left(1 - |\tau|/T_p\right)\right]}{\pi f_D T_p \left(1 - |\tau|/T_p\right)} \right| \\
&= \left[1 - \frac{|\tau|}{T_p}\right] \operatorname{sinc}(x), \ |\tau| < T_p, \\
&= 0, \ |\tau| \ge T_p,
\end{aligned} \tag{4.16}$$

where $x = f_D T_p(1 - |\tau|/T_p)$.

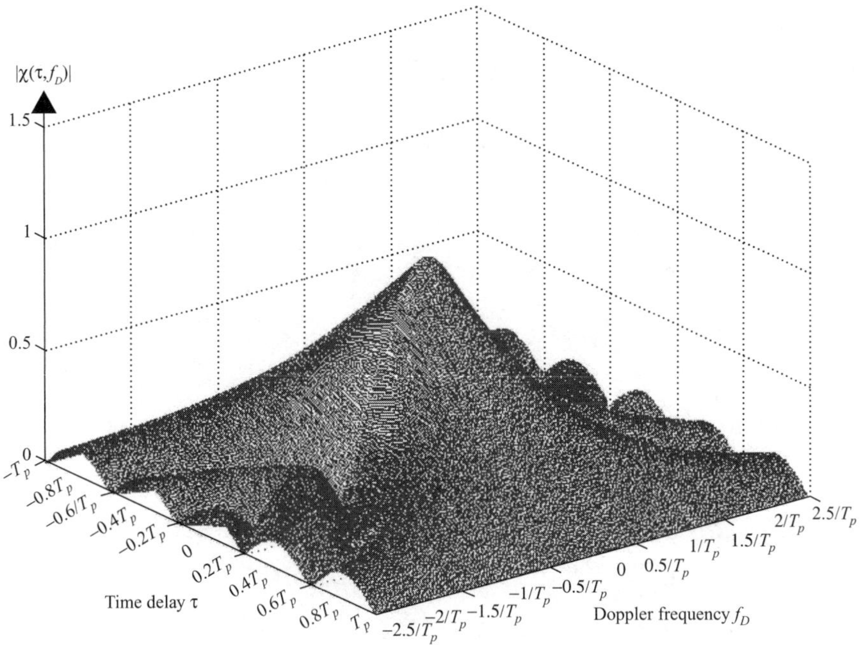

FIGURE 4.14 Ambiguity function of a rectangular pulse in 3D view.

The ambiguity surface $|\chi(\tau, f_D)|$ is a function of both the time delay τ and the Doppler frequency f_D. For $|\tau| < T_p$ and at a given value of τ, the sinc function has its maximum occurring at the origin and zeros located at $x = \pm n$, where $n = 1, 2, 3, \ldots$.

Figure 4.14 is a 3D display of $|\chi(\tau, f_D)|$. The magnitude of the sinc function decreases along the delay-time τ axis and reaches zero when $|\tau|$ equals T_p. The bandwidth of the sinc function along the f_D axis is stretched as the value of $|\tau|$ increases from zero to T_p.

Figure 4.15 displays the cross-sectional view of Fig. 4.14. The waveform in Fig. 4.15a shows the magnitude of $|\chi(\tau = 0, f_D)|$ against the Doppler frequency with $\tau = 0$. The waveform in Fig. 4.15b shows the magnitude of $|\chi(\tau = 0.5T_p, f_D)|$ against the Doppler frequency with $\tau = 0.5T_p$. The 3-dB bandwidth doubles at $\tau = 0.5T_p$ when compared with the one at $\tau = 0$. The Doppler frequency is used to compute the speed of target, which has higher magnitude and narrow bandwidth at $\tau = 0$ than do those at $\tau = 0.5T_p$. This implies that the accuracy of the target's speed measurement is adversely affected when the delay time increases.

Figure 4.16 displays another cross-sectional view of Fig. 4.14. The waveform in Fig. 4.16a shows the magnitude of $|\chi(\tau, f_D = 0)|$ against the time delay with $f_D = 0$. The waveform in Fig. 4.16b shows the magnitude of $|\chi(\tau, f_D = 2.5/T_p)|$ against the time delay with $f_D = 2.5/T_p$. As can be seen, the magnitude of $|\chi(\tau, f_D)|$ equals zero at time delay $\tau = T_p$ and $\tau = -T_p$ in both cases. The function

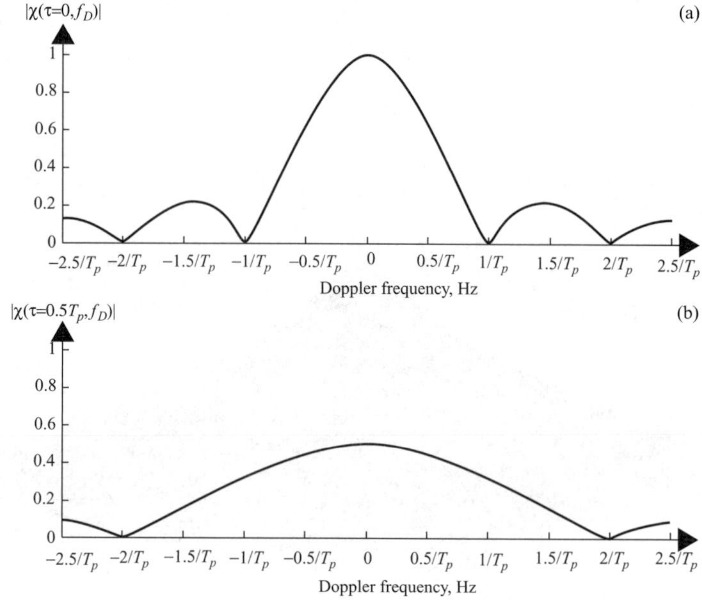

FIGURE 4.15 Cross-sectional view of Fig. 4.14 with $\tau = 0$ (a) and $\tau = 0.5T_p$ (b).

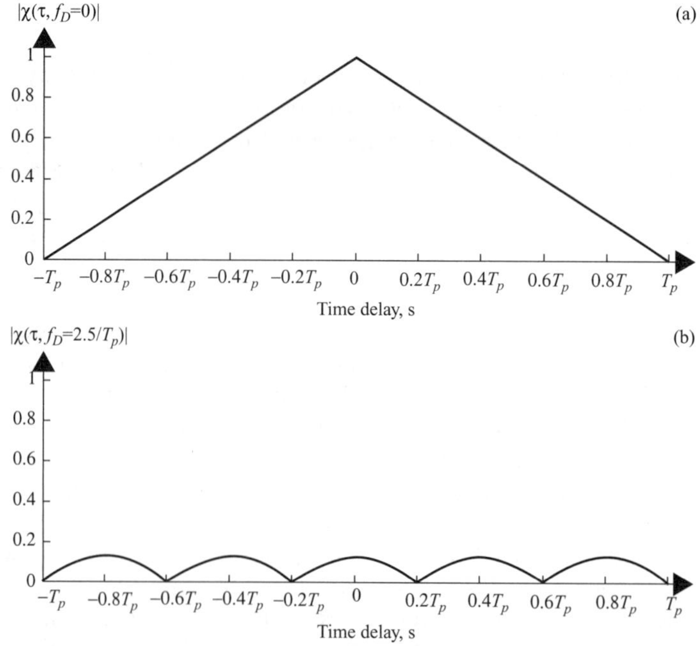

FIGURE 4.16 Cross-sectional view of Fig. 4.14 with $f_D = 0$ (a) and $f_D = 2.5/T_p$ (b).

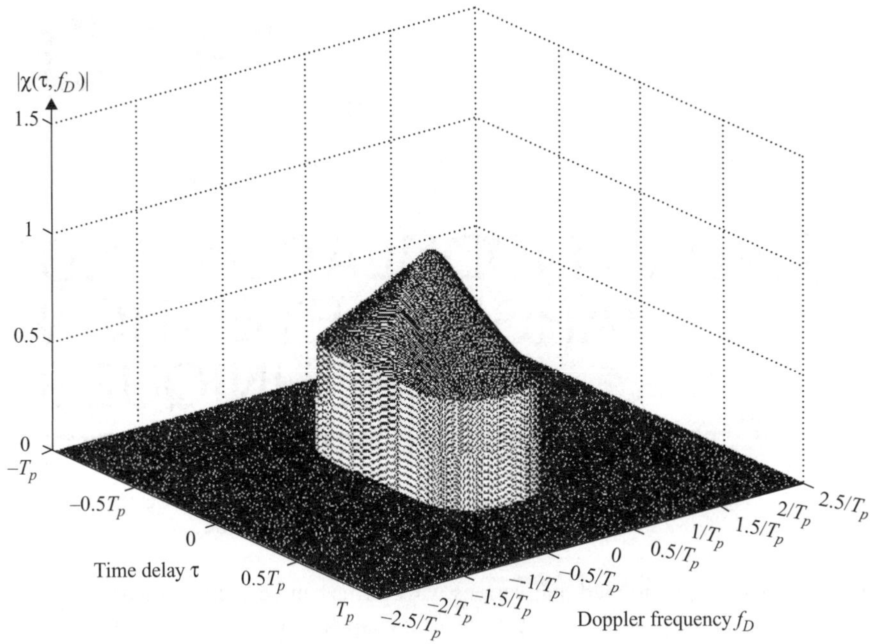

FIGURE 4.17 A 3-dB contour of ambiguity function of a rectangular pulse in 3D view.

$|\chi(\tau, f_D = 0)|$ corresponds to the output of pulse compression, which is used to measure the target's range. As shown in Fig. 4.16b, the outputs of pulse compression values are reduced at increased f_D, and therefore cause errors or ambiguity in target range measurement.

The 3-dB energy contour of the magnitude of $|\chi(\tau, f_D)|$ is shown as a 3D view in Fig. 4.17. This is the region where the target's speed and range should both be measured.

5

RADAR MODULATION AND TARGET DETECTION TECHNIQUES

In Chapter 4, we mentioned that various waveforms can be used by radars to detect the range and/or speed of a target. In this chapter, we discuss several types of these waveforms. Section 5.1 reviews amplitude modulation (AM)-based radars, including continuous–wave radar and pulse–Doppler radar. Section 5.2 describes target detection techniques on AM-based radar. Frequency modulation (FM)-based radars are covered in Section 5.3, including linear frequency modulation (LFM), pulsed LFM, continuous-wave LFM (CWLFM), and stepped frequency modulation (SFM). The target detection techniques for FM-based radar, which utilize the in-phase–quadrature-phase demodulator and matched filtering, are discussed in Section 5.4.

5.1 AMPLITUDE MODULATION (AM) RADAR

Amplitude modulation (AM) is a modulation technique in which the amplitude of the carrier is varied in accordance with some characteristics of the baseband modulating signal. It is the most common form of modulation because of the ease with which the baseband signal can be recovered from the transmitted signal.

An AM signal $p(t)$ can be described in terms of carrier frequency f_c and baseband signal $p_b(t)$ by the following equation:

$$p(t) = A_t [1 + p_b(t)] \cos(2\pi f_c t). \qquad (5.1)$$

Digital Signal Processing Techniques and Applications in Radar Image Processing, by Bu-Chin Wang.

Letting the modulating signal $p_b(t)$ be a sinusoidal signal as $p_b(t) = A_m \cos(2\pi f_a t)$, Eq. (5.1) then becomes

$$p(t) = A_t [1 + A_m \cos(2\pi f_a t)] \cos(2\pi f_c t). \tag{5.2}$$

In general $A_m < 1$; otherwise a phase reversal occurs and demodulation becomes more difficult. The extent to which the carrier's amplitude has been modulated is expressed in terms of a percentage modulation that is calculated by multiplying A_m by 100.

5.1.1 Continuous-Wave (CW) Radar

A radar system that continuously transmits a signal is referred to as *continuous-wave* (CW) *radar*. CW radar is normally used to detect the speed of a moving target. When a CW radar waveform continuously strikes a target that is moving either toward or away from the transmitting radar, the frequency of the reflected waveform is changed; this is known as the *Doppler frequency*. To detect the moving target speed, the CW radar receiver mixes (or homodynes) the received signal with a replica of the transmitted signal. After lowpass filtering, the only remaining component is the Doppler frequency, which can be used to calculate the speed of the target. The continuous-wave radar is the best means of detecting fast-moving objects that do not require range resolution. The disadvantage of the CW Doppler radar system is that it does not determine the range of the object, nor is it able to differentiate between objects when they lie in the same direction and are traveling at the same speed.

5.1.2 Pulse Modulation Radar

The pulse–doppler radar is identical to the CW radar except that the transmitted signal is a sequence of pulses. This technique allows the radar to measure both range as well as velocity of the target. A key requirement for any Doppler radar is coherence; that is, the transmitted signal frequency and the locally generated reference signal frequency must be kept the same with fixed phase difference. This reference signal is then used to detect both the range and the Doppler frequency of the received signal.

The pulse-modulated radar uses a sequence of narrow pulses to serve as a modulating signal and modulates the carrier frequency to form the transmitting waveform. Let $p_N(t)$ be the pulse train, consisting of N pulses with period T, and assume that each pulse has time duration T_p and amplitude A. Then $p_N(t)$ can be expressed as

$$p_N(t) = \sum_{n=0}^{N-1} p_T(t - nT)$$

and

$$\begin{aligned} p_T(t) &= A \quad \text{for} \quad 0 < t < T_p, \\ &= 0 \quad \text{otherwise.} \end{aligned} \tag{5.3}$$

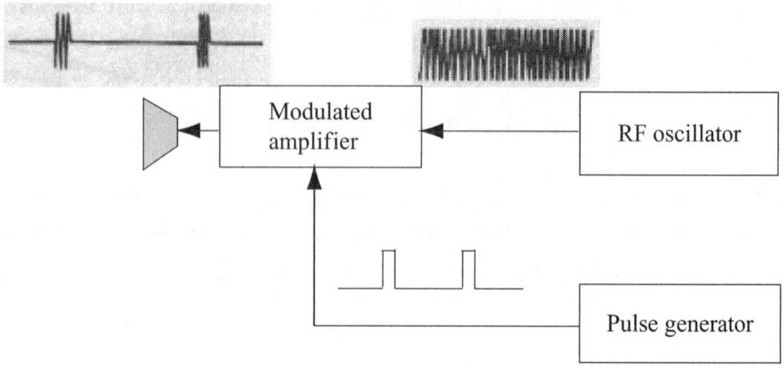

FIGURE 5.1 Transmitter block diagram of a pulse-modulated radar system.

The pulse-modulated signal then becomes

$$p(t) = p_N(t)\cos(2\pi f_c t). \tag{5.4}$$

The pulse modulation radar transmits radiofrequency energy in very short bursts, usually on the order of 0.1 to approximately 50 microseconds (μs). It is sometimes called the *pulse–doppler radar*. The carrier frequency f_c normally ranges from several hundred megahertz to tens of gigahertz. Figure 5.1 shows the transmitter block diagram of a typical pulse-modulated radar system.

The spectrum of a narrow pulse, as discussed in Chapter 1, is a sinc function. The first minimum (or zero) of the sinc function is located at both sides of the peak, which are displaced from the peak by $1/T_p$ with T_p as the pulsewidth. For a pulsed radar signal, the relationship between frequency bandwidth B and T_p can therefore be represented as

$$B \approx \frac{1}{T_p}.$$

The radar range resolution is determined as follows, from Eq. (4.1c):

$$\begin{aligned} \Delta R &= \frac{cT_p}{2} \\ &= \frac{c}{2B}. \end{aligned} \tag{5.5}$$

Figure 5.2 shows the pulse-modulated signal and its corresponding spectrum. As can be seen, the frequency spectrum of the pulse train is discrete and centered on the carrier frequency f_c. The envelope of the spectrum is a sinc function. For a periodic signal of period T, the frequency resolution is equal to $1/T$. Figure 5.2 is plotted with $T = 2T_p$, and the first zero of the sinc function appears at $f = 1/T_p$ away from the peak center f_c.

Another type of high-resolution radar is the impulse or short-pulse radar. The time duration of these radars is usually 0.25–1 nanoseconds (ns). The pulses are transmitted without carriers, so they are often called *carrierless impulses* or *baseband video*

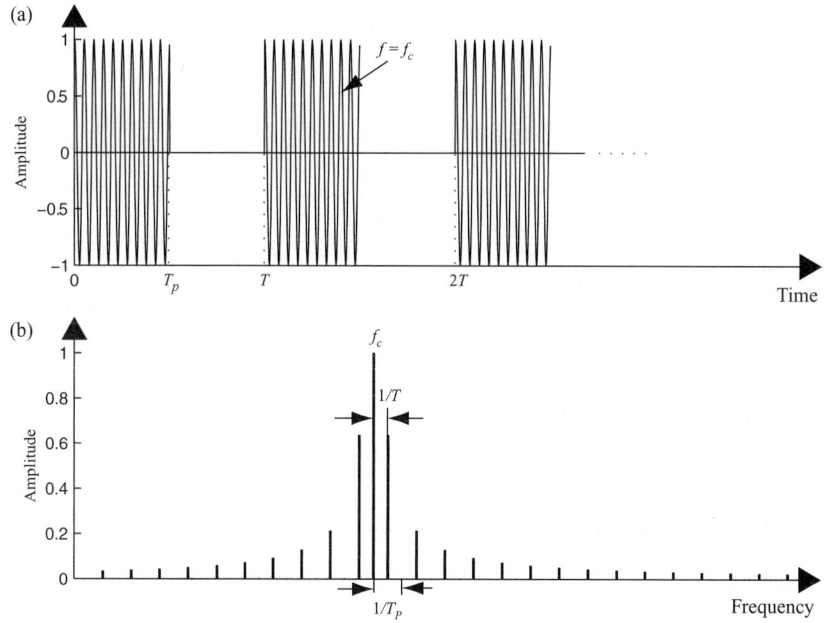

FIGURE 5.2 Time- and frequency-domain waveforms of pulse-modulated radar signal.

pulses. The resulting pulses are often called *monocycle pulses*. These types of radar are used for stationary target ranging only. No radial speed detection is required. Two popular pulses and their corresponding power spectra are shown in Fig. 5.3.

Since radar range resolution depends on the bandwidth of the received signal, and because the bandwidth of a time-gated sinusoid is inversely proportional to the pulse duration, the short pulse provides better range resolution. However, since short-pulse duration requires a high-powered transmitter in order to ensure good reception, this type of radar has the disadvantage of causing hardware problems and safety issues.

5.2 TARGET DETECTION TECHNIQUES OF AM-BASED RADAR

This section presents some techniques for detecting Doppler frequency in a received signal based on amplitude modulation.

5.2.1 Doppler Frequency Extraction

The transmitted signal of an amplitude modulated radar pulse can be represented as

$$p(t) = A_t \cos(2\pi f_c t),$$

where f_c is the carrier frequency. The corresponding received signal from a single moving target will be

$$s(t) = A_r \cos[2\pi(f_c + f_D)t + \phi],$$

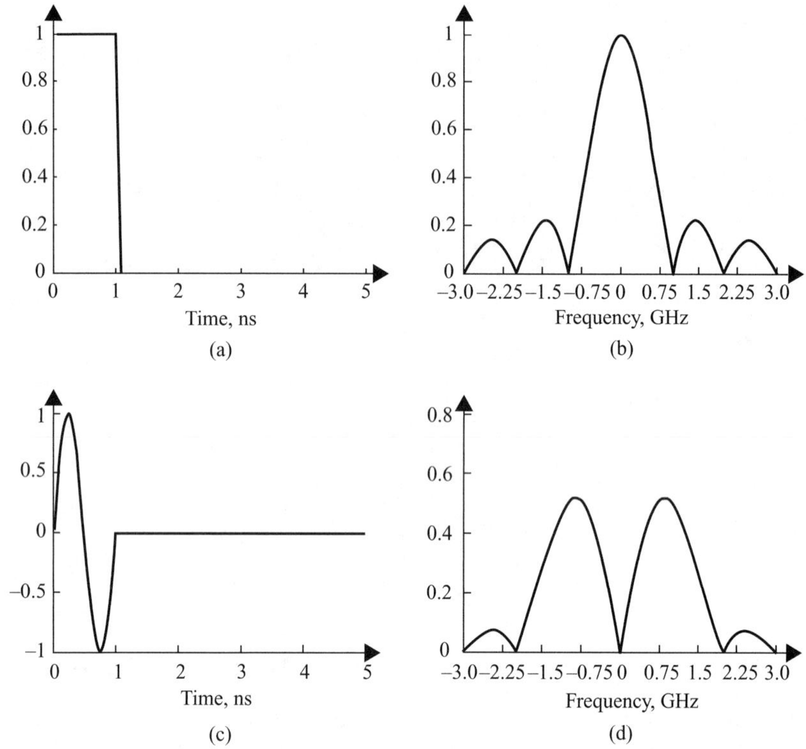

FIGURE 5.3 Time- and frequency-domain waveforms of two video pulses.

where f_D is the Doppler frequency, ϕ is a phase term dependent on the distance to the target, and $A_r < A_t$.

To extract the Doppler frequency f_D, the $p(t)$ and $s(t)$ are multiplied together:

$$\begin{aligned} p(t)s(t) &= A_t A_r \cos(2\pi f_c t)\cos[2\pi(f_c + f_D)t + \phi] \\ &= \frac{A_t A_r}{2}\{\cos(2\pi f_D t + \phi) + \cos[2\pi(2f_c + f_D)t + \phi]\}. \end{aligned}$$

By passing the signal described above through a lowpass filter, one can remove the high-frequency component at $2f_c$ and leave only the Doppler frequency signal

$$s_D(t) = \frac{A_t A_r}{2}\cos(2\pi f_D t + \phi).$$

A block diagram illustrating this Doppler frequency extraction procedure is shown in Fig. 5.4.

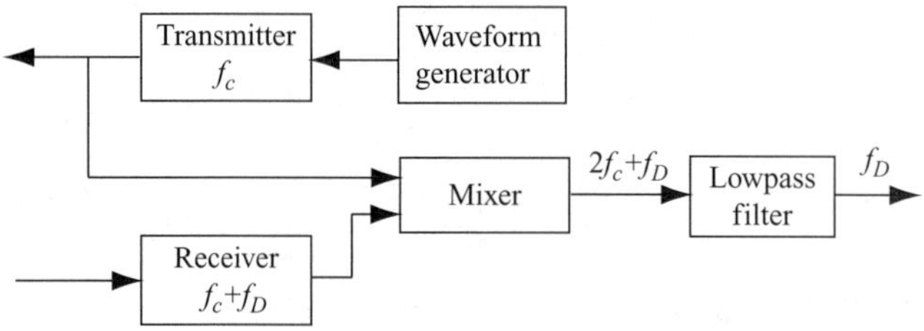

FIGURE 5.4 Block diagram of Doppler frequency extraction.

5.2.2 Motion Direction Detection

The Doppler process discussed above provides an absolute value of frequency difference; it contains no information regarding the direction of target motion. A number of techniques can be applied to preserve this directional information. Among them is a simple offset carrier demodulation method based on the AM technique, described next.

As shown in Eq. (4.8), the Doppler frequency f_D is positive if the target moves toward the receiver, while f_D becomes negative if the target moves away from it. This also implies that with amplitude modulation, the returned echo signal, which consists of frequency component $f_c + f_D$, will be greater than f_c if f_D is positive, and less than f_c if f_D is negative.

Let the received signal $s(t)$ be represented as

$$s(t) = A_r \cos[2\pi(f_c + f_D)t + \phi],$$

where A_r and ϕ are the amplitude and phase of the received signal. Let $p_0(t)$ be a reference signal and represented as

$$p_0(t) = A_0 \cos[2\pi(f_c + f_0)t],$$

where A_0 and f_0 are the amplitude and frequency of the reference signal, and f_0 is chosen to be greater than the maximum possible value of f_D.

The offset carrier demodulation method mixes the received signal $s(t)$ with the reference signal $p_0(t)$, and the resulting signal becomes

$$s(t)p_0(t) = \frac{A_r A_0}{2} \{\cos[2\pi(2f_c + f_0 + f_D)t + \phi] + \cos[2\pi(f_0 - f_D)t + \phi]\}.$$

This signal is then passed through a lowpass filter to remove the high-frequency part of $2f_c + f_0 + f_D$. If the target is approaching the receiver, then f_D is positive and the filter output will have frequency $f_0 - f_D < f_0$. If the target is moving away from the receiver, then f_D is negative and the filter output will have frequency $f_0 - f_D > f_0$. Figure 5.5 is a block diagram illustrating the process discussed above.

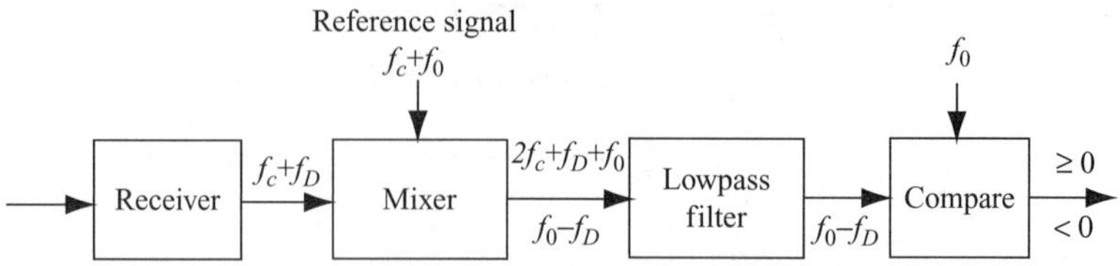

FIGURE 5.5 Block diagram of an offset carrier demodulation.

As an example, Fig. 5.6 shows the block diagram of a pulse–doppler radar, which can be used to detect the range and speed of multiple targets. A 5.0-GHz sinusoid wave is gated by a narrow pulse generator with period $T = 0.1$ ms to produce a pulsed transmitter waveform. The pulse generator is controlled by the "Timing & sync" circuit. The 5.0-GHz sinusoid wave is also used to mix with a 30-MHz coherent oscillator (COHO) to generate a 4.97-GHz signal, which is then mixed with the returned echo signal. The returned echo signal consists of the 5.0-GHz Tx signal plus a Doppler frequency. Multiple targets will cause multiple echoes with multiple Doppler frequencies. The output of the receiver mixer, which consists of the 30-MHz intermediate-frequency (IF) signal with Doppler frequencies, is fed to another mixer. The output of this mixer, which is the baseband target echo signal with Doppler frequency, is then fed to an A/D converter and the range gate. The dashed-line box that

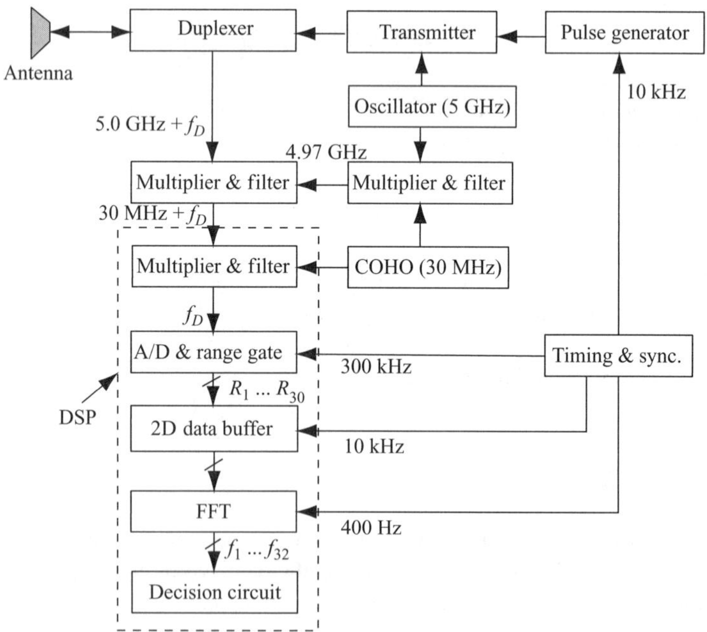

FIGURE 5.6 Block diagram of a pulse–Doppler radar system.

includes the mixer, A/D converter and range gate, 2D data buffer, FFT, and decision circuit can be implemented by a firmware-based DSP processor.

The A/D converter outputs 30 samples per radar pulse period. The magnitude of each sample (or range bin) corresponds to the strength of the returned echo signal at that range. The resolution of each range bin depends on the sampling frequency of the A/D converter. The 30 range samples per period will be stored as one row in the 2D databank. After 25 radar pulse periods, a 25 × 30 data array will be processed by fast Fourier transform. Each column of the 2D array corresponds to one target response. The output of column-based FFT is used to determine the value of the Doppler frequency for the corresponding target. Once the Doppler frequency is found, the moving target's speed can be computed accordingly. The number of samples in each column of the 2D databank represents the number of pulse periods needed for computation. To increase the frequency resolution without increasing the number of radar pulse periods, the time-domain zero-padding method, as described in Chapter 2, can be used. This is a combination of time- and frequency-domain co-processing on a pulse–Doppler radar and is commonly used in many modern radar systems.

A method that has the advantage of both the high energy of a long pulsewidth and the high resolution of a short pulsewidth is discussed next.

5.3 FREQUENCY MODULATION (FM)-RADAR

Frequency modulation (FM) is a technique in which the frequency of the carrier is varied in accordance with some characteristic of the baseband modulating signal $p_b(t)$:

$$p(t) = A_0 \exp\left\{ j \left[\omega_c t + \beta \int_{-\infty}^{t} p_b(t)\, dt \right] \right\}. \tag{5.6}$$

The instantaneous frequency of $p(t)$ can be obtained by differentiating the instantaneous phase of $p(t)$:

$$\begin{aligned} f &= \frac{1}{2\pi}\frac{d}{dt}\left[\omega_c t + \beta \int_{-\infty}^{t} p_b(t)\, dt \right] \\ &= f_c + \frac{1}{2\pi}\beta p_b(t). \end{aligned} \tag{5.7}$$

The baseband-modulating signal $p_b(t)$ normally is a sinusoidal function. Therefore Eq. (5.6) can be expressed as

$$p(t) = A_0 \exp\left[j\left(2\pi f_c t + \beta \sin\omega_a t\right)\right],$$

where f_c is the carrier frequency and β is referred to as the *modulation index* and is the maximum value of phase deviation.

The *linear frequency modulation* (LFM) technique involves a transmitter frequency that is continually increasing or decreasing from a fixed reference frequency. That is, the transmitted FM signal is modified so that the frequency is modulated in a linear manner with time. In other words, from Eq. (5.7), one obtains

$$f = f_c + \alpha t$$

and

$$\beta p_b(t) = 2\pi \alpha t,$$

where α is the frequency changing rate or chirp rate, and is defined as

$$\alpha = \frac{df}{dt}.$$

Substituting these equations into Eq. (5.6), one obtains

$$p(t) = A_0 \exp\left[j\left(2\pi f_c t + \pi \alpha t^2\right)\right]. \tag{5.8}$$

Two types of LFM signals are commonly used in radar applications: the pulsed LFM (also called "chirp") and the continuous-wave LFM (CWLFM). The pulsed LFM signal is described first.

5.3.1 Pulsed Linear Frequency Modulation (LFM) Radar

Let $a_1(t)$ and $a_2(t)$ be defined as

$$\begin{aligned} a_1(t) &= \text{Rect}\left(\frac{t - T_p/2}{T_p}\right), \\ a_2(t) &= A_0 \exp\left[j\left(2\pi f_c t + \pi \alpha t^2\right)\right], \end{aligned} \tag{5.9}$$

where Rect(t) is a rectangular gate function, defined as

$$\begin{aligned} \text{Rect}(t) &= 1 \quad \text{for } -\frac{1}{2} \le t \le \frac{1}{2}, \\ &= 0 \quad \text{otherwise}. \end{aligned} \tag{5.10}$$

A pulsed symmetric LFM signal $a(t)$ with duration T_p can be written as follows, from Eq. (5.8):

$$\begin{aligned} p(t) &= a_1(t)\,a_2\left(t - \frac{T_p}{2}\right) \\ &= \text{Rect}\left(\frac{t - T_p/2}{T_p}\right) \cdot A_0 \exp\left[j\left(2\pi f_c\left(t - \frac{T_p}{2}\right) + \pi\alpha\left(t - \frac{T_p}{2}\right)^2\right)\right]. \end{aligned} \tag{5.11a}$$

The pulsed nonsymmetric LFM signal $p(t)$ with duration T_p can be written as

$$\begin{aligned} p(t) &= a_1(t)\,a_2(t) \\ &= \text{Rect}\left(\frac{t - T_p/2}{T_p}\right) \cdot A_0 \exp\left[j(2\pi f_c t + \pi\alpha t^2)\right]. \end{aligned} \tag{5.11b}$$

A baseband nonsymmetric pulsed LFM waveform with pulse duration T_p and period T is shown in Fig. 5.7a. The corresponding frequency–time relationship is shown in Fig. 5.7b. As can be seen, the LFM signal frequency increases from 0 to $f_{\max}$

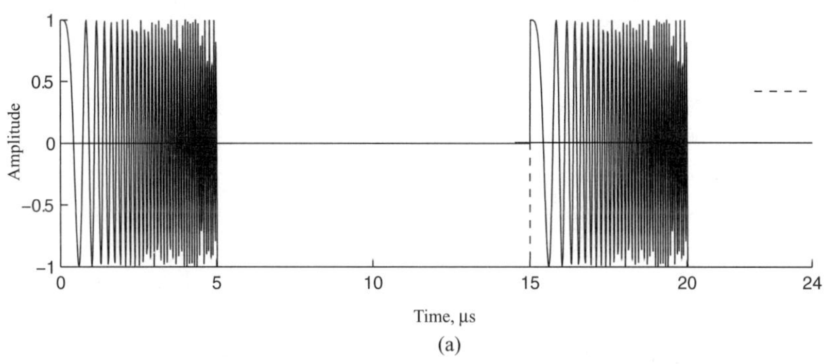

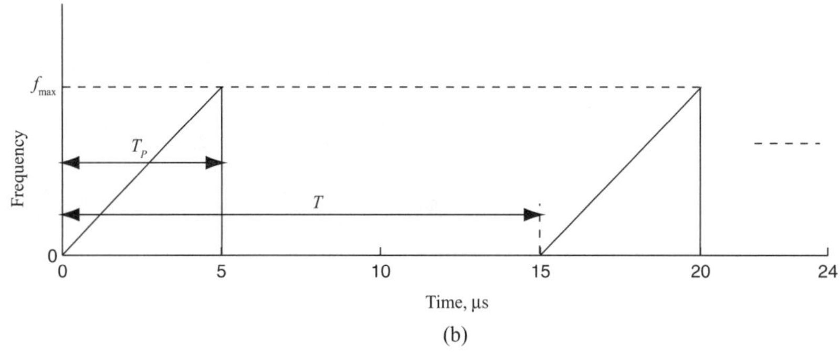

FIGURE 5.7 Time-domain waveform (a) and time–frequency relation (b) of a pulsed LFM signal.

during the pulse duration time T_p, where $f_{\max} = \alpha T_p$ and the frequency bandwidth is $B = f_{\max} = \alpha T_p$.

Analytically, the frequency spectrum of $p(t)$, in the nonsymmetric LFM case, can be considered as the convolution of the two spectra $A_1(f)$ and $A_2(f)$, where $A_1(f)$ and $A_2(f)$ are the Fourier transform of $a_1(t)$ and $a_2(t)$, respectively; that is

$$\begin{aligned}
A_1(f) &= \mathfrak{F}[a_1(t)] \\
&= T_p \frac{\sin(\pi f T_p)}{\pi f T_p} \exp(j\pi f T_p) \\
&= T_p \operatorname{sinc}(f T_{\mathrm{p}}) \exp(j\pi f T_p), \\
A_2(f) &= \mathfrak{F}[a_2(t)] \\
&= \int_{-\infty}^{\infty} \exp\left[j\left(2\pi f_c t + \pi \alpha t^2\right)\right] \cdot \exp(-j2\pi f t)\, dt \\
&= \int_{-\infty}^{\infty} \exp\left\{ j\pi\alpha \left[\left(t - \frac{f - f_c}{\alpha} \right)^2 - \frac{(f - f_c)^2}{\alpha^2} \right] \right\} dt \\
&= \exp\left[-j \frac{\pi (f - f_c)^2}{\alpha} \right] \cdot \frac{1}{\sqrt{\pi}} \int_{-\infty}^{\infty} \exp(j\alpha x^2) dx,
\end{aligned}$$

where

$$x = \sqrt{\pi}\left(t - \frac{f - f_c}{\alpha}\right).$$

Let $-u^2 = j\alpha x^2$, or $u = \sqrt{-j\alpha}x$; then

$$\begin{aligned}
\int_{-\infty}^{\infty} \exp(j\alpha x^2) dx &= \frac{1}{\sqrt{-j\alpha}} \int_{-\infty}^{\infty} \exp(-u^2) du \\
&= \sqrt{\frac{\pi}{\alpha}} \exp\left(j\frac{\pi}{4}\right).
\end{aligned}$$

Therefore

$$A_2(f) = \exp\left(j\frac{\pi}{4}\right) \cdot \sqrt{\frac{1}{\alpha}} \exp\left[-j \frac{\pi (f - f_c)^2}{\alpha} \right],$$

and

$$P(f) = T_p \exp\left(j\pi f T_p\right) \operatorname{sinc}\left(f T_p\right) \otimes \exp\left(j\frac{\pi}{4}\right) \cdot \frac{1}{\sqrt{\alpha}} \exp\left[-j\frac{\pi (f - f_c)^2}{\alpha}\right], \tag{5.12}$$

where the symbol $\otimes$ represents convolution.

In practical applications, the frequency spectrum of LFM-based signals is generated by applying DFT on the signals. As an example, the following parameters are used to generate the pulsed LFM waveforms together with their frequency spectra. Both symmetric and nonsymmetric LFM signals are covered:

Frequency chirp rate $\alpha = 3 \times 10^{12}$
Pulse duration time $T_p = 5 \times 10^{-6}$ s
Period of pulse train $T = 1.024 \times 10^{-5}$ s
Sampling frequency $f_s = 50 \times 10^6$ Hz
Bandwidth: $T_p \times \alpha = 15$ MHz

Figure 5.8 displays the symmetric pulsed LFM signal and its frequency spectrum. The real and imaginary parts of the time-domain pulsed LFM waveform are shown in Figs. 5.8a and 5.8b. Both drawings are symmetric around the center of pulse duration. Figures 5.8c and 5.8d show the real and imaginary parts of the frequency-domain spectrum. Figures 5.8e and 5.8f illustrate the absolute value (or magnitude) and phase of the frequency spectrum. As can be seen, the spectrum is almost flat from frequency bins 175 to 325 for a total of approximately 150 bins. Notice that the frequency spectrum was shown with the origin located at the center of display, or realigned between $-f_s/2$ and $f_s/2$. Given the sampling frequency $f_s = 50$ MHz, which corresponds to frequency bin 512, the bandwidth of the pulsed LFM is approximately $(325 - 175) \times 30/512 \approx 15$ MHz. The "sinc"-like shape at the two band edges of the LFM spectrum is caused by the convolution of the sinc function of $A_1(\omega)$ with $A_2(\omega)$, which is a constant in magnitude and is a band-limited signal.

Figure 5.9 displays the nonsymmetric pulsed LFM signal and its frequency spectrum. Figures 5.9a and 5.9b show the real and imaginary parts of the time-domain pulsed LFM waveform. Both drawings are no longer symmetric around the center pulse duration. Figures 5.9c and 5.9d show the real and imaginary parts of the frequency-domain spectrum; Figs. 5.9e and 5.9f illustrate the magnitude and phase of the frequency spectrum. Again, the spectrum is almost flat between frequency bins 260 and 410, for a total of ~150 bins. The frequency display is centered around the origin, or from $-f_s/2$ to $f_s/2$. The bandwidth of the pulsed nonsymmetric LFM is ~15 MHz. Notice that no negative frequency component exists in this case, which differs from the symmetric LFM case.

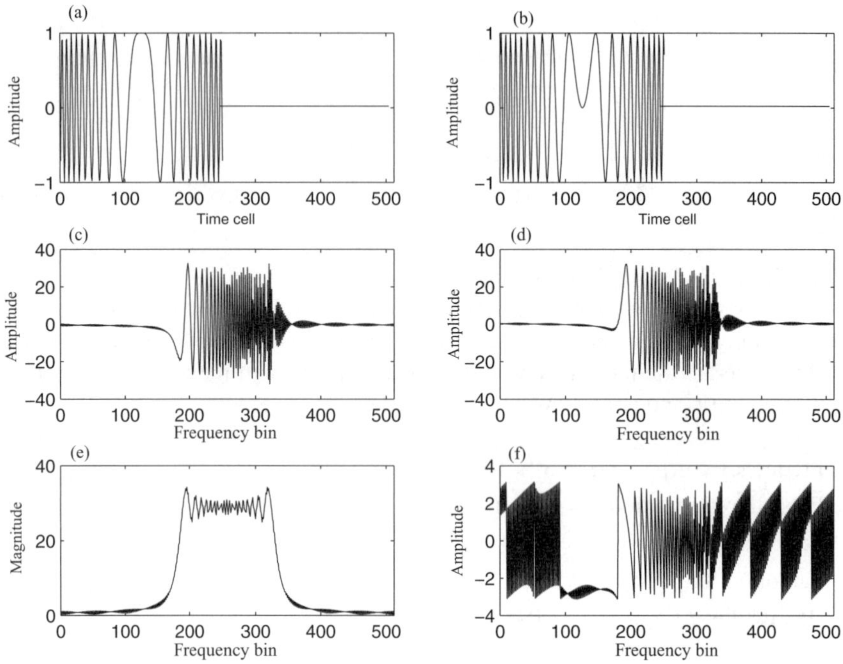

FIGURE 5.8 Time- and frequency-domain waveforms of a pulsed symmetric LFM signal.

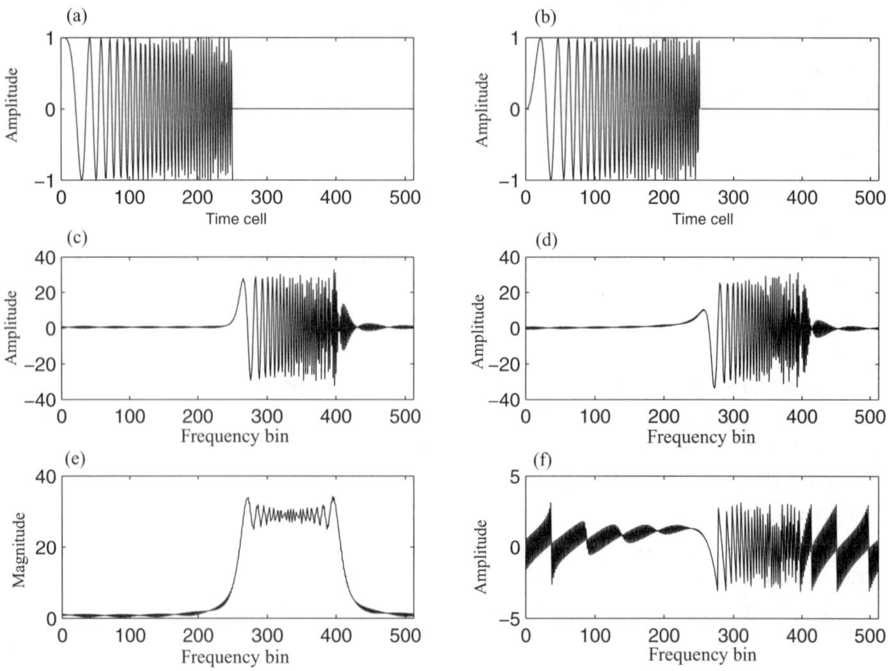

FIGURE 5.9 Time- and frequency-domain waveforms of a pulsed nonsymmetric LFM signal.

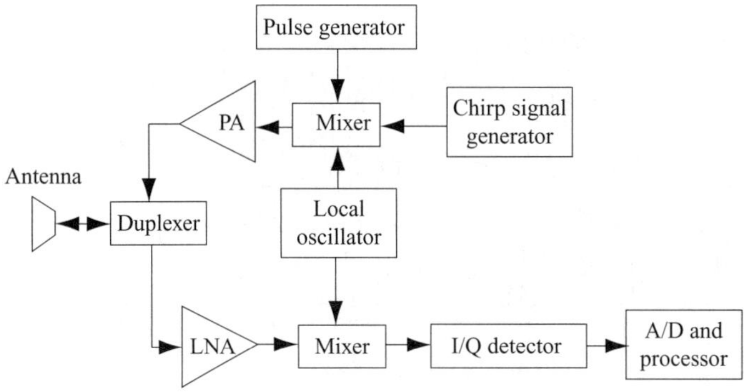

FIGURE 5.10 Block diagram of a PLFM radar system.

A block diagram of pulsed LFM radar system is shown in Fig. 5.10. Here the transmitter consists of a pulse generator, chirp signal generator, mixer, power amplifier, and local oscillator. The receiver consists of a low-noise amplifier (LNA), mixer, in-phase–quadrature-phase detector (or demodulator), local oscillator, analog–digital (A/D) converter, and signal processor. The antenna is connected to either the transmitter or the receiver through a duplexer. The in-phase–quadrature-phase (I–Q) detector and other target-detection-related processing are discussed in later sections.

5.3.2 Continuous-Wave Linear Frequency Modulation Radar

Continuous-wave linear frequency modulation (CWLFM) radar transmits and receives signals continuously. The transmitting frequency is first increased then decreased, and the same pattern is repeated at time interval T. Figure 5.11 is the block diagram of a CWLFM radar system.

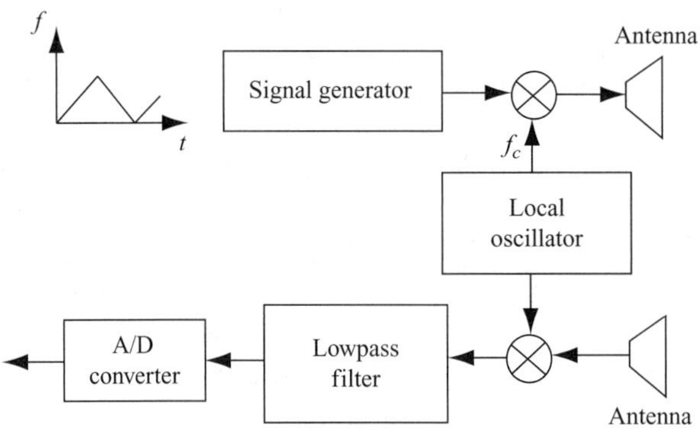

FIGURE 5.11 Block diagram of a CWLFM radar system.

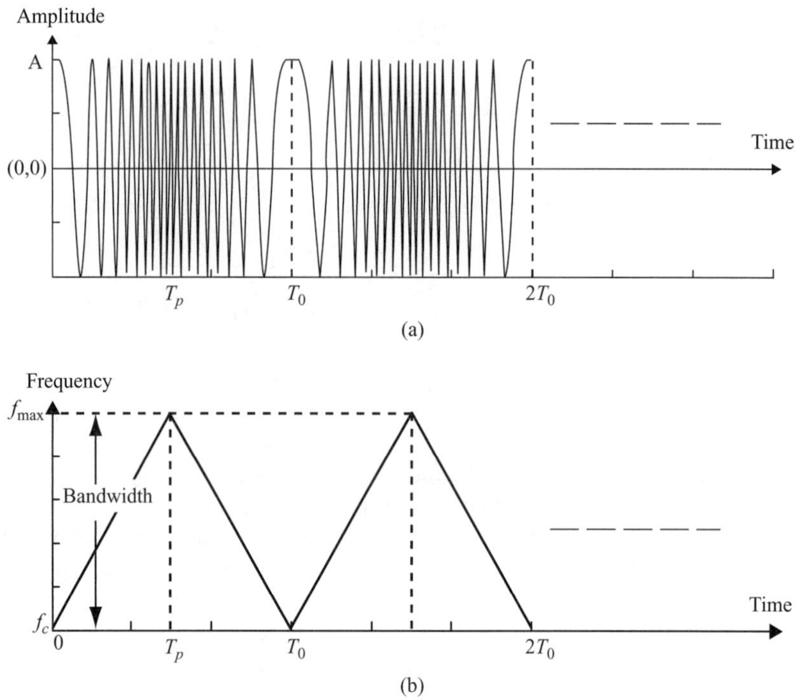

FIGURE 5.12 Time–Frequency relationship of a CWLFM radar signal.

A CWLFM signal has a triangular shape of frequency change with respect to time. The upsweep and downsweep frequencies of the triangle are defined as

$$f = f_c + \alpha t \quad \text{for} \quad 0 < t < T_p,$$
$$f = f_{\max} + \beta(T_p - t) \quad \text{for} \quad T_p < t < T_0,$$

where $f_{\max} = f_c + \alpha T_p$ and α, β are the two chirp rates of the upsweep and downsweep frequencies. If $T_0 = 2T_p$, then the triangle is symmetric and $\alpha = \beta$. In this case, the signal frequency increases from f_c to $f_{\max}$ during the first half of the triangle, and decreases from $f_{\max}$ to f_c for the second half. The sweep frequency bandwidth $B = f_{\max} - f_c = \alpha T_p$. Figure 5.12 displays the time–frequency relationship of a CWLFM signal when $\alpha = \beta$.

A third type of frequency-modulation-based technique for radar application is stepped frequency modulation, discussed next.

5.3.3 Stepped Frequency Modulation Radar

The stepped frequency modulation (SFM) radar has been used extensively in short-range measurements to study the scattering properties of geophysical surfaces and

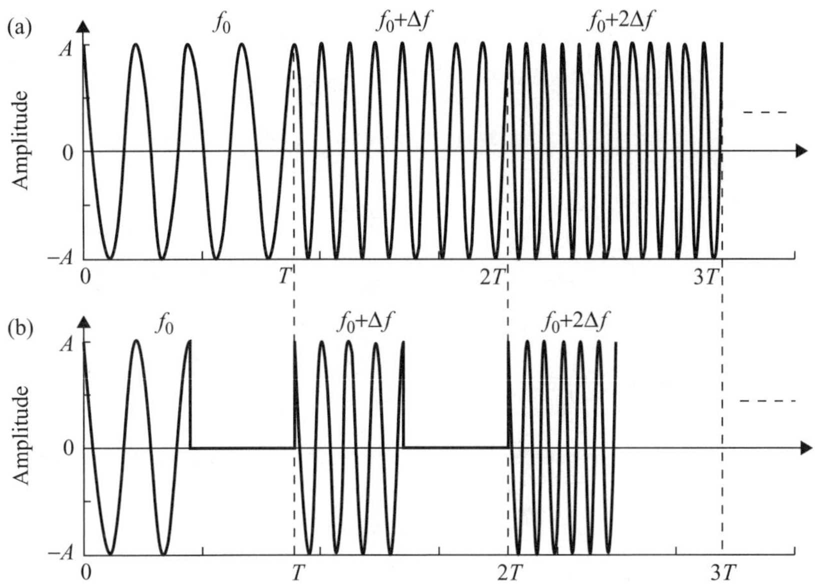

FIGURE 5.13 Waveforms of (a) a CWSFM radar signal and (b) a pulsed SFM radar signal.

other nontransparent media. It computes the target range by measuring the magnitude and phase response over a number of stepped frequencies (N) within a given frequency band.

There are two types of stepped-frequency-based radar: the continuous-wave stepped frequency (CWSFM) radar and the pulsed SFM (PSFM) radar. Both the CWSFM radar and PSFM radar transmit and receive a sequence of N-frequency waveforms changed by the frequency step Δf. However, each section of the N-frequency waveforms is consecutive for the CWSFM radar, but in a pulsed form for the PSFM radar. Figure 5.13 shows the SFM waveforms in the time domain; Fig. 5.13a displays the CWSFM signal, while Fig. 5.13b shows the PSFM signal.

The starting frequency of the CWSFM or PSFM signal called the *base frequency* f_0 is the lowest frequency transmitted. The frequency change or step is a constant increment from the preceding frequency, and is denoted by Δf. The bandwidth B of a stepped frequency radar is $f_U - f_0 = (N-1)\Delta f$, where f_U is the highest frequency and N is the number of frequency steps. Each single stepped frequency signal lasts for a period of time T, which is also considered as the pulse repetition interval (PRI), which serves as the time window where both transmitting and receiving signals are used to detect the targets. The N-frequency section group is considered as one burst, and each burst lasts for a time period NT.

Figure 5.14 displays the time–frequency relationship for the N-frequency CWSFM waveform (Fig. 5.14a) and PSFM waveform (Fig. 5.14b), respectively. The target detection and range computation schemes for both CWSFM and PSFM signals are quite similar; only the hardware implementation is different.

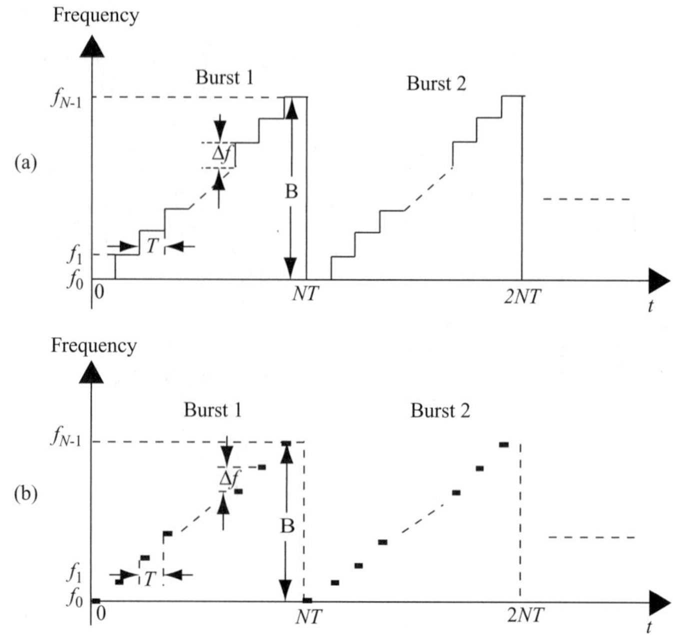

FIGURE 5.14 Time–frequency relationship of (a) CWSFM radar signal and (b) a PSFM radar signal.

The block diagram of a stepped-frequency-based radar system is shown in Fig. 5.15. The receiver box normally consists of an intermediate-frequency (IF) demodulator as shown in Fig. 5.6. Compare this with the CWLFM case shown in Fig. 5.11, where an extra system box, namely, the quadrature mixer, or the I–Q demodulator, is used. The I–Q demodulator generates the in-phase and quadrature-phase signals from the received signal, discussed next.

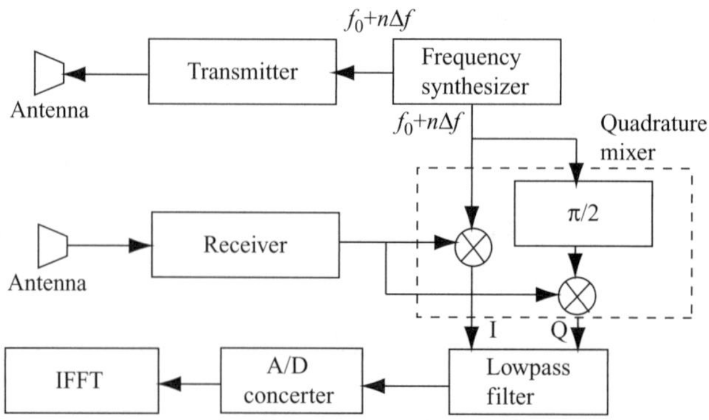

FIGURE 5.15 Block diagram of a stepped frequency modulation radar.

5.4 TARGET DETECTION TECHNIQUES OF FM-BASED RADAR

5.4.1 In-Phase Quadrature-Phase Demodulator

Consider the LFM signal represented by Eq. (5.8):

$$p(t) = A_0 \exp\left[j\left(2\pi f_c t + \pi\alpha t^2\right)\right].$$

The observed and real-valued signal of this equation is represented as

$$\begin{aligned} p(t) &= \cos(2\pi f_c t + \pi\alpha t^2) \\ &= \cos[2\pi f_c t + \phi(t))]. \end{aligned}$$

In general, the carrier frequency f_c has a high value and αt has a relative low value. The phase term $\phi(t) = \pi\alpha t^2$ is presented for illustration purposes. In a digital communication system, a complex-valued signal is needed to perform functions such as phase detection of a QAM signal, adaptive equalization, and many other functions. The complex-valued signal is required to transmit the time-varying phase information, and an in-phase–quadrature-phase (I–Q) demodulator is used to receive the correct phase information.

Figure 5.16 displays a typical I–Q demodulator where the input signal $v(t)$ is a general LFM signal as described above, and the output are the real and imaginary parts of the phase signal $\phi(t)$. The two carrier signals, $\cos(2\pi f_c t)$ and $\sin(2\pi f_c t)$, are locally generated. The lowpass filter is designed to retain the low-frequency component $\phi(t)$, but will filter out the high-frequency component $4\pi f_c t + \phi(t)$. The A/D converter is used to digitize the real and imaginary parts of the low-frequency phase signal $\phi(t)$.

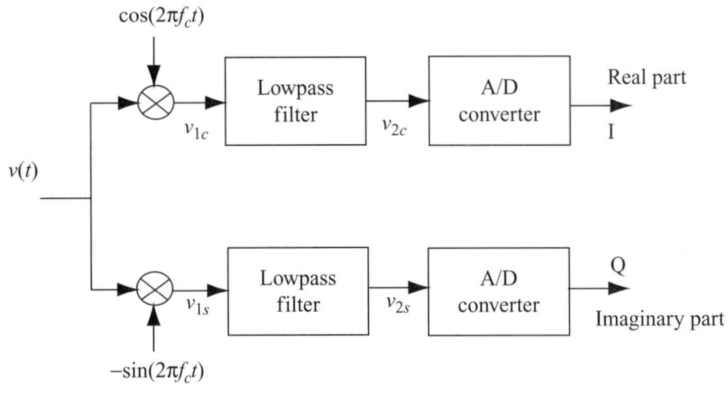

FIGURE 5.16 In-phase–quadrature-phase (I–Q) demodulator.

The intermediate signals $v_{1c}(t)$, $v_{1s}(t)$, $v_{2c}(t)$, and $v_{2s}(t)$ can be derived as follows:

$$v_{1c}(t) = \frac{\{\cos[4\pi f_c t + \phi(t)] + \cos[\phi(t)]\}}{2},$$

$$v_{1s}(t) = \frac{\{\sin[4\pi f_c t + \phi(t)] + \sin[\phi(t)]\}}{2},$$

$$v_{2c}(t) = \frac{\cos[\phi(t)]}{2},$$

$$v_{2s}(t) = \frac{\sin[\phi(t)]}{2}.$$

The phase signal $\phi(t)$ can then be computed as follows:

$$\phi(t) = \tan^{-1}\left[\frac{v_{2s}(t)}{v_{2c}(t)}\right].$$

5.4.2 Matched Filter and Pulse Compression

The matched filter technique is widely used in digital communication fields to recover a signal that is corrupted by additive white Gaussian noise. In radar applications, a matched filter is applied on received signals to identify a target by determining whether the filter output, which is a compressed pulse, exceeds a specific threshold. Multiple targets will produce multiple pulses with varying magnitudes. The magnitude of a compressed pulse depends on the duration time of the receiving pulse (T_p), the reflectivity of target (σ), and the distance from radar to target (D). The effective width of a compressed pulse is determined by the duration time and frequency bandwidth (B) of the receiving pulse, or the time–bandwidth (T_pB) product. In the following discussion, a target is assumed to be an ideal target with reflectivity $\sigma = 1$, with no signal attenuation occurring between radar and target.

Given an input signal $s(t)$ and its Fourier transform $S(f)$, the matched filter $h(t)$ and its Fourier transform $H(f)$ are defined as

$$h(t) = s^*(-t) \tag{5.13a}$$

and

$$H(f) = S^*(f). \tag{5.13b}$$

Let the output of a matched filter be $f(t)$; then its Fourier transform satisfies the relation

$$\begin{aligned} F(f) &= S(f)H(f) \\ &= |S(f)|^2. \end{aligned} \tag{5.14a}$$

Therefore

$$\begin{aligned} f(t) &= \mathfrak{F}^{-1}[|S(f)|^2] \\ &= s(t) * h(t) \\ &= \int s(\tau) h(t-\tau) d\tau. \end{aligned} \tag{5.14b}$$

Since $h(t) = s^*(-t)$, it follows that

$$f(t) = \int s(\tau) s^*(t+\tau) d\tau, \tag{5.14c}$$

which is the definition of autocorrelation of $s(t)$. In other words, the matched filtering of $s(t)$ is equivalent to the autocorrelation of $s(t)$ with itself.

Let the echo signal from a single target be the pulsed nonsymmetric LFM signal described in Eq. (5.11b):

$$\begin{aligned} p(t) &= a_1(t) a_2(t) \\ &= \text{Rect}\left(\frac{t - T_p/2}{T_p}\right) \cdot A_0 \exp\left[j\left(2\pi f_c t + \pi\alpha t^2\right)\right]. \end{aligned}$$

With $A_0 = 1$ and $h(t) = p^*(-t)$, the matched filter output becomes

$$\begin{aligned} f(t) &= \int_0^{T_p} p(\tau) \cdot p^*(\tau + t))d\tau \\ &= \int_0^{T_p} \exp\left[j\left(2\pi f_c \tau + \pi\alpha\tau^2\right)\right] \cdot \exp\left\{-j\left[2\pi f_c(t+\tau) + \pi\alpha(t+\tau)^2\right]\right\} d\tau \\ &= \exp\left[-j\left(2\pi f_c t + \pi\alpha t T_p + \pi\alpha t^2\right)\right] \cdot T_p \cdot \frac{\sin\left(\pi\alpha t T_p\right)}{\pi\alpha t T_p}. \end{aligned} \tag{5.15}$$

Let $x = \alpha t T_p = Bt$; then the absolute value of $f(t)$ becomes

$$\begin{aligned} |f(x)| &= T_p \cdot \frac{\sin\pi x}{\pi x} \\ &= T_p \cdot \text{sinc}(x). \end{aligned} \tag{5.16}$$

The function $|f(x)|$ has a maximum value at $x = 0$, and its magnitude equals T_p. From the discussion in Chapter 2, the first x value of $\text{sinc}(x) = 0$ occurs at $x = Bt = 1$, or $t = 1/B$. It is also true that the 3-dB mainlobe width of $\text{sinc}(x)$ equals the first x value of $\text{sinc}(x) = 0$. Therefore, the 3-dB mainlobe width or the compressed pulsewidth is $t = 1/B$, where $B = \alpha T_p$ is the bandwidth of an LFM signal with time duration T_p. The pulse compression ratio is therefore

$$\frac{\text{Original LFM pulse duration}}{\text{Compressed pulse duration}} = \frac{T_p}{(1/B)} = T_p B. \tag{5.17}$$

Notice that T_pB is the time–bandwidth product.

Since the time-domain convolution of two signals is equivalent to the multiplication of two signals in the frequency domain, the matched filter can be implemented by transforming both the received signal and the matched filter function into a frequency domain using FFT. The two frequency-domain functions are then multiplied together and followed by IFFT to convert back to the time domain. Because the matched filter function is normally shorter than the received signal (due to multiple targets at different ranges), zero padding on the matched filter function is required when computing the FFT on the matched filter function.

Figure 5.17 illustrates the process of DFT-based pulse compression (or matched filtering). Here the reference chirp signal is the matched filter function $h(t)$, which equals $p^*(-t)$, with $p(t)$ as the transmitting signal.

Examples 5.1 and 5.2 (below) illustrate the implementation of pulse compression in two approaches; one through direct convolution in time domain and the other through DFT in the frequency domain. The first example is based on a symmetric pulsed LFM with a single target, while the second example is based on a nonsymmetric pulsed LFM with dual targets. The signal and target related parameters are listed here:

Frequency chirp rate $\alpha = 2 \times 10^{12}$

Pulse duration time $T_p = 4 \times 10^{-6}$ s

Frequency bandwidth of pulsed LFM $B = \alpha\, T_p = 8$ MHz

Sampling frequency $f_s = 20$ MHz

Number of samples in $T_p = 80$ (81 was used in simulation)

Time bandwidth product $T_pB = 32$

Delay time between Tx pulse and Rx signal:

Example 5.1: 5×10^{-6} s (or 100-sample delay)

Example 5.2: First target: 5×10^{-6} s (or 100-sample delay)

Second target: 7.5×10^{-6} s (or 150-samples delay)

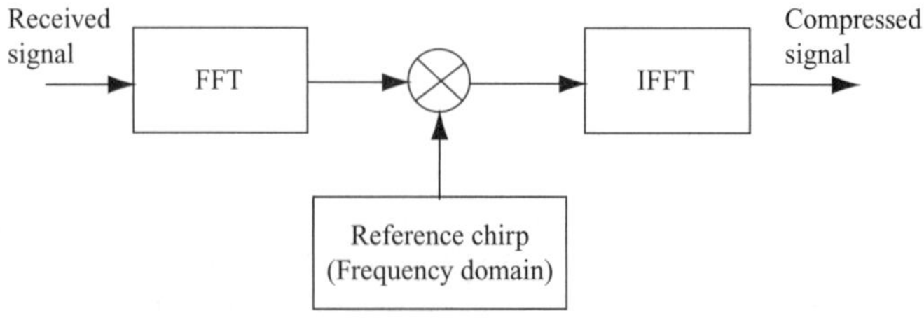

FIGURE 5.17 DFT-based processing of chirp signal.

Example 5.1 Symmetric Pulsed LFM with Single-Target Response Figure 5.18 displays the waveforms of both the transmitter (Tx) signal and the matched filter (MF) function. Since the Tx signal is symmetric, the real part of the MF function is identical to that of the time reversed Tx signal, while the imaginary part of the MF function becomes negative of that of the time reversed Tx signal.

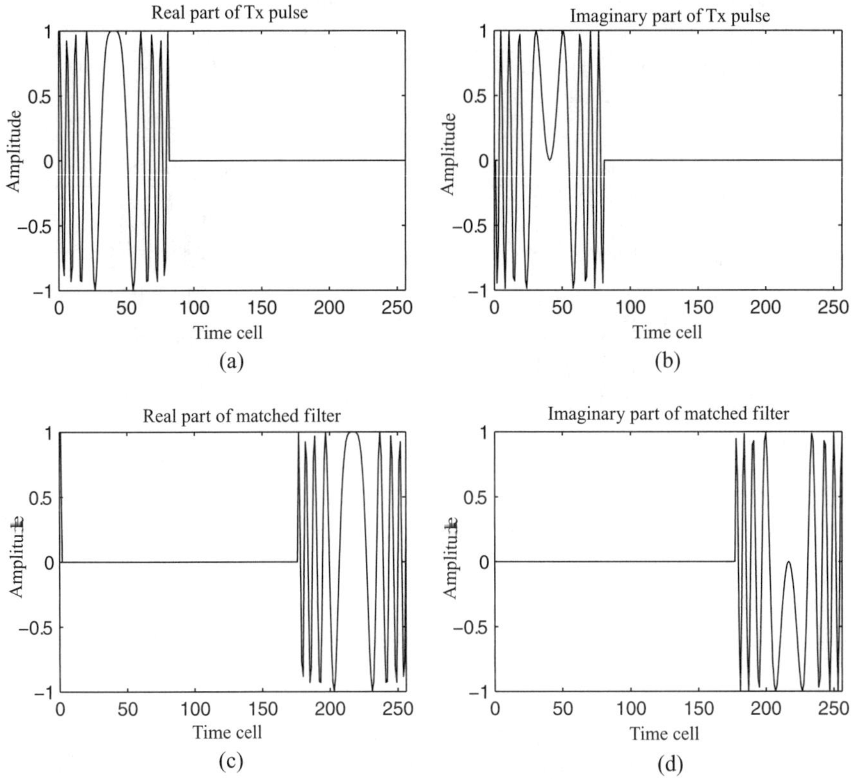

FIGURE 5.18 Waveforms of Tx signal and matched filter function.

Figure 5.19 displays the relationship between the transmitter signal and the received signal. The Rx and Tx waveforms are identical except that the former is delayed by 100 samples relative to the Tx. Here an ideal target is assumed, and the received signal has the same magnitude as the transmitter.

The frequency spectrum of a transmitter signal is shown in Fig. 5.20. Figures 5.20a and 5.20b display the real and imaginary parts of the spectrum, while Figs. 5.20c and 5.20d display the spectrum in terms of magnitude and phase. A 256-point FFT is used to compute the frequency spectrum with bin 128 as the center or origin of the spectrum. As can be seen, the frequency spectrum is symmetric around the origin (bin 128). Since the sampling frequency is 20 MHz, which corresponds to 256 frequency bins, the bandwidth is ~8 MHz (from bins ~80 to 180).

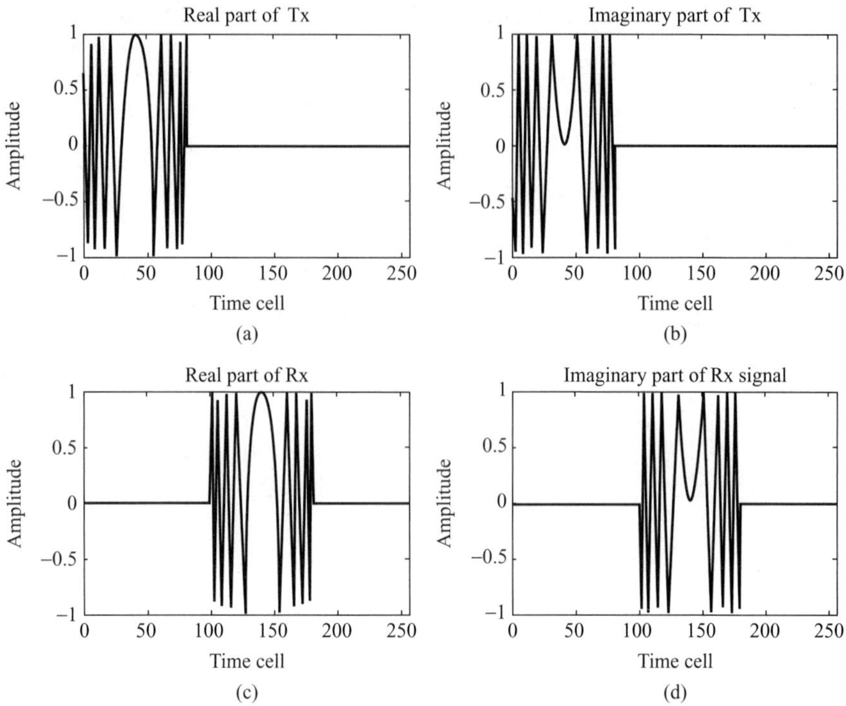

FIGURE 5.19 Waveforms of Tx signal and Rx signal.

The frequency spectrum of a matched filter is shown in Fig. 5.21. Figures 5.21a and 5.21b display the real and imaginary parts of the spectrum; Figs. 5.21c and 5.21d display the spectrum in terms of magnitude and phase. When comparing the spectrum of a matched filter with that of the Tx signal, one can see that the real part is identical to that of the Tx signal, and the imaginary part is negative of that of the Tx signal. The magnitude is the same, but the phase is negative of that of the Tx signal. Again, a 256-point FFT is used to compute the frequency spectrum with bin 128 as the origin of the spectrum.

Figure 5.22 displays the frequency spectrum of the received signal Rx. Although the received signal is a delayed version of the transmitter signal, the real and imaginary parts of the frequency spectrum are quite different from those of the Tx signal. The phase of the Rx spectrum also differs from that of Tx, but the magnitude is identical to that of Tx.

Figure 5.23 displays the compressed receiving signal, or matched filter output. Figures 5.23a and 5.23b show results based on time-domain processing; this is the direct convolution of received signal Rx with the MF function. Figures 5.23c and 5.23d show results based on frequency-domain processing, where the received signal and the MF function are first transformed into the frequency domain. The two frequency spectra are then multiplied together, followed by inverse DFT to obtain the final result as shown in Figs. 5.23c and 5.23d. Both time-domain and frequency-domain processing of pulse compression yield the same results.

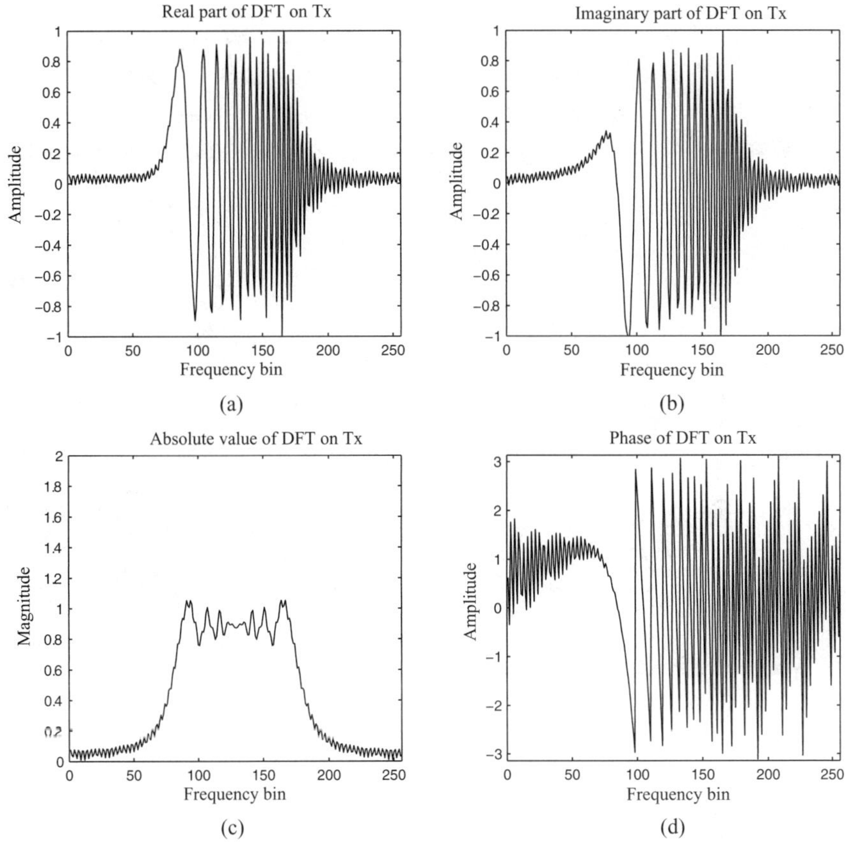

FIGURE 5.20 Frequency spectrum of Tx signal.

Notice that the magnitude of the compressed pulse is 80, which equals the sample number of the LFM pulse. The peak of the compressed pulse occurs at time cell 180. This number corresponds to the sum of the Rx signal sample delay (100 in this example) and the sample number of LFM pulse (80 in this example).

The sample number in the compressed pulse, which equals the number between the peak of the pulse and the first zero crossing, is about 2.5 samples. Compared with the sample number in Tx pulse duration time, which equals 80, the pulse compression ratio is 80/2.5 = 32. This is the same number as the time–bandwidth product, or $T_pB = 32$.

Example 5.2 Nonsymmetric LFM Pulse with Dual-Target Response Figure 5.24 displays the time-domain waveforms of a pulsed LFM transmitter signal and the matched filter (MF). Figures 5.24a and 5.24b show the real and imaginary parts of the Tx signal. The waveform is nonsymmetric, and there are 80 samples within the pulse duration time T_p. Figures 5.24c and 5.24d show the real and imaginary parts of the MF. Both the real and imaginary part of MF waveforms are time-reversed relative

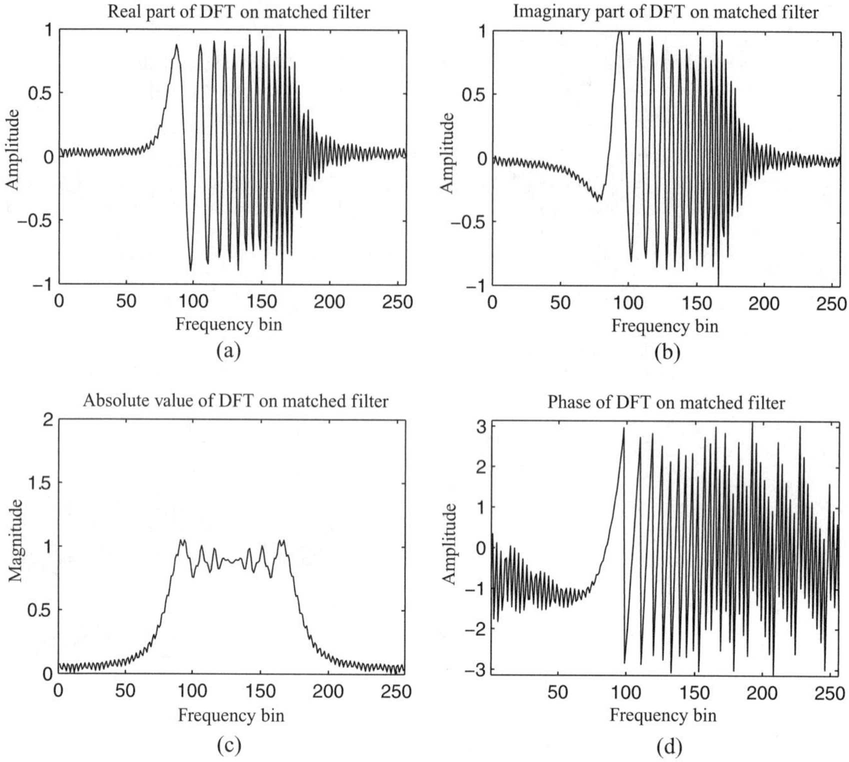

FIGURE 5.21 Frequency spectrum of matched filter (MF) function.

to those of the Tx signal. In addition, the magnitude is negated for the imaginary part of the MF. Similar to the symmetric pulsed LFM matched filter, the nonsymmetric LFM MF is flipped around the origin (time cell 1), and appears from time cells 178 to 256 and cell 1 for a 256-point DFT computation.

Figure 5.25 displays the frequency spectra of a pulsed LFM Tx signal and the MF signal. Figures 5.25a and 5.25b show the magnitude and phase parts of the Tx signal, while Figs. 5.25c and 5.25d show the magnitude and phase parts of the MF signal. As can be seen, the magnitude of the MF spectrum is identical to that of Tx, while the phase of the MF spectrum is the negated version of the Tx spectrum. Given the sampling frequency $f_s = 20$ MHz and the 256-point DFT used, the Tx bandwidth is approximately 8 MHz (from about bins ~130 to 230).

The received signal consists of two target responses. One occurs at time cell 100 and the other, at time cell 150. These two target-reflected signals are added together and shown in Figs. 5.26a and 5.26b. The magnitude and phase of the frequency spectrum corresponding to the Rx signal are shown in Figs. 5.26c and 5.26d. A 512-point DFT is used to compute the spectrum of the received signal. Given the sampling frequency $f_s = 20$ MHz, the bandwidth of Tx is approximately 8 MHz (from about bins ~260 to 460).

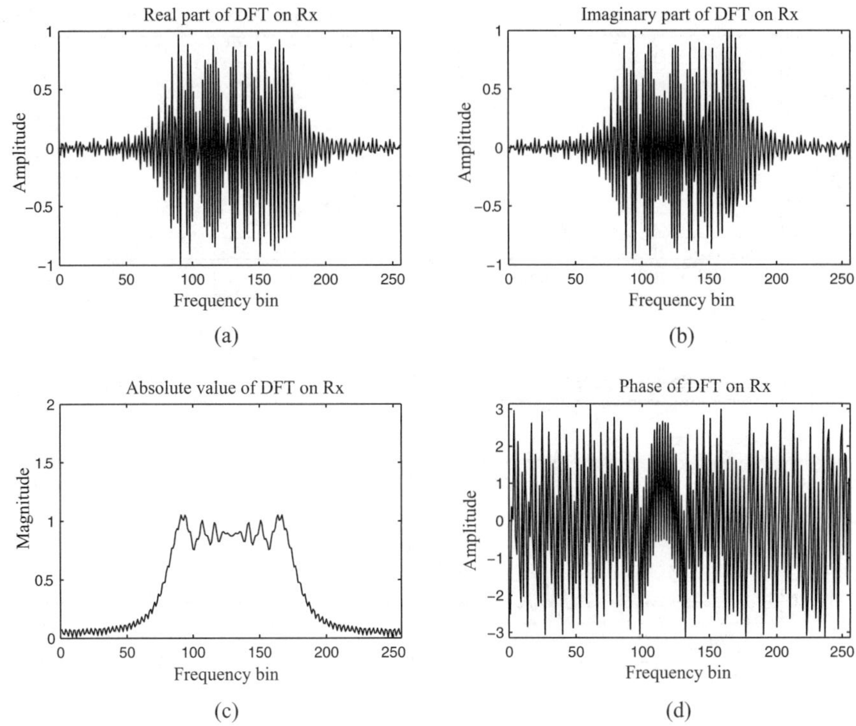

FIGURE 5.22 Frequency spectrum of Rx signal.

Figure 5.27 displays the received signal after compression; Figs. 5.27a and 5.27b show results based on convolution of the Rx signal with the MF function, while Figs. 5.27c and 5.27d show results based on frequency-domain processing. Both time-domain and frequency-domain processes yield the same results.

Similar to the case of symmetric LFM, the magnitude of the compressed pulse is 80. There are two pulses appearing at time cells 180 and 230, respectively. Time cells 180 and 230 correspond with respect to the sum of the sample numbers of the LFM pulse (80) and the sample delay of the first Rx target (100); and also to the sum of the sample delay of the second target (150) and the sample number of the LFM pulse (80).

Similar to the case in Example 5.1, the sample number in the compressed pulse is ~2.5 samples and the pulse compression ratio is 80/2.5 = 32.

5.4.3 Target Detection Techniques of LFM Radar

A pulsed LFM signal with duration T_p can be represented as follows from Eq. (5.8):

$$p(t) = A_0 \exp\left[j(2\pi f_c t + \pi \alpha t^2)\right].$$

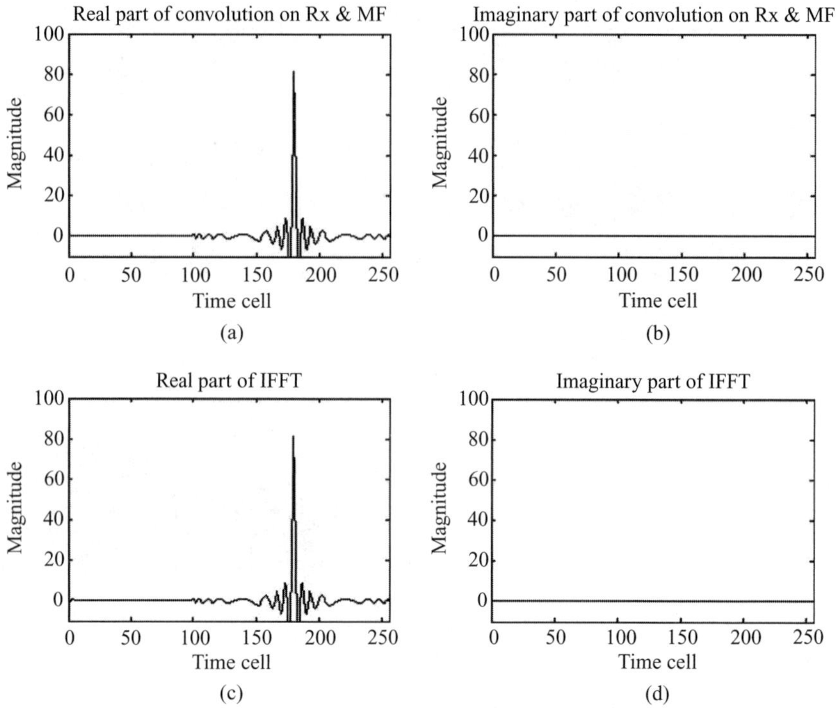

FIGURE 5.23 Comparison of pulse compression based on convolution and DFT.

The frequency spectrum of the signal is bounded as

$$
\begin{aligned}
f &\in [f_c, f_{\max}] \\
&= [f_c, f_c + \alpha T_p],
\end{aligned}
$$

and its frequency bandwidth is $B = \alpha T_p$.

Assuming ideal targets and $A_0 = 1$, the received signal can then be expressed as

$$
\begin{aligned}
s(t) &= \sum_n \exp\left[j2\pi f_c (t - t_n) + j\pi\alpha (t - t_n)^2\right] \\
&= \sum_n \exp\left[j2\pi f_c (t - t_n)\right] \exp\left[\left(\pi\alpha t^2 - 2\pi\alpha t_n t + \pi\alpha t_n^2\right)\right],
\end{aligned}
\tag{5.18}
$$

where $t_n = 2R_n/c$ is the round-trip echo delay from the nth target.

To dechirp the received signal, the complex conjugate of the echo signal $s(t)$ is multiplied with a locally generated chirp signal, which is the same as the transmitting chirp signal, except that it is a continuous LFM signal with a much greater

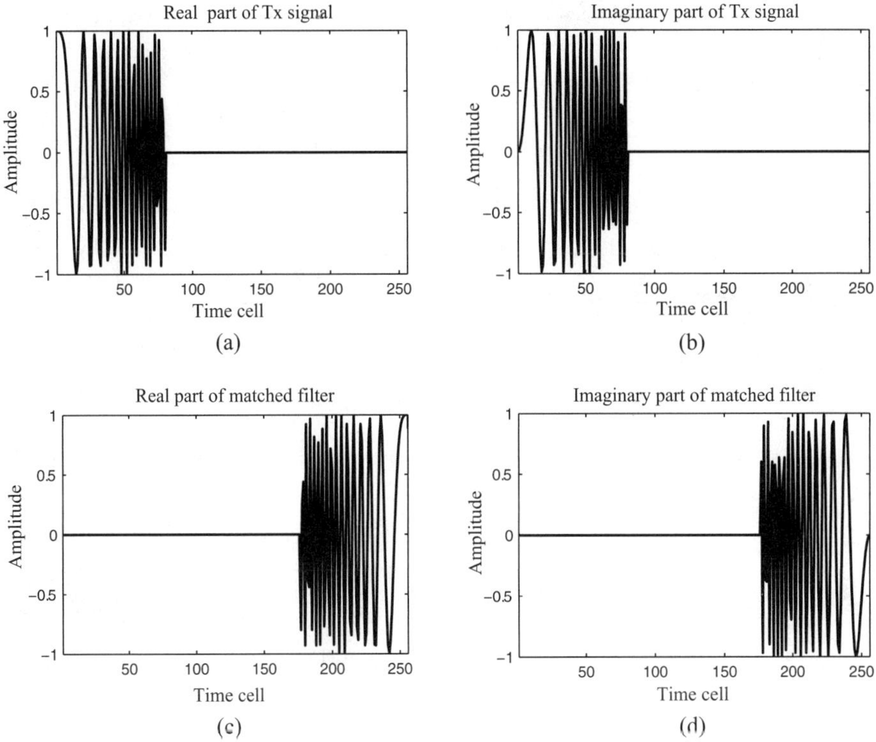

FIGURE 5.24 Waveforms of Tx signal and MF function.

bandwidth:

$$
\begin{aligned}
r(t) &= s^*(t)\exp\left(j2\pi f_c t + j\pi\alpha t^2\right) \\
&= \sum_n \exp\left(j2\pi f_c t_n - j\pi\alpha t_n^2\right)\exp\left(j2\pi\alpha t_n t\right).
\end{aligned}
\tag{5.19}
$$

The dechirped signal $r(t)$ has a frequency component αt_n, which carries information on the round-trip delay t_n, and is generally called the *beat frequency* f_{bn}. As an example, Figure 5.28 shows the time–frequency relationship of both the local chirp signal and the echo signals from two targets. Figure 5.28a shows the instantaneous frequency versus time, while Fig. 5.28b show beat frequencies of two targets versus time.

As can be seen in Fig. 5.28, the transmitted chirp signal with pulse duration T_p and period T has two echoes returned from targets at t_1 and t_2. After dechirp processing, the beat frequencies of the two targets become f_{b1} and f_{b2}, respectively. The target range R_n can then be derived from the beat frequency f_{bn} and the delay time t_n as

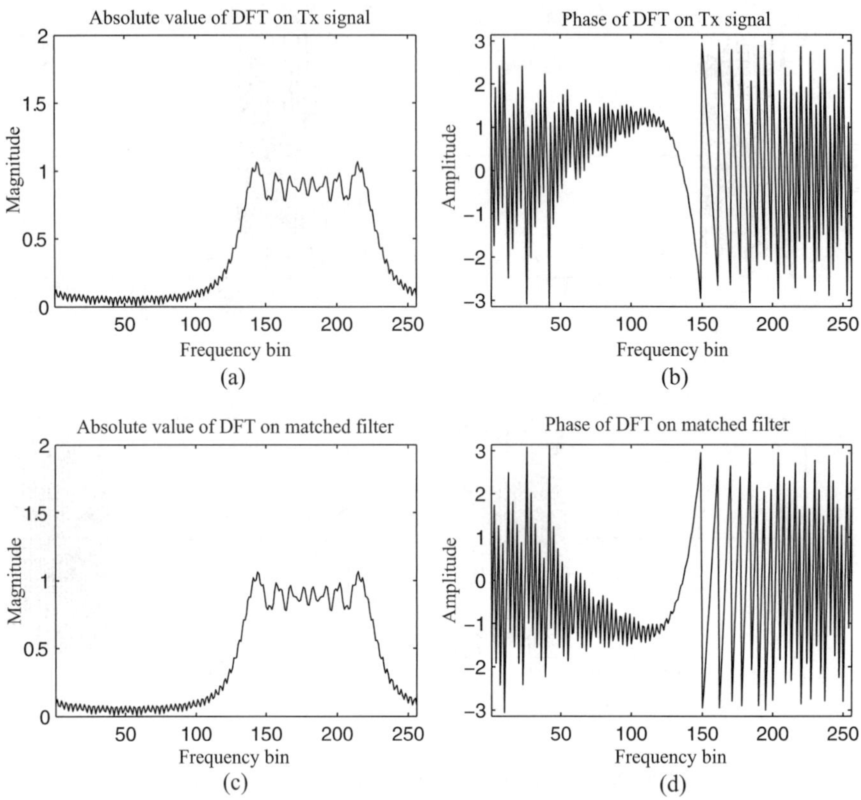

FIGURE 5.25 Frequency spectra of Tx signal and MF function.

follows:

$$\frac{df}{dt} = \frac{f_{\max} - f_c}{T_p} = \frac{\alpha T_p}{T_p} = \alpha,$$

$$\frac{df}{dt} = \frac{f_{bn}}{t_n} = \frac{f_{bn}}{(2R_n)/c}.$$

Therefore

$$R_n = \frac{c \cdot f_{bn}}{2\alpha}. \tag{5.20}$$

The minimum beat frequency can be computed, from Fig. 5.28, as $f_{b\min} = f_{\max} - f_c = \alpha T_p$, which corresponds to the minimum detectable range. The maximum beat frequency is $f_{b\max} = \alpha(T - T_p)$, which corresponds to the maximum detectable range. The bandwidth of the reference chirp signal is then αT. The block diagram of dechirp processing is shown in Fig. 5.29.

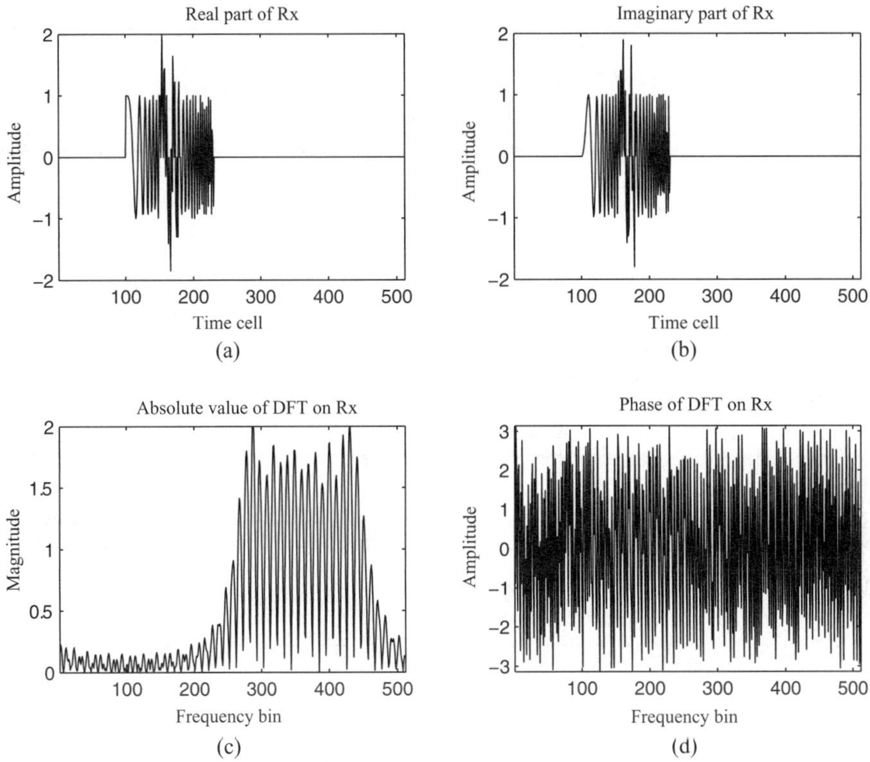

FIGURE 5.26 Time- and frequency-domain waveforms of Rx signal.

Since the maximum beat frequency is $\alpha(T - T_p)$, the Nyquist sampling frequency for the A/D converter is therefore $2\alpha(T - T_p)$, which could be much smaller than the $2(f_c + \alpha T_p)$ required for direct sampling on received signals. The output of the A/D converter is then processed through FFT to obtain the beat frequencies f_{bn}, and the target range can be computed using Eq. (5.20).

The pulsed LFM radar has a cost-effective way to detect the target range, yet it has a relatively lengthy pulsewidth and shares the same antenna for transmitting and receiving; therefore, it has poor minimum/blind range detection capability.

The CWLFM radar uses separate antennas for transmitting and receiving. It consumes less power than that of the pulsed LFM radar, yet still maintains the same signal-to-noise ratio as that of the pulsed LFM radar. It operates at a lower sampling frequency and therefore at a lower system cost. In addition, it can also detect target velocity, described next.

Consider the CWLFM waveform shown in Fig. 5.12 with $T_p = T_0/2$. Let $p(t)$ be the transmitted signal of CWLFM radar:

$$p(t) = \exp\left[j\left(2\pi f_c t + \pi\alpha t^2\right)\right]. \tag{5.21}$$

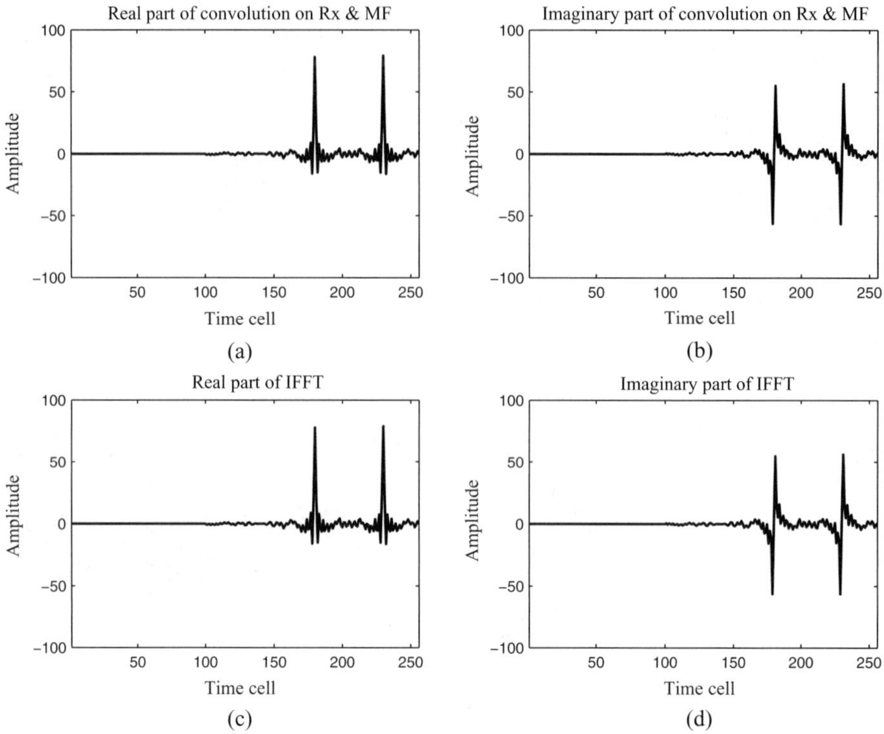

FIGURE 5.27 Comparison of pulse compression based on convolution and DFT.

The returned echo signal from a single target with a time delay τ will have the following form:

$$p(t-\tau) = \exp\left\{j\left[2\pi f_c(t-\tau) + \pi\alpha(t-\tau)^2\right]\right\}. \tag{5.22}$$

By multiplying the transmitted signal with the complex conjugate of the echo signal, one obtains

$$\begin{aligned} r(t) &= p(t)\cdot p^*(t-\tau) \\ &= \exp[j(2\pi f_c t + \pi\alpha t^2)]\cdot\exp\{-j[2\pi f_c(t-\tau) + \pi\alpha(t-\tau)^2]\} \\ &= \exp\left[j\left(2\pi f_c\tau - \pi\alpha\tau^2\right)\right]\cdot\exp\left[j2\pi\alpha\tau t\right]. \end{aligned} \tag{5.23}$$

The first item is a constant phase that depends on the delay time only, while the second item is a time-varying signal with a beat frequency

$$f_b = \alpha\tau. \tag{5.24}$$

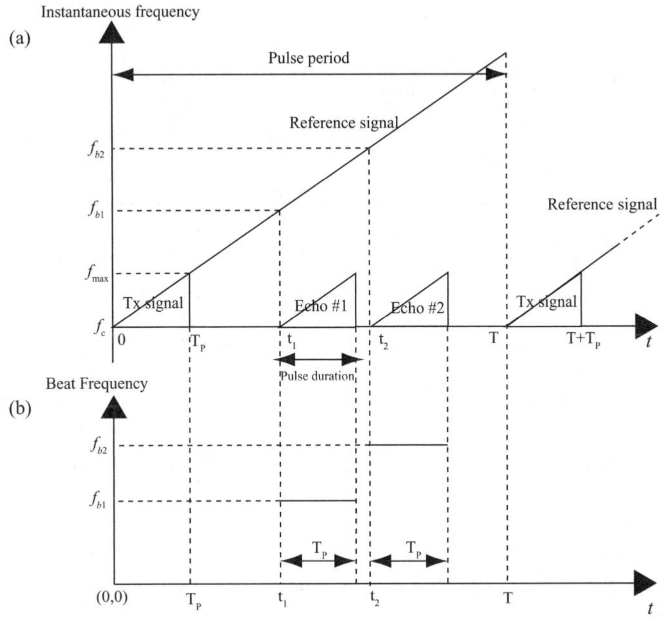

FIGURE 5.28 Time–frequency relationship of Tx, reference, and echo signals.

The maximum beat frequency occurs at $\tau = T_p$ and equals to $f_{b\max} = \alpha T_p = \alpha T_0/2$. The dechirped signal $r(t)$ can then be sampled by an A/D converter with sampling frequency $f_s \geq 2f_{b\max} = \alpha T_0$. Since $\tau = 2R/c$, the target range R can then be computed as

$$R = \frac{cf_b}{2\alpha}. \tag{5.25}$$

Figure 5.30 shows an example of dechirping with two stationary targets; Fig. 5.30a shows the time–frequency relationship corresponding to the transmitting signal and two echo signals, while Fig. 5.30b shows the time–frequency relationship of the two dechirped or beat frequencies. The beat frequencies f_{b1} and f_{b2}, corresponding to delay times t_1 and t_2, respectively, are positive values during the

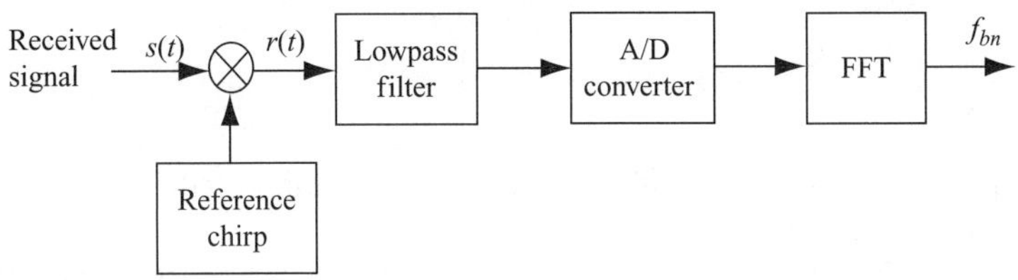

FIGURE 5.29 Block diagram of dechirp processing.

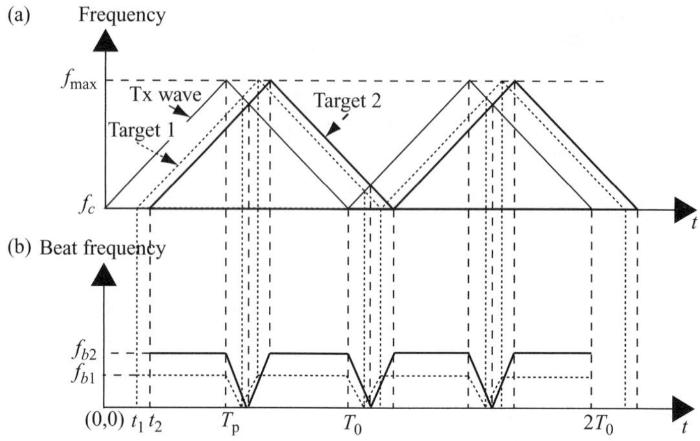

FIGURE 5.30 Time–frequency relationship of Tx and echo signals from two stationary targets.

upsweep and the downsweep periods. However, during the crossover between the Tx signal and the echo signal, the echo's beat frequency value reduces to zero first and then increases to the original beat frequency. This is graphically represented in Fig. 5.30. Since $T_0 = 2T_p$, the beat frequency values in the upsweep and downsweep periods are identical.

As discussed in Chapter 4, when a target approaches the radar, the received signal frequency is increased by an amount of the Doppler frequency f_D. This increment in f_D reduces the beat frequency in the upsweep region and increases the beat frequency in the downsweep region, that is

$$\text{In upsweep region:} \quad f_{bu} = f_b - f_D, \tag{5.26a}$$

$$\text{In downsweep region:} \quad f_{bd} = f_b + f_D, \tag{5.26b}$$

where f_{bu} and f_{bd} are the observed or measured frequencies. The true range-related beat frequency and Doppler frequency could then be computed as follows:

$$f_b = \frac{f_{bu} + f_{bd}}{2}. \tag{5.27a}$$

$$f_D = \frac{f_{bd} - f_{bu}}{2}. \tag{5.27b}$$

A similar situation applies if the target is moving away from the radar. Once the beat frequency f_b and the Doppler frequency f_D are known, the target range and velocity can be computed accordingly.

As an example, Fig. 5.31 displays the time–frequency relationship of Tx and echo signals from two moving targets. Figure 5.31a displays three triangular-shape plots,

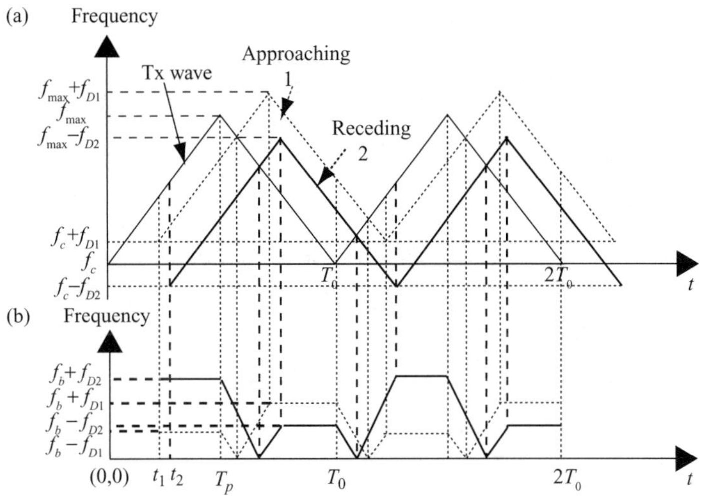

FIGURE 5.31 Time–frequency relationship of Tx and echo signals from two moving targets.

corresponding to the time–frequency relationship of the transmitting signal, the echo signal with target approaching, and the echo signal with target receding. The lighter (dotted) line shows the target approaching the radar while the darker (solid) line shows the target moving away from the radar. Figure 5.31b shows the beat frequencies corresponding to the targets both approaching and receding. Again, the dotted line refers to the approaching target and the solid line refers to the receding target.

5.4.4 Target Detection Techniques of SFM Radar

The SFM-based radar is used primarily for short-range measurement. This is because its detection range is limited to $R \leq (cT/2)$, where T is the time interval between each stepped frequency and T is normally a small number. Consider a target with distance R from a PSFM radar with stepped frequency f_k, where $k = 0, 1, \ldots, N-1$. The transmitted PSFM signal $p_k(2\pi f_k t)$ and the returned echo signal $s_k(2\pi f_k t, \tau)$, with time delay $\tau = 2R/c$, can be expressed as follows:

$$p_k(2\pi f_k t) = A_k \exp(j2\pi f_k t),$$
$$s_k(2\pi f_k t, \tau) = B_k \exp[j2\pi f_k(t-\tau)].$$

The returned signal $s_k(2\pi f_k t, \tau)$ is then coherently downconverted by multiplying it by a portion of the transmitted signal, follows by a lowpass filter:

$$\begin{aligned} s_k(2\pi f_k t, \tau) p_k(2\pi f_k t) &= A_k B_k \exp[j2\pi f_k(t-\tau)]\exp(j2\pi f_k t) \\ &= A_k B_k \exp(j4\pi f_k t)\exp(-j2\pi f_k \tau). \end{aligned}$$

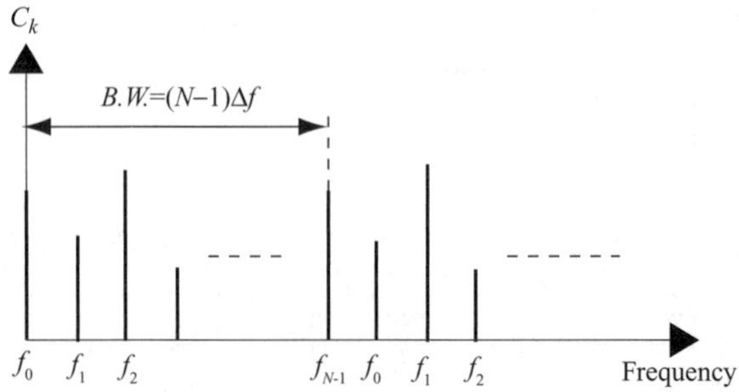

FIGURE 5.32 Baseband echo response from PSFM signal.

Let C_k be the lowpass filter output, which can be expressed as

$$\begin{aligned} C_k &= A_k B_k \exp(-j2\pi f_k \tau). \\ &= A_k B_k \exp\left(-j2\pi f_k \frac{2R}{c}\right). \end{aligned} \tag{5.28}$$

Figure 5.32 shows an example of the digitized and normalized baseband echo signal C_k.

For simplicity, we let $A_k B_k = 1$ in the following discussion. The frequency-domain baseband signal $C_k = \exp(-j2\pi f_k \tau)$ is then transformed into the time domain, using IFFT, as a range profile; that is

$$\begin{aligned} y_n &= \frac{1}{N} \sum_{k=0}^{N-1} C_k \exp\left[j \frac{2\pi nk}{N}\right] \\ &= \frac{1}{N} \sum_{k=0}^{N-1} \exp\left[j \frac{2\pi nk}{N} - j2\pi f_k \tau\right], \end{aligned} \tag{5.29}$$

where $n = 0, 1, \ldots, N-1$. With $f_k = f_0 + k\Delta f$ and $\tau = 2R/c$, y_n becomes

$$\begin{aligned} y_n &= \frac{1}{N} \sum_{k=0}^{N-1} \exp\left[j \frac{2\pi nk}{N} - j2\pi\left(f_0 + k\Delta f\right) \frac{2R}{c}\right] \\ &= \frac{\exp\left(-j4\pi f_0 R/c\right)}{N} \frac{\exp\left[j2\pi n - j4\pi N \Delta f R/c\right] - 1}{\exp\left[j\left(2\pi n/N\right) - j\left(4\pi \Delta f R/c\right)\right] - 1}. \end{aligned}$$

Let $a = (n/N) - (2\Delta f R/c)$; then

$$y_n = \frac{\exp\left(-j4\pi f_0 R/c\right)}{N} \exp[j(N-1)\pi a] \frac{\sin\left(\pi N a\right)}{\sin\left(\pi a\right)}. \tag{5.30}$$

The magnitude response of the IFFT becomes

$$|y_n| = \frac{1}{N}\left|\frac{\sin(\pi N a)}{\sin(\pi a)}\right|. \tag{5.31}$$

Note that the magnitude response of the IFFT, called the *synthetic pulse*, consists of N components and is repeated for each N time interval. The peaks of the mainlobes of Eq. (5.31) occur at $a = \pm\, l$, or $n_p = 2NR\Delta f/c + lN$, $l = 0,1,2,\ldots$.

The range of target R, which corresponds to the peak of Eq. (5.31), can then be computed as follows:

$$R = \frac{cn_p}{2N\Delta f}. \tag{5.32}$$

As an example, Fig. 5.33 shows the simulated plot of $|y_n|$, where $N = 64$, $\Delta f = 20 \times 10^3$ Hz, and $R = 1.5 \times 10^3$ m were used. According to this plot, the peak of $|y_n|$ appears between time cells 13 and 14, which corresponds to $n \approx 12.8 = 13$. (Notice that n is an integer and $n = 0,1,\ldots, N\!-\!1$.) The range R can be calculated from Eq. (5.32), which is 1.52×10^3 m. This result is very close to the data R used to plot Fig. 5.33.

The range resolution can be computed as follows, with $n_p = 1$ in Eq. (5.32):

$$\Delta R = \frac{c}{2N\Delta f}. \tag{5.33}$$

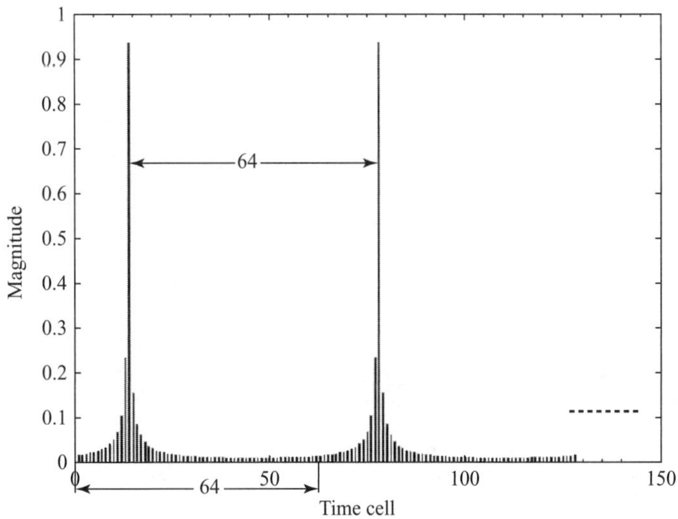

FIGURE 5.33 A single-target range profile based on PSFM signal.

It is clear that the range resolution of step frequency depends on both the frequency step size and the number of steps that are used. This is different from the range resolution of single-frequency pulse radar with pulse duration T_p, which is

$$\Delta R = \frac{cT_p}{2}.$$

If a number of closely spaced targets are present, but the distance between each target is greater than ΔR, then each target will have its own unique frequency-dependent phase and baseband signal C_i. The final target function (or range profile) is the superposition of the synthetic pulses obtained from each target.

Figure 5.34 displays N-section stepped frequency radar pulses, together with three echo signals corresponding to each frequency pulse. The group consists of N stepped frequencies, starting from f_0 and ending at f_{N-1}, with each frequency separated by time interval T, shown in the top row of the figure. For each stepped frequency f_i, there exist three echo responses, $C_{i,1}$, $C_{i,2}$, and $C_{i,3}$. The first subscript i of echo response $C_{i,j}$ denotes the ith stepped frequency. The second subscript j refers to the jth target. The individual frequency component f_i and its associated target echoes are shown under the top row. They were displayed for a period of T without being synchronized with the transmitted pulses.

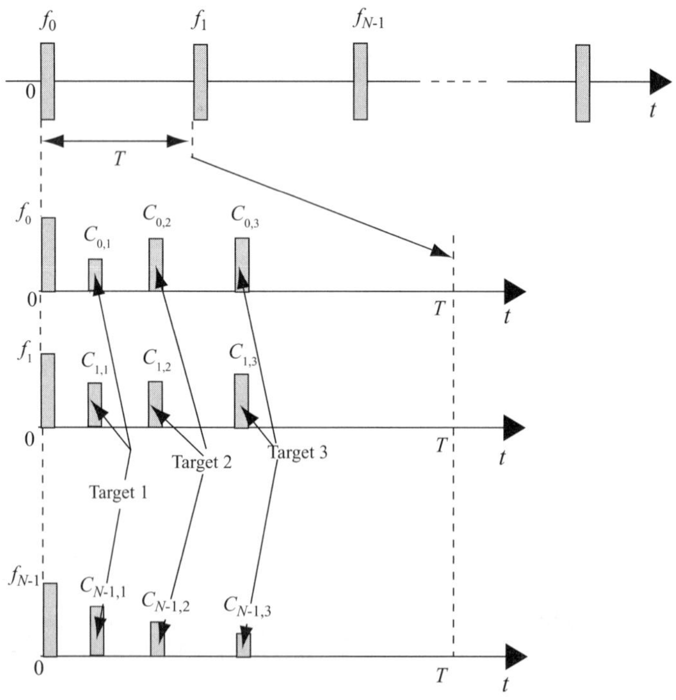

FIGURE 5.34 Stepped frequency pulse train and echoes returned in one pulse period.

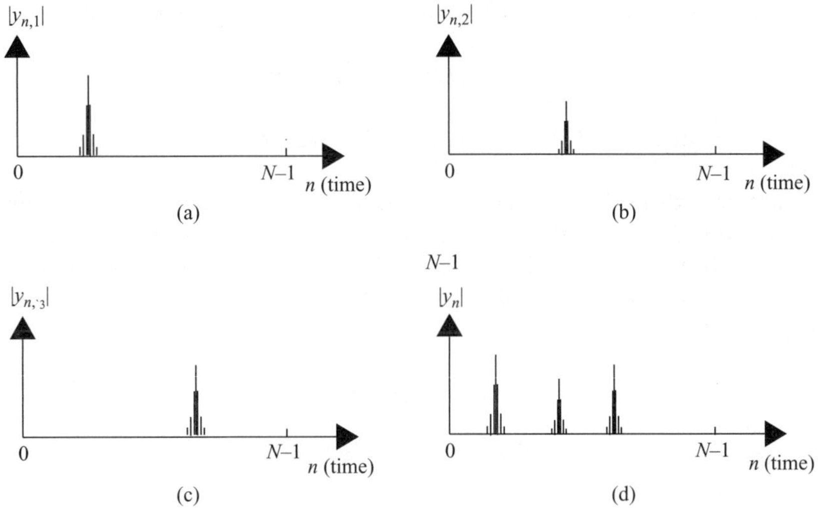

FIGURE 5.35 A multiple-target range profile based on PSFM.

Figure 5.35 displays the synthesized range profile, which consists of three targets. Figure 5.35a shows the target 1 range profile $|y_{n,1}|$, which is generated by taking IFFT on $\{C_{k,1}\}$ for $k = 0,1, \ldots, N–1$. Figure 5.35b shows the target 2 range profile $|y_{n,2}|$, and Fig. 5.35c shows the target 3 range profile $|y_{n,3}|$. Both $|y_{n,2}|$ and $|y_{n,3}|$ are obtained by taking IFFT on $\{C_{k,2}\}$ and $\{C_{k,3}\}$, respectively, for $k = 0,1, \ldots, N–1$. The three range profiles are then superimposed to become the complete range profile $|y_n|$, which is shown in Fig. 5.35d.

In summary, the following steps are used by the stepped frequency radar to obtain a high-resolution synthetic range profile (SRP) based on stationary targets:

1. Transmit a burst of N pulses (or sections), with each pulse shifted in frequency by a fixed step size Δf. (Assume that N is a number the power of 2.)
2. Collect the in-phase (I) and quadrature-phase (Q) complex value samples of the target's baseband echo response in each coarse range bin for every transmitted pulse. These samples are considered as the frequency-domain measurements of the target's spectral profile.
3. Apply IFFT to the N-complex samples of each coarse range bin to obtain an N-element SRP of the target in the range bin.
4. Repeat steps 1–3 for each multiple target.
5. The final range profile is the superposition of the SRP from each target.

In case the target is moving at speed $v_t(n)$, $n = 1,2,3$, the target will be at a slightly different position with displacement Δr_i on a pulse-to-pulse basis. This position shift will generate an extra phase factor on the frequency response signal C_i; that is

$$C_i\,(2\pi f_i) \Rightarrow C_i\,(2\pi f_i)\exp\left(-j2\pi f_i\,\Delta t_i\right),$$

where $\Delta t_i = 2\Delta r_i/c$ and $\Delta r_i = r_i - r_0$ is the target displacement at the stepped frequency f_i.

The target radial velocity $v_t(n)$, $n = 1,2,3$, can be computed as follows. By applying M bursts of N-section SFM signal on the three targets described above, let $r_{1N}(n)$ and $r_{MN}\,(n)$ be the range profiles corresponding to the first and the Mth bursts of the N-section signal, and let $\Delta r_{MN}(n)$, $n = 1,2,3$ be the range difference corresponding to targets 1, 2, and 3. Then

$$\begin{aligned}\Delta r_{MN}(n) &= r_{MN}(n) - r_{1N}(n)\\ &= (M-1)NT\,v_{tn}\end{aligned}$$

and

$$v_t(n) = \Delta r_{MN}(n)/[(M-1)NT] \quad \text{for} \quad n = 1,2,3.$$

6

BASICS OF RADAR IMAGING

In this chapter, Section 6.1 surveys several popular imaging radars and possible applications. Section 6.2 covers the geometry of stripmap SAR, which consists of broadside SAR and squint SAR. Section 6.3 describes the relationship between Doppler frequency and radar image processing, where both broadside SAR and squint SAR are covered. A key problem of radar image processing is range migration, which is described in Section 6.4. Radar image distortion and resolution are covered in Sections 6.5 and 6.6. Most hand drawings in this chapter are for illustration purposes and are not to scale.

6.1 BACKGROUND

Imaging radar has been used to measure the motion of the earth's surface to help us better understand earthquakes and volcanoes. It can be used to study the movements and changing size of glaciers and ice floes to understand long-term climate variability. The development of highly detailed and accurate elevation maps is based on imaging radar. Imaging radar is also used to find oil or other natural resources, to study land cover and land use change, to assess the health of crops and forests, and to plan for urban development.

In early 1950, scientist Carl Wiley found that the use of Doppler frequency analysis could improve radar image resolution of side-looking radar. This new finding led to the development of the SAR (synthetic aperture radar) technique. The SAR technique derives from the along-track (or azimuth) processing of the signal data

Digital Signal Processing Techniques and Applications in Radar Image Processing, by Bu-Chin Wang.

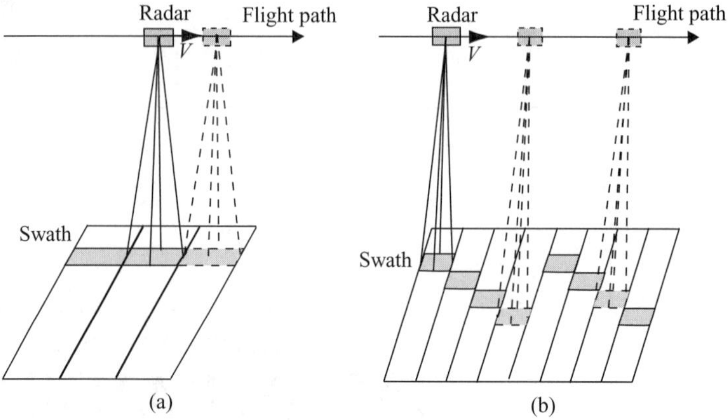

FIGURE 6.1 Configurations of (a) a stripmap SAR and (b) a scan SAR.

by synthesizing an aperture that is longer than the actual physical antenna to yield a higher resolution. Depending on the system's configuration, SAR has been used to acquire data in three major modes: (1) *stripmap SAR*, which images a long strip terrain during the full transit distance; (2) *scan SAR*, which scans the ground in wider swaths by varying the elevation angle of the antenna beam along the flight path; and (3) *spotlight SAR*, which images a scene with finer resolution at multiple viewing angles during a single pass.

Figure 6.1 displays two configurations of the SAR system: the stripmap SAR and the scan SAR. Figure 6.1a shows that with fixed antenna angle, the radar scans through the ground with a fixed width of strip along the flight path. Figure 6.1b shows that the antenna varies the angle along the flight path and scans through a wider swath on the ground.

Figure 6.2a shows the spotlight SAR, which achieves finer resolution on the target by rotating the antenna and aiming at a target along the flight path. In addition to the stripmap, scan, and spotlight modes, other techniques are also used for radar image processing. Interferometric SAR (InSAR) has been widely used in many applications

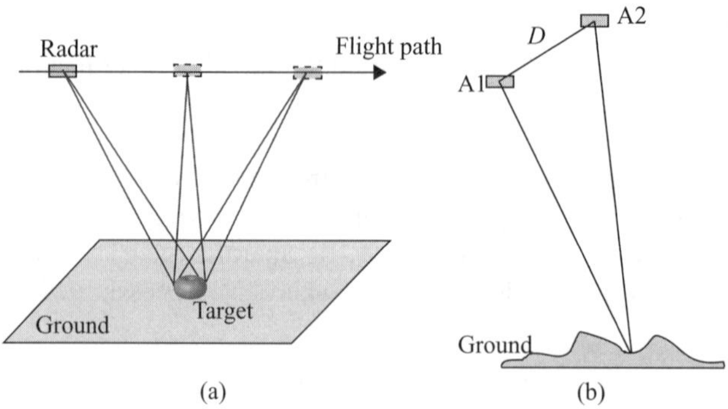

FIGURE 6.2 Imaging radar for (a) a spotlight SAR and (b) an interferometric SAR.

by applying multiple radar antennas and/or different acquisition geometries. InSAR may also achieve the same purpose by scanning the same ground with a single antenna but at different flight paths. Figure 6.2b displays an example of InSAR, where A1 and A2 are two radars separated by distance D. The inverse SAR (ISAR) is an imaging system that deals with moving targets while the radar is stationary.

The following chapters focus only on discussion of the stripmap SAR, which consists of the broadside-mode SAR and the squint-mode SAR.

6.2 GEOMETRY OF IMAGING RADAR

Figure 6.3 shows an imaging radar that operates on a flying vehicle, which could be a satellite or an airplane. The radar antenna is oriented parallel to the flight direction; that is, it is looking sideward to the ground. The radar is moving along the flight path above the earth with height H at velocity $\overline{V}$. The radar antenna, which is assumed to be a phased array, has dimensions of length L and width W. The ground surface area from which the radar pulse is reflected is called the *footprint*. *Swath* is the ground

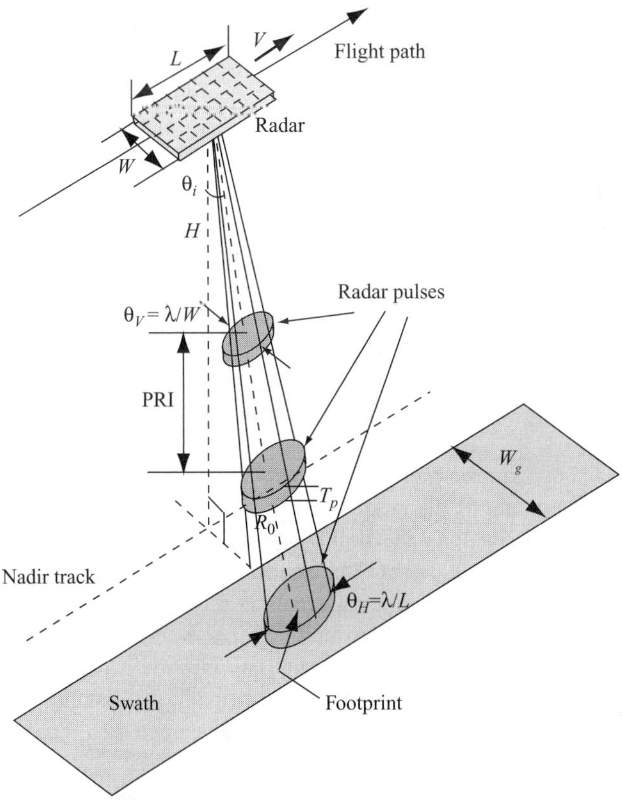

FIGURE 6.3 Geometry of stripmap imaging radar.

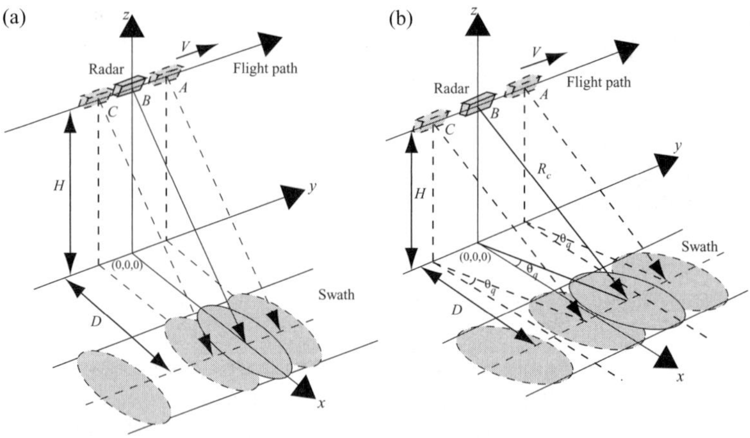

FIGURE 6.4 Geometry of (a) a broadside SAR and (b) a squint SAR.

surface area covered by the consecutive radar pulses. The radar transmits short pulses with duration T_p and repeats at period PRI $= T = 1/f_{\text{PRF}}$ to the ground, where f_{PRF} is the pulse repetition frequency. The 3-dB beamwidth along the track is $\theta_H = \lambda/L$, while across the track it is $\theta_V = \lambda/W$ and the wavelength of the transmitted signal is λ. The pulse is directed at some angle off nadir (directly below the radar) called the *look angle* or *incident angle* θ_ι. The distance from radar antenna to center of footprint is represented by R_0.

Imaging radar is extremely sensitive to the ground surface roughness of the area being imaged. It does not detect the visible color of the surface, but detects the moisture and electrical properties of the surface. Ground targets are illuminated numerous times by the radar beams. The time interval during which the target is illuminated depends on the beamwidth of the radar antenna and the speed of the flying vehicle, namely, $\Delta T = R_0\theta_H/V$. The returned signals from ground targets at the same slant range will arrive at the same time. These equal-slant-range targets are separable only in Doppler frequency, because of the radial velocity difference of the radar relative to the target.

Figure 6.4a shows the geometry of a broadside-mode side-looking aperture radar system, where the center of the radar beams is perpendicular to the flight path. The radar moves at speed V along the flight path, which is parallel to the y axis. The three radar positions A, B, and C are the locations where the radar transmits a signal and receives an echo from the ground targets at three pulse periods. The length of the center beam between radar and ground target is R_0. The radar–target distance is called *slant range*; however, when the target is under the center beam of the radar and the center beam is perpendicular to the flight path, the distance is called *range*. The radar beam covers the ground with an elliptical shaped area. The two axes of the elliptical shape are determined by $R_0\theta_H = R_0\lambda/L$ and $R_0\theta_V = R_0\lambda/W$, respectively, where θ_H and θ_V are the radar beamwidths and L and W are the antenna sizes. The consecutive radar beams cover the ground and form a rectangular area, which can be processed to form a rectangular-shape radar image.

Figure 6.4b shows the geometry of a forward-looking aperture radar system with nonzero squint angle, where the center of the radar beam forms an angle θ_q with the axis perpendicular to the flight path. All other parameters are the same as in Fig. 6.4a, except that the radar beams cover the ground with a rhombus area, which can be processed to form a rhombus-shape radar image.

6.3 DOPPLER FREQUENCY AND RADAR IMAGE PROCESSING

The radar image is formed by processing the 2D raw data collected by range radar. Each row of 2D raw data is formed by the target echo returned from every radar pulse, which is transmitted at the rate of f_{PRF}. The 2D radar image data are represented in complex numbers and normally can be processed separately by processing range data first, followed by azimuth data. Various algorithms are available to process the 2D raw data to generate a radar image, and they are discussed in later chapters.

Figure 6.5 shows the configuration of imaging radar and the corresponding radar transmitter and receiver signal. Figure 6.5a shows a radar pulse that covers an elliptical-shape ground area. The two axes of the elliptical shape are $R_0\lambda/L$, and $R_0\lambda/W$, respectively, where L and W are the length and width of radar antenna. The two slant ranges from the radar to the closest target and farthest target along the x axis are represented as R_N and R_F, respectively. R_0 is the distance from the radar center beam to the target. The radar moves at speed V in the y direction at $x = 0$ and height $z = H$. Figure 6.5b shows the relationship between the transmitted pulse and the returned echoes from all targets within the elliptical shaped area in one pulse period. All targets with equal distance from the radar will superimpose on each other and share the same data from the range viewpoint. Delay time t and slant range r are related by $t = 2r/c$, where c is the speed of light; therefore, t and r are used

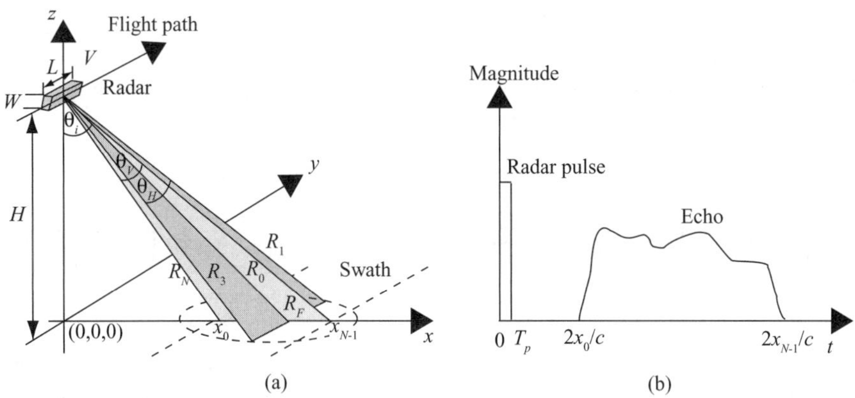

FIGURE 6.5 (a) Imaging radar and (b) radar pulse and received echo.

interchangeably for convenience. Notice that the parameter x in Fig. 6.5b refers to the slant range, while in Fig. 6.5a it refers to the ground range.

The transmitter signal from a pulsed LFM radar can be written as follows, from Eq. (5.8):

$$p(t) = a(t)\,\mathrm{Rect}\left(\frac{t}{T_p}\right)\exp\left(j2\pi f_c t + j\pi\alpha t^2\right).$$

Here $a(t)$ is the amplitude of the signal, $\mathrm{Rect}\left(t/T_p\right)$ is a rectangular function with $|t| \le (T_p/2)$, and T_p is the pulse duration time. The symbol f_c is the carrier frequency, and α is the LFM pulse chirp rate.

The echo signal from a single target, with distance R from the radar and target reflection coefficient σ, can be represented as

$$s(t) = \sigma a\left(t - \frac{2R}{c}\right)\mathrm{Rect}\left(\frac{t - 2R/c}{T_p}\right)\exp\left[j2\pi f_c\left(t - \frac{2R}{c}\right) + j\pi\alpha\left(t - \frac{2R}{c}\right)^2\right]. \tag{6.1}$$

The baseband signal $s_b(t)$ can be obtained by removing the carrier frequency followed by a lowpass filter through the quadrature demodulation process. The demodulated baseband signal can be represented as:

$$s_b(t) = \sigma\left|a\left(t - \frac{2R}{c}\right)\right|^2\mathrm{Rect}\left(\frac{t - 2R/c}{T_p}\right)\exp\left[-j2\pi f_c\frac{2R}{c} + j\pi\alpha\left(t - \frac{2R}{c}\right)^2\right]. \tag{6.2}$$

Let the single target and the imaging radar be located at $(X_0,0,0)$ and $(0,u,H)$ respectively, where X_0 and H are constant but the radar position u is a variable. The slant range R can be represented as follows:

$$R(u) = \sqrt{X_0^2 + u^2 + H^2}. \tag{6.3}$$

The slant range $R(u)$ is a function of both target location and radar position; therefore the 2D received signal can be represented as follows:

$$s_b(t,u) = \sigma\left|a\left(t - \frac{2R(u)}{c}\right)\right|^2\mathrm{Rect}\left(\frac{t - 2R(u)/c}{T_p}\right)\exp\left[-j2\pi f_c\frac{2R(u)}{c} + j\pi\alpha\left(t - \frac{2R(u)}{c}\right)^2\right]. \tag{6.4}$$

The data collected along the flight path, or y axis, are discrete, with sampling frequency equal to f_{PRF}. The sample spacing along the y axis can be computed as $\Delta y = V/f_{\mathrm{PRF}}$.

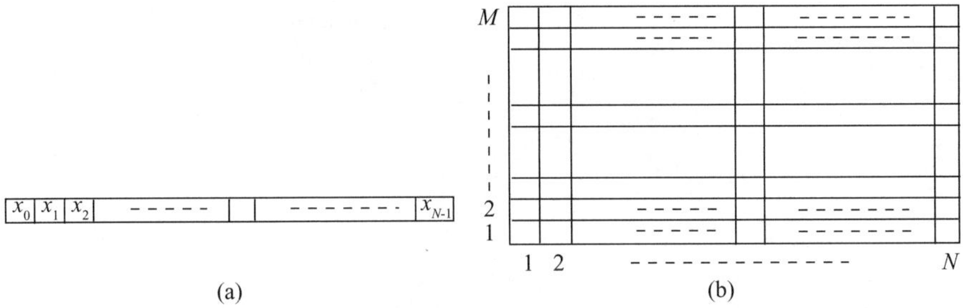

FIGURE 6.6 (a) Single channel radar range data; (b) $M \times N$ radar imaging data array.

Figure 6.6 displays the formation of raw data for the radar image. Figure 6.6a represents a single channel of slant range data, which corresponds to a given radar position $u = u_0$ from an echo signal as shown in Fig. 6.5b. This is a 1D data digitized with N samples starting from ground location x_0 to x_{N-1}. Figure 6.6b shows an $M \times N$ radar image data array, which corresponds to M radar positions of $u = u_i$ for $i = 1,2,\ldots,M$. At each position u_i, the radar receives a returned signal with N samples. A more detailed analysis of radar image data is addressed later.

The principle of using radar to determine target reflectivity and range distance has been addressed in Chapters 4 and 5. The following sections discuss the key parameters of Doppler frequency, which serves as the basis to synthesize a LFM signal along the azimuth direction. Broadside SAR is addressed first, followed by squint SAR later.

6.3.1 Broadside SAR

In a broadside SAR system, the radar beam is perpendicular to its direction of movement. Figure 6.7 displays a broadside SAR system with three radar positions, A, B, C, together with a ground point target located at $(D,0,0)$. At radar position A, the radar beam begins to illuminate the target. At position B the center of the radar beam is on the target. At position C the radar beam ends the illumination on the target. Position B is chosen to be at $(0,0,H)$, with beamwidth $\theta_H = \lambda / L$. The distance between positions A and C is Ls, which is the synthetic aperture length, and $Ls = R_0\theta_H = R_0\lambda/L$.

A simplified version of Fig. 6.7 is shown in Fig. 6.8a, where the plane covered by points A,B,C and the point target on the ground forms a new x'–u coordinate system. The distance D from the target to the y axis in Fig. 6.7 is replaced by the slant range D' in Fig. 6.8, where $D' = (D^2 + H^2)^{1/2} = R_0$, and R_0 is the closest distance between target and radar. Radar positions A and C are the two ends of radar's 3-dB beamwidth, and the slant range r between position A (or C) and the point target equals $R = \sqrt{R_0^2 + (Ls^2/4)}$.

Figure 6.8b shows the relationship between the slant range r and the point target. Here radar position B is chosen to be the origin of the x'–u coordinate system, and θ is the angle between the slant range r and the x' axis. In practical applications

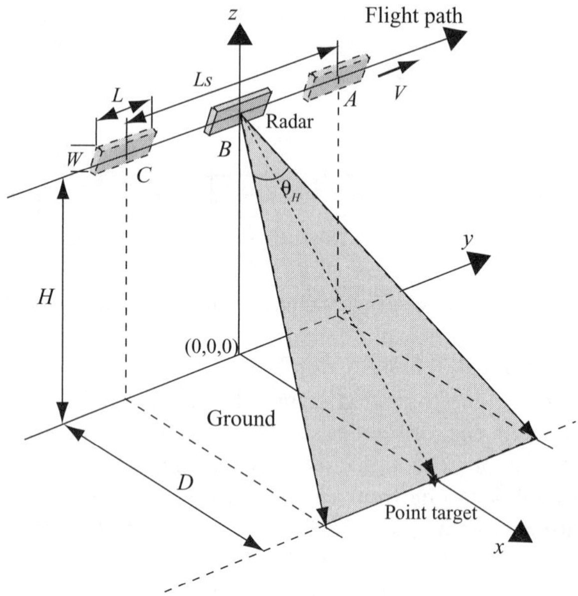

FIGURE 6.7 Configuration of a broadside SAR system.

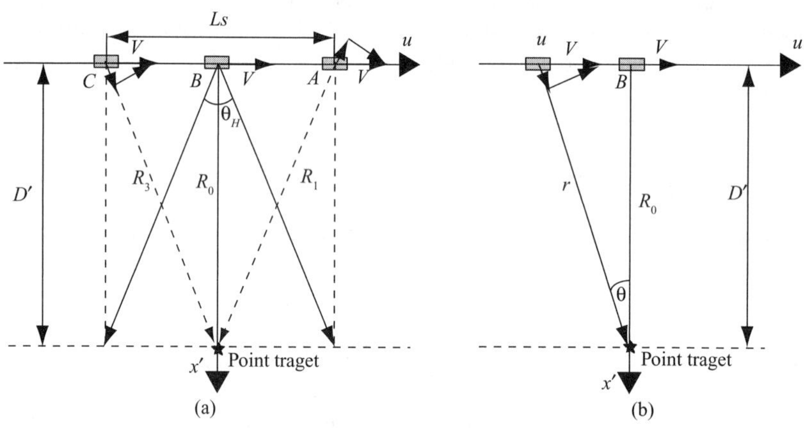

FIGURE 6.8 A simplified broadside SAR system.

$|u| \ll R_0$, the slant range r can be computed as

$$
\begin{aligned}
r &= \sqrt{R_0^2 + u^2} \\
&= R_0\sqrt{1 + \frac{u^2}{R_0^2}} \\
&\approx R_0 + \frac{u^2}{2R_0}
\end{aligned}
\tag{6.5a}
$$

or

$$r - R_0 \approx \frac{u^2}{2R_0}. \tag{6.5b}$$

Equation (6.5b) expresses a parabolic function, with the vertex located at $r = R_0$ and $u = 0$. Figure 6.9 shows the slant range r versus the radar movement along the flight path. The origin of the r–u coordinate is set at $r = R_0$. The plot is based on Eq. (6.5b) with the assumption that both slant range r and radar position u are normalized with respect to R_0, and the transmitted signal is a pulsed signal. For illustration purposes, we set $\theta_H = 0.0349$ (2°) and $Ls = 0.0349R_0$. Notice that the x' axis in Fig. 6.8 is a reference axis connecting point B and the ground target, which is perpendicular to the flight path. The r axis of Fig. 6.9 is a slant range axis connecting the radar and the ground target. For a radar pulse with finite duration, the returned echo signal will last for the same time interval. Figure 6.9a shows a pulse duration equivalent to $R_0/200$. Figure 6.9b shows the range-compressed version of Fig. 6.9a. Only range-compressed signals are considered in the following discussion,

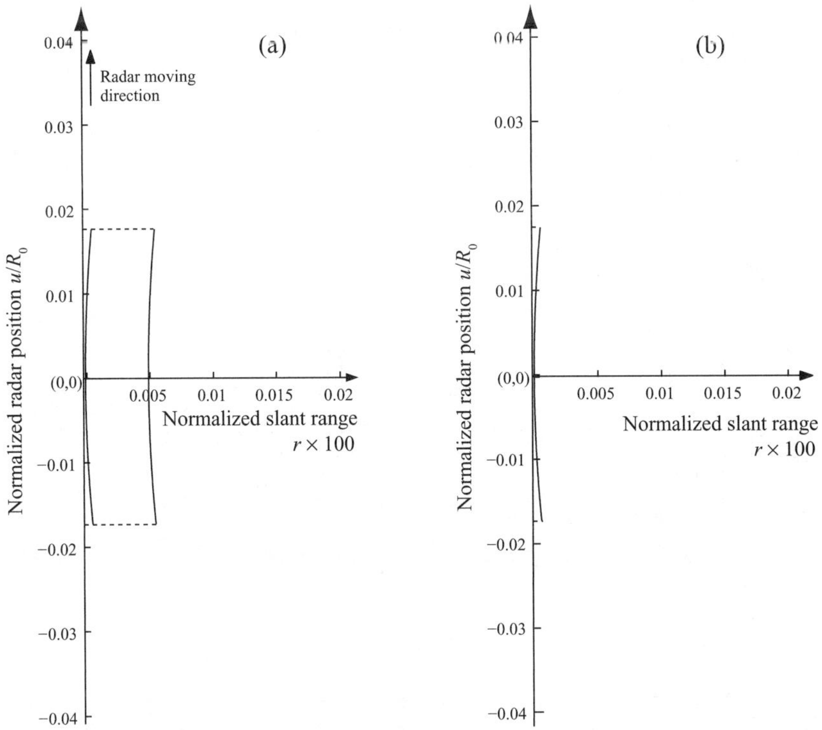

FIGURE 6.9 Echo signal from the point target before (a) and after (b) range compression.

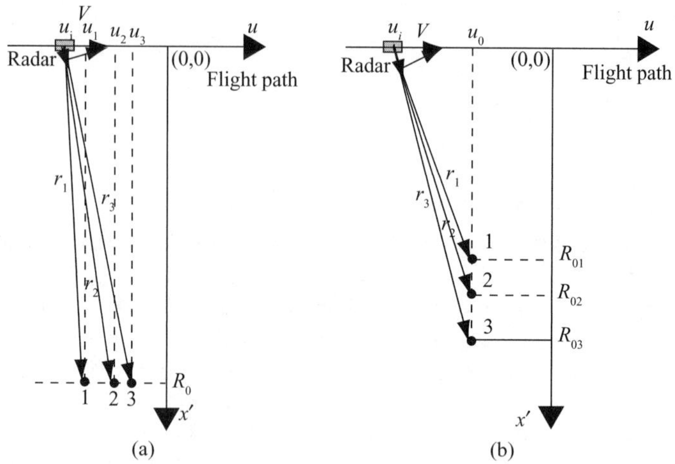

FIGURE 6.10 Broadside SAR with multiple targets.

Figure 6.10 displays the geometry of two cases of broadside SAR with three targets on the ground. The corresponding slant ranges between radar and targets are shown as r_1, r_2, and r_3, respectively.

Figure 6.10a shows the three targets, labeled as 1, 2, and 3, and located at (u_1, R_0), (u_2, R_0), and (u_3, R_0), respectively, where R_0 is the closest distance between targets and flight path. Figure 6.10b displays the three targets aligned along the x' axis with the same cross-range u_0, and located at (u_0, R_{01}), (u_0, R_{02}), and (u_0, R_{03}), respectively. Each target corresponds to one of the three targets to flight path distances R_{01}, R_{02}, and R_{03}, respectively. The trajectories of echo signals in the u–r or azimuth–slant range domain corresponding to target 1–3 of Fig. 6.10 are shown in Fig. 6.11, where $\theta_H = 10°$ is assumed.

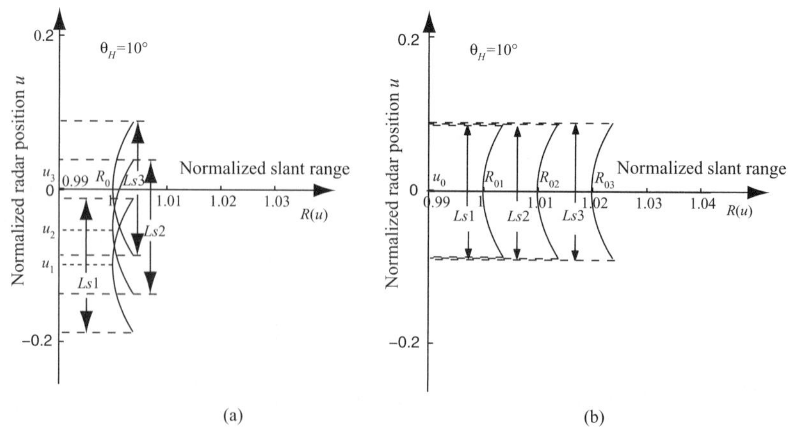

FIGURE 6.11 Slant range $R(u)$ versus radar position u for three targets at equal (a) and different (b) ranges.

Figure 6.11a displays the trajectories of echo signals due to three targets, which share the same distance R_0 but are located at three different positions, u_1, u_2, and u_3, along the u axis. For simplicity, we have made $u_3 = 0$. The peaks of the three curves appear at u_1, u_2, and u_3 where the radar has the closest distance to the ground targets, respectively. Since the three targets have the same closest distance R_0 from the radar, the parabolic trajectories appear to have the same curvature for all three targets. In addition, the synthesized aperture lengths $Ls1 = Ls2 = Ls3$.

Figure 6.11b shows the trajectories of echo signals due to three targets, which line up in the direction perpendicular to the flight path (u axis) but have different distances R_{0i} from the radar. The parabolic trajectories appear to have a different curvature for different targets. The greater the distances R_{0i} is, the flatter the parabolic trajectory becomes and the wider the synthetic aperture length is. The peaks of the three parabolic trajectories are all aligned along the r axis with $u = u_0$. For simplicity, we have made $u_0 = 0$.

In general, the ground targets are randomly distributed; therefore the trajectories of echo signals will be a combination of those in both Figs. 6.11a and 6.11b.

In the SAR system described above, the radar moves along the flight path, or u axis. The Doppler frequency therefore occurs and can be expressed in terms of the radar position u, the slow time s (along the u axis), and the slant range r. Discussion of the Doppler frequency in terms of u or s will lead to synthesis of pulsed LFM along the azimuth direction, while the representation of Doppler frequency in terms of slant range r will serve as the basis for "range cell migration correction."

Figure 6.12 shows a broadside SAR with radar moving at velocity $\overline{V} = V\hat{u}$, where $\hat{u}$ is the unit vector along the u axis. The radar position B is set to be the origin of the x'–u coordinate, and has the shortest range R_0 from the target. This is also the position where the target is under the center beam of radar. The radar positions A and

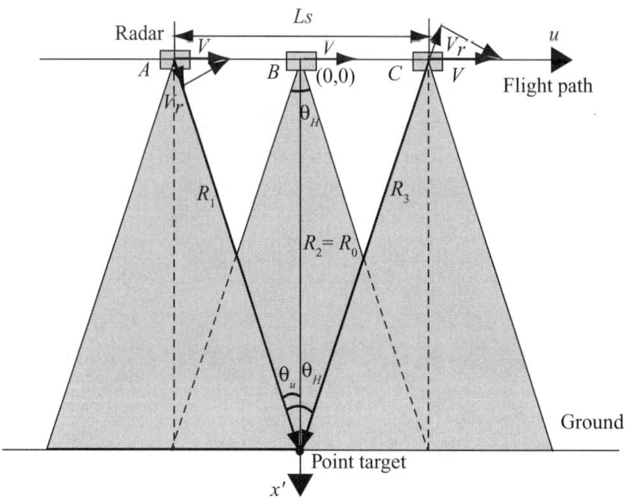

FIGURE 6.12 Broadside SAR with single point target.

C are two endpoints of the 3-dB beamwidth. The angle between radar position u and the x' axis is represented as θ_u. At radar position A or C, the angle θ_u equals $\theta_H/2$, where θ_H is the radar's 3-dB beamwidth and the synthetic aperture length $Ls = R_0\theta_H$.

At any radar position $\overline{u} = u\hat{u}$, the radial velocity of radar is defined as $\overline{V_r} = V_r\hat{r}$. The symbol $\hat{r}$ is the unit vector from radar position to the target. The dot product of $\hat{u}$ and $\hat{r}$ is defined as $\hat{u} \cdot \hat{r} = \sin\theta_u$. At position B the radar's radial velocity $\overline{V_r} = 0$.

Given the radar position u, where $|u| \leq Ls/2$, the relationship between V_r and u can be expressed as follows:

$$\begin{aligned} V_r &= \overline{V} \cdot \hat{r} \\ &= V \sin\theta_u \\ &= -V \frac{u}{\sqrt{R_0^2 + u^2}} \operatorname{sgn}(u). \end{aligned} \tag{6.6a}$$

Here sgn (u) denotes "sign of u." The value of V_r is positive when the radar is located at the left side of the origin (or u is negative) and becomes negative when the radar is at the right side of the origin (or u is positive).

Normally $R_0 \gg |u|$; therefore Eq. (6.6a) becomes

$$\begin{aligned} V_r &\approx -V \frac{u}{R_0}\left(1 - \frac{u^2}{2R_0^2}\right) \operatorname{sgn}(u) \\ &\approx -\frac{V}{R_0} u \operatorname{sgn}(u). \end{aligned} \tag{6.6b}$$

As discussed in Chapter 4, the relationship between the radial velocity $\overline{V_r}$ and the Doppler frequency is $f_D = 2V_r/\lambda$. The Doppler frequency f_D is positive for inward velocity $\overline{V_r}$ and becomes negative for outward velocity $\overline{V_r}$. Therefore, at any radar position u, the Doppler frequency can be expressed as

$$\begin{aligned} f_{Du} &= \frac{2V_r}{\lambda} \\ &= -\frac{2Vu}{\lambda\sqrt{R_0^2 + u^2}} \operatorname{sgn}(u) \\ &= -\frac{2V}{\lambda} \sin\theta_u \operatorname{sgn}(u) \\ &\approx -\frac{2Vu}{\lambda R_0} \operatorname{sgn}(u). \end{aligned} \tag{6.7a}$$

The change rate of f_D with respect to the radar position u is

$$\begin{aligned} \dot{f}_{Du} &= \frac{\partial f_{Du}}{\partial u} \\ &= -\frac{2V}{\lambda R_0}\cos^3\theta_u \\ &\approx -\frac{2V}{\lambda R_0}. \end{aligned} \tag{6.7b}$$

Given the parameters V and λ, Eq. (6.7b) states that the change rate of f_{Du} is always negative and its magnitude decreases when R_0 increases. The subscript u used in f_{Du} and $\dot{f}_{Du}$ indicates that f_{Du} and $\dot{f}_{Du}$ are associated with variable u. Here $\theta_u \approx 0$ for small beamwidth.

Equation (6.7a) states that the Doppler frequency $f_{Du} = 0$ when u equals zero. The maximum value of Doppler frequency occurs at both ends of the 3-dB radar beamwidth, which implies $u_m = \pm Ls/2$; that is

$$\begin{aligned} f_{Dum} &= f_{Du}\big|_{u=u_m} \\ &= \pm\frac{VLs}{\lambda R_0} \\ &= \pm\frac{V}{L}, \end{aligned} \tag{6.7c}$$

where $Ls = R_0\lambda/L$.

The bandwidth of the Doppler frequency therefore becomes

$$B_{Du} = \frac{2V}{L}. \tag{6.7d}$$

Since the slow time $s = u/V$, the Doppler frequency f_D and associated parameters can also be represented in terms of s as follows, from Eq. (6.7a):

$$\begin{aligned} f_{Ds} &= -\frac{2V\sin\theta_u}{\lambda}\,\mathrm{sgn}\,(u) \\ &= -\frac{2V^2 s}{\lambda R_0}\,\mathrm{sgn}\,(u). \end{aligned} \tag{6.8a}$$

$$\dot{f}_{Ds} \approx -\frac{2V^2}{\lambda R_0}. \tag{6.8b}$$

$$\begin{aligned} f_{Dsm} &= 2V^2\frac{\pm(Ls/2V)}{\lambda R_0} \\ &= \pm\frac{V}{L}. \end{aligned} \tag{6.8c}$$

$$B_{Ds} = \frac{2V}{L}. \tag{6.8d}$$

The total time duration during which the Doppler frequency is observable along the flight path can be computed as

$$T_a = \frac{Ls}{V}. \tag{6.9}$$

From Eqs. (6.8a) and (6.8b), it is clear that for a single target with a small radar beamwidth, the following conditions apply:

1. The Doppler frequency f_{Ds} is linearly propositional to the slow time s.
2. The rate of Doppler frequency change $\dot{f}_{Ds}$ is independent of the slow time s.

These two characteristics of Doppler frequency associated with slow time s fit the characteristics of the linear frequency modulation waveform discussed in Chapter 5. In other words, a new signal along the flight path can be derived, and it can be considered as a synthesized pulsed linear frequency-modulated (LFM) signal in terms of slow time s. This LFM signal has the following characteristics:

$$\begin{aligned}
\text{Center of frequency:} f_{Dc} &= 0 \\
\text{Frequency change rate:} \alpha &= \dot{f}_D \\
&= -\frac{2V^2}{\lambda R_0} \\
\text{Maximum frequency:} f_{Dm} &= \pm\frac{V}{L} \\
\text{Pulse duration:} T_a &= \frac{Ls}{V} \\
\text{Bandwidth:} B_{\text{Dop}} &= \dot{f}_D T_a \\
&= \frac{2V}{L}.
\end{aligned}$$

Here the subscript s is dropped from f_D and related parameters for convenience.

Derivation of the synthesized LFM waveform is based on the 3-dB beamwidth of the radar antenna. Therefore it is amplitude-modulated by the antenna's 3-dB radiation pattern. Figure 6.13 shows a typical radiation pattern of antenna array and a pulsed LFM waveform along the azimuth direction.

Figure 6.14 displays the 3-dB beamwidth of the antenna radiation pattern shown in Fig. 6.13a. This radiation pattern serves as a built-in windowing function on the synthesized LFM waveform. By multiplying the data of Fig. 6.14a by those of Fig. 6.13b, one obtains the synthesized and symmetric LFM waveform along the azimuth axis, which is shown in Fig. 6.14b. The synthesized LFM signal has a duration time $T_a = Ls/V$, and its maximum frequency $f_{Dm} = \pm(V/L)$ occurs at two ends of the synthetic aperture length Ls.

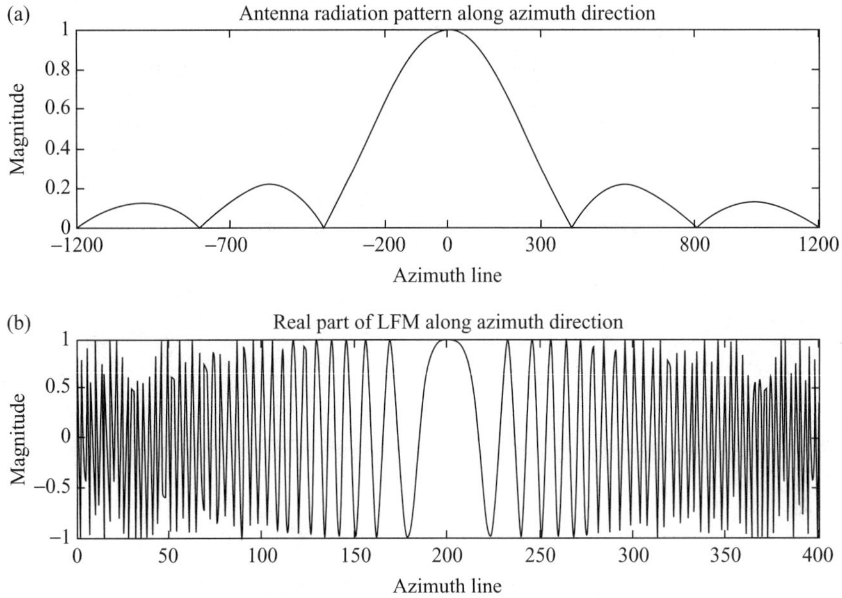

FIGURE 6.13 (a) Radiation pattern from a typical antenna array; (b) real part of a LFM signal.

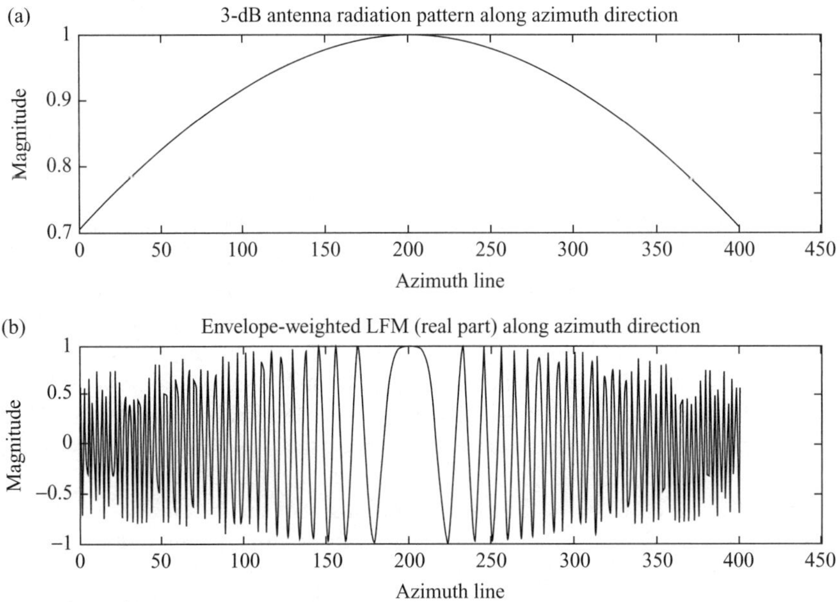

FIGURE 6.14 (a) 3-dB beamwidth of a radiation pattern from a typical antenna array; (b) real part of amplitude-weighted LFM signal.

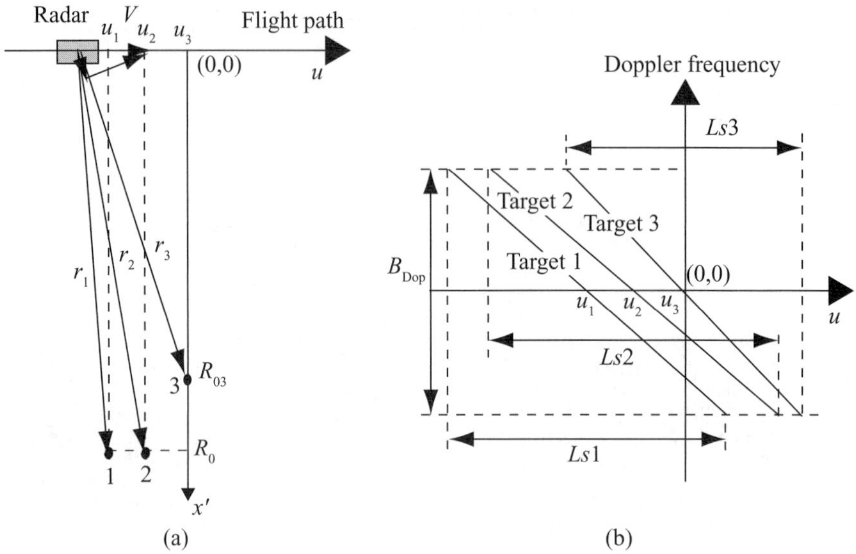

FIGURE 6.15 Doppler frequency and multiple targets

Figure 6.15 displays the relationship between the Doppler frequency and the radar displacement u under multiple target situations. Figure 6.15a shows the geometry of the SAR system with three targets located at (u_1, R_0), (u_2, R_0), and (u_3, R_{03}), respectively. The slant ranges r_1, r_2, and r_3 are the respective target–radar distances. The origin of the u–x' coordinate system is chosen with radar position $u_3 = 0$ and $x' = 0$. The radar moves at velocity $\overline{V} = V\hat{u}$ along the u axis, and the x'axis is perpendicular to the u axis.

Figure 6.15b plots the Doppler frequency f_D against the radar position along the u axis or azimuth direction. Equations (6.7a), (6.7b), and (6.7c) are used for this illustration. Targets 1 and 2 have identical slopes, namely, $-2V/(\lambda R_0)$. Their frequency bandwidth and the synthetic aperture length are also the same: $B_{D1} = B_{D2} = (2VLs/\lambda R_0) = (2V/L)$ and $Ls1 = Ls2 = (R_0\lambda/L)$. However, the zero-frequency positions of targets 1 and 2 are different and located at u_1 and u_2,, respectively. Target 3, has a shorter closest distance to radar R_{03}, and its aperture duration time is shorter than those of targets 1 and 2. The frequency bandwidth of target 3 is identical to those of targets 1 and 2. The zero frequency occurs at $u_3 = 0$. The center frequency f_{Dc} is zero for all three targets.

Since the slow time $s = u/V$, Fig. 6.15b can also be considered as f_D versus slow time s by dividing u with V. All parameters discussed above will apply, except the slope of f_D becomes $-(2V^2/\lambda R_0)$.

The preceding descriptions are based on a Doppler frequency related to the radar position u, or slow time s, along the azimuth direction. The correlation between Doppler frequency and slant range r is discussed next.

Consider the point target case and use Fig. 6.12 for illustration. Given the radar position u and the slant range r, Eq. (6.6a) can be rewritten as follows:

$$\begin{aligned} V_r &= V \sin\theta_u \\ &= -V\frac{u}{r}\,\text{sgn}\,(u) \\ &= -V\frac{1}{r}\sqrt{r^2 - R_0^2}\,\text{sgn}\,(u)\,. \end{aligned} \tag{6.10a}$$

Equation (6.7) then becomes

$$\begin{aligned} f_{Dr} &= \frac{2V_r}{\lambda} \\ &= -\frac{2V}{\lambda r}\sqrt{r^2 - R_0^2}\,\text{sgn}\,(u), \end{aligned} \tag{6.10b}$$

$$\begin{aligned} \dot{f}_{Dr} &= \frac{\partial f_{Dr}}{\partial r} \\ &= -\frac{2V}{\lambda}\frac{R_0^2}{r^2\sqrt{r^2 - R_0^2}}\text{sgn}\,(u), \end{aligned} \tag{6.10c}$$

$$\begin{aligned} f_{Drm} &= \pm\frac{2V}{\lambda}\sqrt{\frac{Ls^2}{4R_0^2 + Ls^2}} \\ &\approx \pm\frac{VL_s}{\lambda R_0} \\ &= \pm\frac{V}{L}. \end{aligned} \tag{6.10d}$$

Let f_{DU} and f_{DL} represent the upper and lower bounds of the Doppler frequency spectrum. These two frequencies occur at both ends of the radar 3-dB beamwidth where a target is under illumination. For a forward-looking SAR, the Doppler frequency f_{DU} occurs at the position where the radar beam begins to illuminate the target, while f_{DL} occurs at the position where the radar beam ends the illumination on the target. Their values are $f_{DU} = V/L$ and $f_{DL} = -V/L$ in the broadside case.

The bandwidth of Doppler frequency for broadside SAR can then be computed as

$$\begin{aligned} B_D &= f_{DU} - f_{DL} \\ &= \frac{2V}{L}. \end{aligned} \tag{6.10e}$$

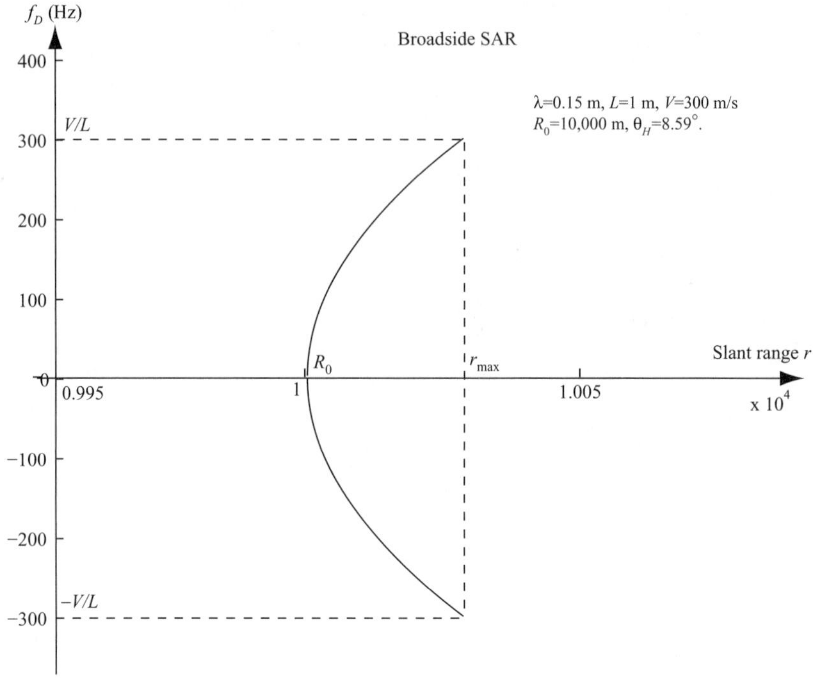

FIGURE 6.16 Doppler frequency versus slant range for single target.

Figure 6.16 displays the correlation between the Doppler frequency and the slant range r. The plotted is based on the single-target model shown in Fig. 6.12 and with the help of Eq. (6.10b). The following parameters are used for the plot: $f_c = 2.0$ GHz, $V = 300$ m/s, $R_0 = 10{,}000$ m, $L = 1$ m, and $\theta_{\rm H} = 0.15$ radian or 8.59°. As can be seen, both ends of the curve have values $V/L = 300$ Hz and $-V/L = -\ 300$ Hz, respectively. The center of the curve is located at range $R_0 = 10{,}000$ m, where the Doppler frequency $f_{Dc} = 0$.

The plot in Fig. 6.17 is similar to that of Fig. 6.16 except with multiple targets. The plot reflects the three-target model shown in Fig. 6.15a. The left curve represents target 3, and the right one represents targets 1 and 2. The maximum slant range occurs at both ends of the curve, which correspond to the two ends of the Doppler frequency. Both curves have the same Doppler frequency center $f_{Dc} = 0$. The start and end of the Doppler frequencies are $f_{DU} = V/L = 300$ Hz and $f_{DL} = -\ V/L = -300$ Hz, respectively, and are the same for both curves. The bandwidth $B_D = 2V/L = 600$ Hz are also identical for both curves. The slant ranges $R_{01} = 10{,}000$ m and $R_{02} = 10{,}025$ m are used to plot the drawing.

Up to this point, we have shown that $f_D, \dot{f}_D, f_{DU}, f_{DL}$, and B_D can be represented in terms of spatial position u, slow time s, and slant range r. Although f_D and $\dot{f}_D$ are in different forms when expressed in terms of u, s, or r, they all share the same results of f_{DU}, f_{DL}, and B_D.

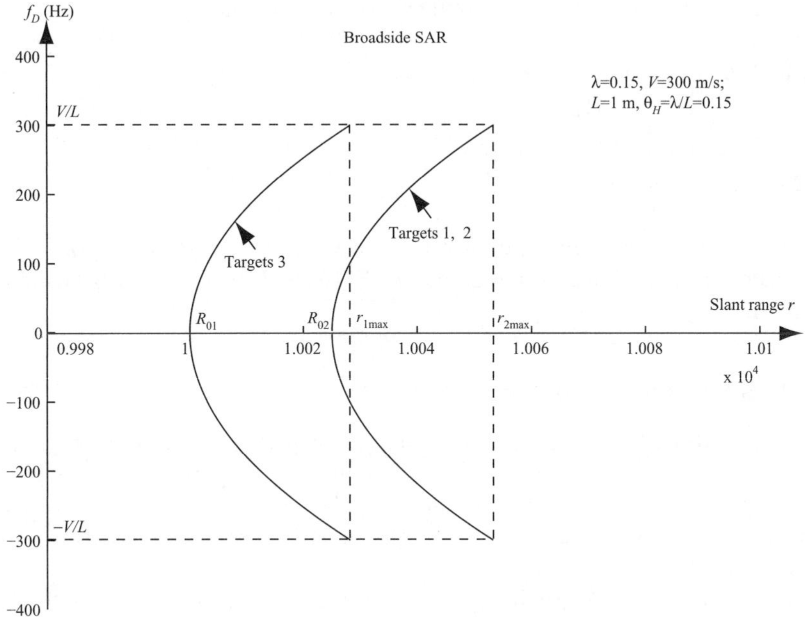

FIGURE 6.17 Doppler frequency versus slant range for multiple targets.

The following are part of the technical specifications of the ERS-1/2 satellite systems. They will be used to compute the parameters related to Doppler frequency discussed in this section.

Antenna size ($L \times W$): 10 × 1 m
Velocity (V): 7125 m/s
Sampling rate (f_s): 18.96 MHz
LFM chirp rate ($\dot{f}$): 4.1778×10^{11} Hz/s
Radar wavelength (λ): 0.0566 m
Pulse duration time (T): 37 μs
f_{PRF}: 1680 Hz

The bandwidth and resolution of range imaging can be computed as

$$
\begin{aligned}
B_t &= \dot{f} \times T \\
&= 4.1778 \times 10^{11} \times 37 \times 10^{-6} \\
&= 15.46 \text{ MHz} \\
\Delta x &= c/(2B) \\
&\approx 10 \text{ m}
\end{aligned}
$$

The related parameters of Doppler frequency can be computed as

$$f_{DU}\colon V/L = 712 \text{ Hz}$$
$$f_{DL}\colon -V/L = -712 \text{ Hz}$$
$$B_D\colon 2V/L = 1424 \text{ Hz}$$
$$\Delta u\colon V/f_{\text{PRF}} = 4.24 \text{ m}$$

Notice that the cross-range Doppler frequency bandwidth of 1424 Hz is much less than that of the range (time-domain) LFM signal, which has a bandwidth of 15.46 MHz.

6.3.2 SAR with Squint Angle

For the broadside SAR systems, the radar beam is perpendicular to the flight path. For satellite SAR systems, the earth's rotation induces an effective squint angle that varies by about $\pm 3^\circ$ for each orbit. Some applications require the radar beam to look forward or aft on the ground targets. A nonzero squint angle SAR system is discussed next.

Figure 6.18 shows the geometry of a typical forward-looking radar with squint angle θ_q and a point target located at $(0, R_0)$ in a u–x' coordinate system. A small-squint SAR system is considered as $\theta_q < 0.5\theta_H$, while a low-squint SAR system is considered as $0.5\theta_H < \theta_q < 10^\circ$. The three radar positions of u_1, u_2, and u_3 correspond to the start, center, and end of the radar beam illuminated on the target. Narrow beamwidth θ_H is assumed in our discussion.

Figure 6.18a shows a small-squint SAR; Fig. 6.18b, a low-squint SAR. The slant ranges R_1, R_2, and R_3 correspond to the three radar positions u_1, u_2, and u_3, respectively. The three shaded areas in Fig. 6.18 represent the 3-dB radar beamwidth corresponding to positions u_1, u_2, and u_3, respectively. The slant range R_2, corresponding to the center beam of radar at position u_2, is also represented as R_c. The distance

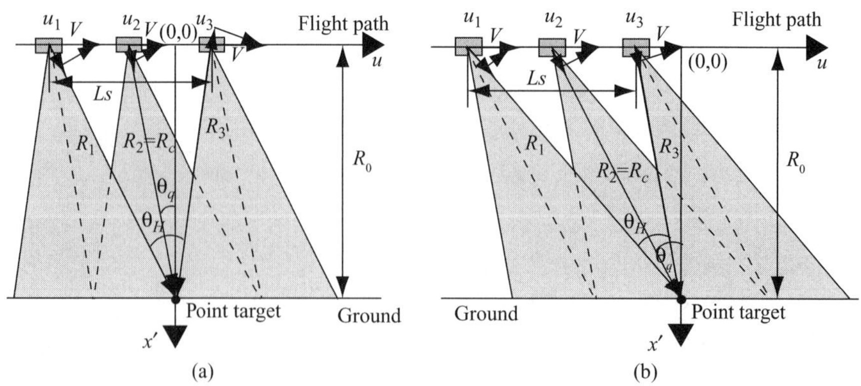

FIGURE 6.18 Geometry of a forward-looking radar system with nonzero squint angle.

between u_1 and u_3 equals the synthetic aperture length Ls. The closest distance between the radar and the target is $R_0 = R_c \cos\theta_q$. The radar beamwidth is shown as θ_H, and the squint angle θ_q is the angle between the radar center beam at position u_2 and the x' axis. The radar moves from left to right at speed V along the u axis. If the radar moves from right to left, the same geometry becomes a backward-looking radar. Forward looking radar is the major configuration discussed in the following chapters.

For a squint angle that satisfies $\theta_q < \theta_H/2$, as shown in Fig. 6.18a, the radar moves from u_1 to u_2, then from u_2 to $u = 0$. The slant range r decreases from R_1 to $R_2 = R_c$ and then to R_0. When the radar moves from $u = 0$ to u_3, the slant range r increases from R_0 to R_3. The radar positions u_1, u_2, and u_3, can be represented as follows:

$$\begin{aligned} u_1 &= -R_0 \tan(0.5\theta_H + \theta_q), \\ u_2 &= -R_0 \tan(\theta_q), \\ u_3 &= R_0 \tan(0.5\theta_H - \theta_q). \end{aligned} \tag{6.11a}$$

The corresponding slant range R_1, R_2 ($=R_c$), and R_3 can be represented as

$$\begin{aligned} R_1 &= \frac{R_0}{\cos(\theta_H/2 + \theta_q)}, \\ R_c &= \sqrt{R_0^2 + u_2^2} \\ &= \frac{R_0}{\cos\theta_q}, \\ R_3 &= \frac{R_0}{\cos(\theta_H/2 - \theta_q)}. \end{aligned} \tag{6.11b}$$

In the case of $\theta_q > \theta_H/2$, as shown in Fig. 6.18b, Eq. (6.11b) remains the same, but u_3 in Eq. (6.11a) changes to become

$$u_3 = -R_0 \tan(0.5\,\theta_H - \theta_q). \tag{6.11c}$$

In the broadside case, we have shown that the Doppler frequency f_D and the related parameters can be expressed in terms of the radar position u, slow time s, or slant range r. Those equations used in the broadside case are now extended to the squint angle case by factoring in Eq. (6.11). Since $u = sV$, the radar position u will be dropped in the following discussion.

The small-squint-angle SAR and the low-squint-angle SAR are considered separately in the following sections to derive the Doppler frequency f_D and other associated parameters.

6.3.2.1 SAR with a Small Squint Angle. Figure 6.18a serves as a small-squint-angle SAR that satisfies $\theta_q \le 0.5\theta_H$. The Doppler frequency and related parameters f_D, $\dot{f}_D$, and B_D can be derived as follows:

1. When radar moves from s_1 to s_3, from Eq. (6.7), the Doppler frequency and the change rate of f_D with respect to the slow time s can be derived as

$$\begin{aligned} f_{Ds} &= \frac{2V_r}{\lambda} \\ &= -\frac{2V^2 s}{\lambda\sqrt{R_0^2 + s^2V^2}} \operatorname{sgn}(u) \\ &= -\frac{2V}{\lambda}\sin\theta_u \operatorname{sgn}(u), \\ \dot{f}_{Ds} &= \frac{\partial f_{Ds}}{\partial s} \\ &= -\frac{2V^2 R_0^2}{\lambda\sqrt[3]{R_0^2 + s^2V^2}} \\ &= -\frac{2V^2}{\lambda R_0}\cos^3\theta_u. \end{aligned} \tag{6.12a}$$

From Eq. (6.10), the Doppler frequency and the change rate or slope of f_D with respect to slant range r can be derived as follows:

$$\begin{aligned} f_{Dr} &= \frac{2V_r}{\lambda} \\ &= -\frac{2V\sqrt{r^2 - R_0^2}}{\lambda r} \operatorname{sgn}(u), \\ \dot{f}_{Dr} &= \frac{\partial f_{Dr}}{\partial r} \\ &= -\frac{2V}{\lambda}\frac{R_0^2}{r^2\sqrt{r^2 - R_0^2}} \operatorname{sgn}(u). \end{aligned} \tag{6.12b}$$

Although f_{Ds} and f_{Dr} appear to differ, one can observe that $f_{Ds} = f_{Dr} = f_D$, but $\dot{f}_{Ds} \ne \dot{f}_{Dr}$. Notice that $\dot{f}_{Dr}$ is dependent on sgn (u), but $\dot{f}_{Ds}$ is not.

2. When $s = 0$, $r = R_0$; therefore

$$f_{Ds} = 0$$

$$\dot{f}_{Ds} = -\frac{2V^2}{\lambda R_0}$$

and

$$f_{Dr} = 0$$

$$\dot{f}_{Dr} = \infty.$$

3. The cenroid of the Doppler frequency f_{Dc} occurs when $s = s_2$, $r = r_2 = R_c$, and $\theta_u = \theta_q$:

$$\begin{aligned} f_{Dcs} &= -\frac{2V \sin\theta_q}{\lambda}\operatorname{sgn}(u_2) \\ &= \frac{2V}{\lambda}\sin\theta_q, \end{aligned} \tag{6.13a}$$

$$\dot{f}_{Dcs} = -\frac{2V^2}{\lambda R_0}\cos^3\theta_q,$$

$$\begin{aligned} f_{Dcr} &= -\frac{2V \sin\theta_q}{\lambda}\operatorname{sgn}(u_2) \\ &= \frac{2V}{\lambda}\sin\theta_q \end{aligned} \tag{6.13b}$$

$$\dot{f}_{Dcr} = \frac{2V}{\lambda R_c}\frac{\cos^2\theta_q}{\sin\theta_q}.$$

Here, $\operatorname{sgn}(u_2) = -1$ is applied, as can be seen from Fig. 6.18a. Therefore, the center of the Doppler frequency $f_{Dc} = f_{Dcs} = f_{Dcr}$.

4. The upper bound of the Doppler frequency occurs at $s = s_1$ and $r = R_1$, and the lower bound of the Doppler frequency occurs at $s = s_3$ and $r = R_3$; therefore

$$\begin{aligned} f_{DUs} &= f_{Ds}\big|_{s=s_1} \\ &= -\frac{2V\sin(0.5\theta_H + \theta_q)}{\lambda}\operatorname{sgn}(u_1) \\ &= \frac{2V}{\lambda}\sin(0.5\theta_H + \theta_q), \\ f_{DLs} &= f_{Ds}\big|_{s=s_3} \\ &= -\frac{2V\sin(0.5\theta_H - \theta_q)}{\lambda}\operatorname{sgn}(u_3) \\ &= -\frac{2V}{\lambda}\sin(0.5\theta_H - \theta_q), \end{aligned} \tag{6.14a}$$

and

$$\begin{aligned} f_{DUr} &= f_{Dr}\big|_{r=R_1} \\ &= -\frac{2V\ \sin(0.5\theta_H + \theta_q)}{\lambda}\operatorname{sgn}(u_1) \\ &= \frac{2V}{\lambda}\sin(0.5\theta_H + \theta_q), \\ f_{DLr} &= f_{Dr}\big|_{r=R_3} \\ &= -\frac{2V\ \sin(0.5\theta_H - \theta_q)}{\lambda}\operatorname{sgn}(u_3) \\ &= -\frac{2V}{\lambda}\sin(0.5\theta_H - \theta_q). \end{aligned} \tag{6.14b}$$

Here, $\operatorname{sgn}(u_1) = -1$ and $\operatorname{sgn}(u_3) = 1$ are applied as can be seen from Fig. 6.18a. Therefore the upper bound and the lower bound of f_D satisfy the relation $f_{DU} = f_{DUs} = f_{DUr}$ and $f_{DL} = f_{DLs} = f_{DLr}$.

5. The bandwidth of the Doppler frequency is then

$$\begin{aligned} B_D &= f_{DU} - f_{DL} \\ &= \frac{2V}{\lambda}\sin(0.5\theta_H + \theta_q) + \frac{2V}{\lambda}\sin\left(0.5\theta_H - \theta_q\right) \\ &= \frac{4V}{\lambda}\sin(0.5\theta_H)\cos\theta_q. \end{aligned} \tag{6.15a}$$

For a small beamwidth θ_H, $\cos(0.5\theta_H) \approx 1$, $\sin(0.5\theta_H) \approx 0.5\theta_H$, and $\theta_H = \lambda/L$; therefore

$$\begin{aligned} f_{DU} &\approx \frac{2V}{\lambda}\left[0.5\theta_H\cos\theta_q + \sin\theta_q\right] \\ &= \frac{2V}{\lambda}\sin\theta_q + \frac{V}{L}\cos\theta_q \\ &= f_{Dc} + \frac{V}{L}\cos\theta_q, \\ f_{DL} &\approx -\frac{2V}{\lambda}\left[0.5\theta_H\cos\theta_q - \sin\theta_q\right] \\ &= \frac{2V}{\lambda}\sin\theta_q - \frac{V}{L}\cos\theta_q \\ &= f_{Dc} - \frac{V}{L}\cos\theta_q, \\ B_D &= f_{DU} - f_{DL} \\ &= \frac{2V}{L}\cos\theta_q. \end{aligned} \tag{6.15b}$$

In the case where the small radar beamwidth is used and the squint angle is zero, the bandwidth B_D of Eq. (6.15b) becomes identical to that of the broadside case as shown in Eq. (6.10e).

The synthetic aperture length Ls can be computed as

$$Ls = R_1 \sin(0.5\theta_H + \theta_q) + R_3 \sin(0.5\theta_H - \theta_q).$$

The pulse duration time is computed as follows:

$$T_a = \frac{Ls}{V}.$$

The correlation between Doppler frequency and slow time s and slant range r is shown in Fig. 6.19. Both plots are based on the small-squint-angle SAR with a relatively large radar beamwidth under a single-target situation for illustration purposes. The following parameters are used to generate the drawings: $V = 300$ m/s, $L \times W = 1 \times 1$ m, $f_c = 2.0$ GHz, $R_0 = 10{,}000$ m, and $\theta_q = 3°$ and $\theta_H = \lambda/L = 0.15$ radian or 8.59°.

Figure 6.19a shows the Doppler frequency distribution in terms of the slow time s, which decreases linearly from slow times s_1 to s_3. The linear relation reflects the

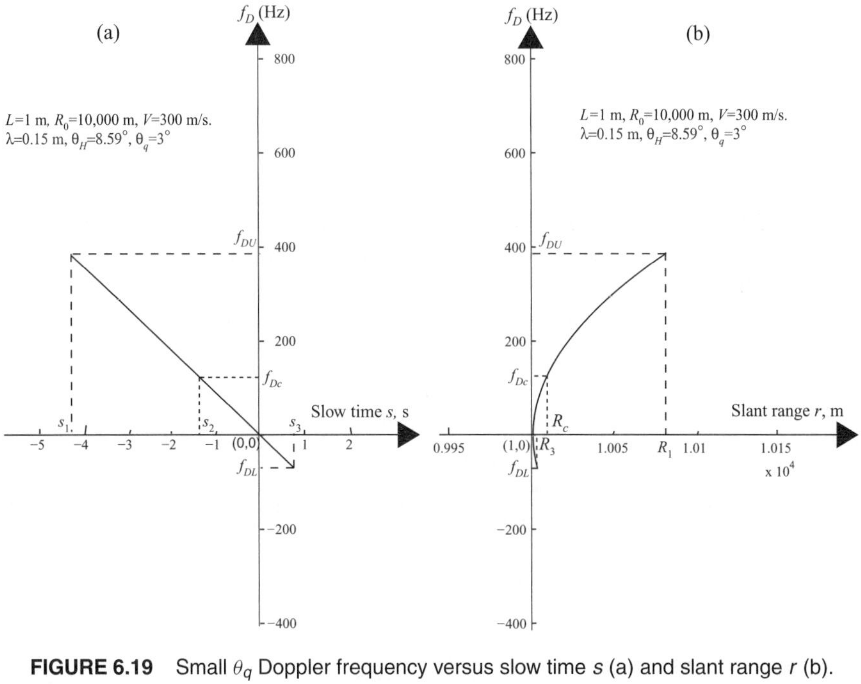

FIGURE 6.19 Small θ_q Doppler frequency versus slow time s (a) and slant range r (b).

linear FM characteristics under the squint-mode radar in terms of slow time s. Figure 6.19b, based on Eq. (6.12c), illustrates Doppler Frequency distribution in terms of slant range r and appears to be a hyperbolic format. The two plots shown in Fig. 6.19 share the same parameters of the Doppler frequency spectrum, namely, the same f_{DU}, f_{DL}, f_{Dc}, and B_D values.

6.3.2.2 SAR with a Low Squint Angle. A low-squint-angle SAR that satisfies $0.5\theta_H < \theta_q < 10°$ is shown in Fig. 6.18b. For such a configuration, the Doppler frequency and related parameters f_D, $\dot{f}_D$, and B_D can be derived as follows:

1. At any radar position u, where $u_1 < u < u_3$ and $R_1 < r < R_3$, one obtains

$$\begin{aligned} f_{Ds} &= \frac{2V}{\lambda}\sin\theta_u \\ &= \frac{2V^2 s}{\lambda\sqrt{R_0^2 + (sV)^2}}\mathrm{Sgn}(u) \\ \dot{f}_{Ds} &= -\frac{2V^2 R_0^2}{\lambda\sqrt[3]{R_0^2 + s^2V^2}} \\ &= -\frac{2V^2}{\lambda R_0}\cos^3\theta_u, \end{aligned} \tag{6.16a}$$

$$\begin{aligned} f_{Dr} &= \frac{2V\sqrt{r^2 - R_0^2}}{\lambda r}, \\ \dot{f}_{Dr} &= \frac{2V}{\lambda}\frac{\cos^2\theta_u}{\sqrt{r^2 - R_0^2}}. \end{aligned} \tag{6.16b}$$

   Equations (6.16a) and (6.16b) show that $f_{Ds} = f_{Dr} = f_D$, but $\dot{f}_{Ds} \neq \dot{f}_{Dr}$. Both f_D and $\dot{f}_{Dr}$ are positive because the value u, shown in Fig. 6.18b, is always negative.

2. The centroid of the Doppler frequency, which occurs at $s = s_2$ and $r = R_2 = R_c$, can be computed as follows:

$$\begin{aligned} f_{Dcs} &= \frac{2V}{\lambda}\sin\theta_q, \\ f_{Dcr} &= \frac{2V}{\lambda}\sqrt{1 - \left(\frac{R_0}{R_c}\right)^2} \\ &= \frac{2V}{\lambda}\sin\theta_q. \end{aligned} \tag{6.16c}$$

   Therefore $f_{Dcs} = f_{Dcr} = f_{Dc}$.

3. The upper bound of the Doppler frequency occurs at $s = s_1$ and $r = R_1$ and the lower bound, at $s = s_3$ and $r = R_3$; therefore

$$
\begin{aligned}
f_{DUs} &= -\frac{2V^2 s_1}{\lambda\sqrt{R_0^2 + V^2 s_1^2}}\operatorname{sgn}(u_1) \\
&= \frac{2V}{\lambda}\sin(0.5\theta_H + \theta_q), \\
f_{DLs} &= -\frac{2V^2 s_3}{\lambda\sqrt{R_0^2 + V^2 s_3^2}}\operatorname{sgn}(u_3) \\
&= \frac{2V}{\lambda}\sin(\theta_q - 0.5\theta_H),
\end{aligned}
\tag{6.16d}
$$

and

$$
\begin{aligned}
f_{DUr} &= \frac{2V}{\lambda}\sin(0.5\theta_H + \theta_q), \\
f_{DLr} &= \frac{2V}{\lambda}\sin(\theta_q - 0.5\theta_H).
\end{aligned}
\tag{6.16e}
$$

Therefore $f_{DUu} = f_{DUr} = f_{DU}$ and $f_{DLu} = f_{DLr} = f_{DL}$.

4. The bandwidth of the Doppler frequency can be computed as follows:

$$
\begin{aligned}
B_D &= f_{DU} - f_{DL} \\
&= \frac{2V}{\lambda}\sin\left(0.5\theta_H + \theta_q\right) + \frac{2V}{\lambda}\sin(0.5\theta_H - \theta_q) \\
&= \frac{4V}{\lambda}\sin(0.5\theta_H)\cos\theta_q.
\end{aligned}
\tag{6.17a}
$$

Comparing Eqs. (6.15a) with (6.17a), one can see that the Doppler frequency bandwidth B_D is identical for both small- and low-squint-angle SAR radars.

For small radar beamwidth θ_H, the following approximations can be derived:

$$\begin{aligned}
f_{DU} &\approx \frac{2V}{\lambda}\left[\sin\theta_q + 0.5\theta_H\cos\theta_q\right] \\
&= \frac{2V}{\lambda}\sin\theta_q + \frac{V}{L}\cos\theta_q \\
&= f_{Dc} + \frac{V}{L}\cos\,\theta_q, \\
f_{DL} &\approx \frac{2V}{\lambda}\left[\sin\theta_q - 0.5\theta_H\cos\theta_q\right] \\
&= \frac{2V}{\lambda}\sin\theta_q - \frac{V}{L}\cos\theta_q \\
&= f_{Dc} - \frac{V}{L}\cos\,\theta_q \\
B_D &= f_{DU} - f_{DL} \\
&= \frac{2V}{L}\cos\,\theta_q.
\end{aligned} \tag{6.17b}$$

Again, the bandwidth B_D shown in Eq. (6.17b) is identical to that of Eq. (6.15b).

The synthetic aperture length, or equivalently the beamwidth, can be computed as

$$Ls = R_1\,\sin(\theta_q + 0.5\theta_H) - R_3\,\sin(\theta_q - 0.5\theta_H)$$

and the pulse duration time, as $T_a = Ls/V$.

Figure 6.20 displays two drawings of the Doppler frequency based on the low-squint-angle SAR with a relatively small radar beamwidth. Figure 6.20a is plotted in terms of slow time s and Fig. 6.20b, in terms of slant range r. A single target is assumed, and the following parameters are used to generate the drawings: $V = 300$ m/s, $L \times W = 1 \times 1$ m, $f_c = 6$ GHz, $R_0 = 10{,}000$ m, and $\theta_q = 9^\circ$ and $\theta_H = \lambda/L = 0.05$ radian or 2.865°.

Figure 6.20a, based on Eq. (6.16a), shows that the Doppler frequency linearly decreases from slow times s_1 to s_3. Figure 6.20b, based on Eq. (6.16b), also appears to be in linear format. This is because the slant range R_0 is a large value relative to the movement of radar. Unlike the f_D shown in Fig. 6.16, where both positive and negative f_D exist, only positive f_D exist for both plots of Fig. 6.20. The Doppler centroid f_{Dc} is located in the middle of the bandwidth of f_D. Again, the

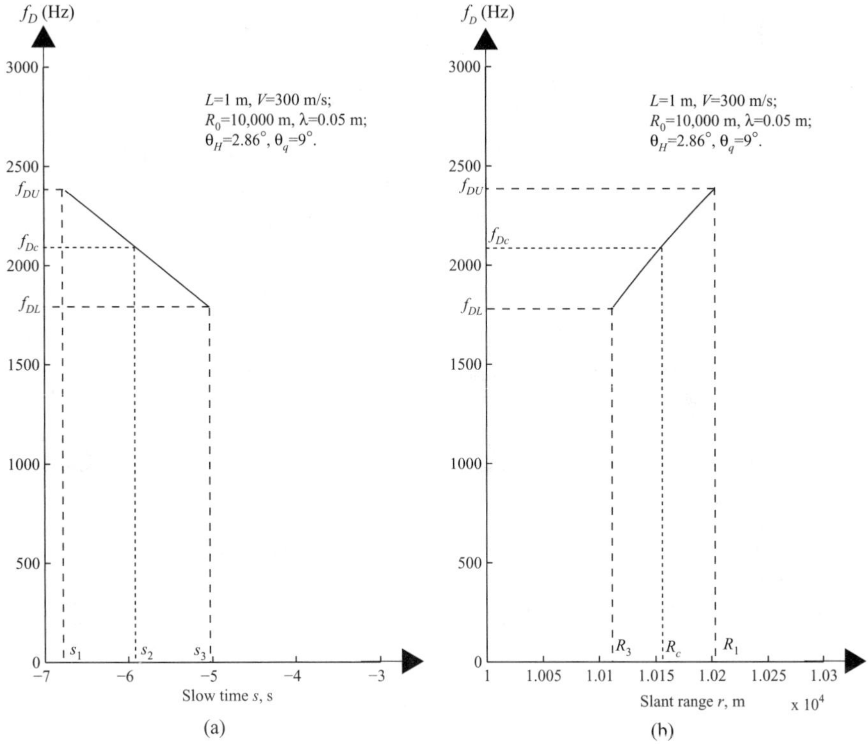

FIGURE 6.20 Low θ_q Doppler frequency versus slow time s (a) and slant range r (b).

two plots shown in Fig. 6.20 share the same parameters of f_D, namely, f_{DU}, f_{DL}, and B_D.

For comparison purposes, three drawings of the Doppler frequency are displayed in Fig. 6.21. The drawings are in terms of the slant range r and squint angles θ_q. Figure 6.21a corresponds to the broadside SAR; Fig. 6.21b, a small-squint-angle SAR; Fig. 6.21c, a large-squint-angle SAR. The parameters used to generate the drawings are listed below:

Radar carrier frequency $f_c = 2$ GHz; (wavelength $\lambda = 0.15$ m)

Antenna array size L × W = 1 × 1 m

Radar beamwidth $\theta_H = 8.6°$

Small squint angle $\theta_q = 3°$

Large squint angle $\theta_q = 15°$

Shortest target–radar distance, $R_0 = 10{,}000$ m

Radar direction of movement (orientation) = from left to right

Radar speed (velocity) $V = 300$ m/s

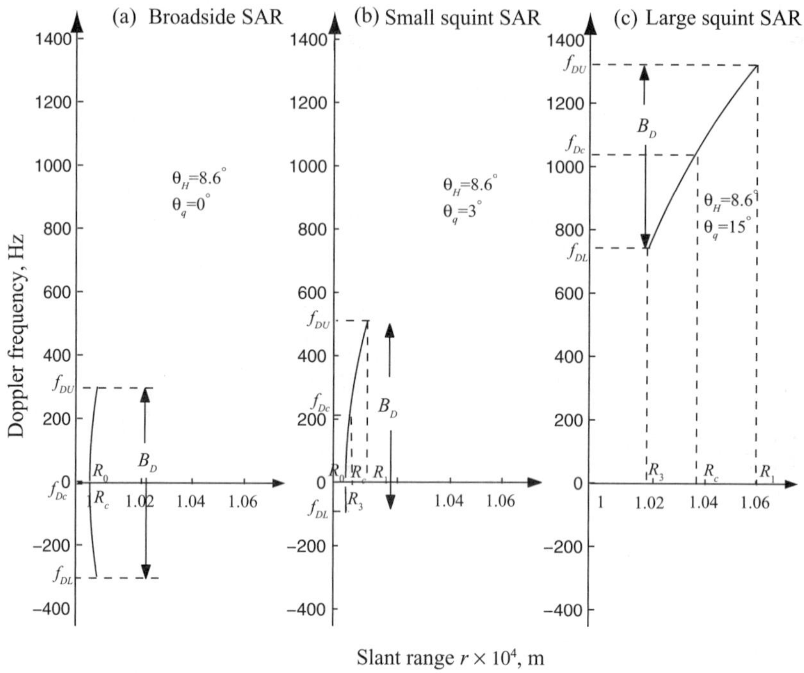

FIGURE 6.21 Comparison of Doppler frequencies for different SAR systems.

The key Doppler frequency–related parameters for the three cases shown in Fig. 6.21 can be computed and are listed below:

θ_q	0°	3°	15°
f_{Dc}, Hz	0	209.3	1035.28
f_{DU}, Hz	300	508.3	1322.1
f_{DL}, Hz	−300	−90.7	742.7
B_D, Hz	600	599	579.4

Although the three cases from Fig. 6.21 share the same radar beamwidth $\theta_H = 8.6°$, the frequency bandwidth is different in all three cases. Since the Doppler frequency bandwidth is squint-angle-dependent, the larger the squint angle θ_q, the smaller the Doppler frequency bandwidth B_D becomes.

Figure 6.22 displays the relationship of the Doppler frequency of a forward-looking squint SAR with respect to radar positions u. Figure 6.22a shows the geometry of a squint radar with three ground targets located at (u_1, R_{01}), (u_2, R_{02}), and (u_3, R_{03}), respectively, with $R_{01} = R_{02} = R_0$ and $u_3 = 0$. The slant ranges between radar and targets 1, 2, and 3 are r_1, r_2, and r_3, respectively.

Figure 6.22b shows that the Doppler frequency f_D varies linearly along the u axis. The two lines on the left, which correspond to targets 1 and 2, have the same slope.

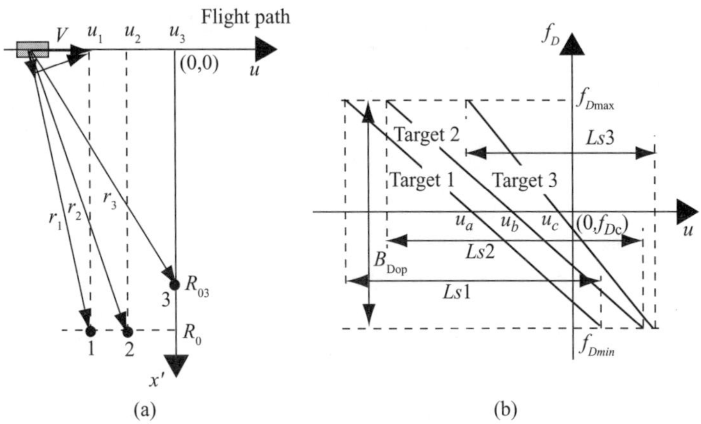

FIGURE 6.22 (a) Multiple-target squint SAR system; (b) plot of Doppler frequency f_D versus radar displacement u.

The third line has a steeper slope and is related to target 3. The centers of all three lines are identical, and serve as the centroid of Doppler frequency f_{Dc}. The intersections of the three lines with the u axis are u_a, u_b, and u_c, which are the radar positions when the targets are under the center beam of the radar. Because of the nonzero squint angle, u_a, u_b, and u_c are different from u_1, u_2, and u_3. The Doppler frequency bandwidths of all three lines are identical. Targets 1 and 2 have the same synthetic aperture length, while target 3 has a shorter one.

6.4 RANGE MIGRATION AND CURVATURE

To help the reader better understand range migration and curvature, Fig. 6.23 is presented as a simplified version of Fig. 6.18. Notice that radar position u is at the left side of the point of origin and is a negative value. Leting $\theta = \pi/2 - \theta_q$, one can obtain the following equation:

$$\begin{aligned} R^2(u) &= R_c^2 + (u_c - u)^2 - 2R_c(u_c - u)\cos\theta \\ &= R_c^2 + (u - u_c)^2 + 2R_c(u - u_c)\sin\theta_q. \end{aligned}$$

Therefore

$$\begin{aligned} R(u) &= R_c\sqrt{1 + \frac{2\,(u - u_c)\sin\theta_q}{R_c} + \frac{(u - u_c)^2}{R_c^2}} \\ &= R_c + (u - u_c)\sin\theta_q + \frac{(u - u_c)^2}{2R_c} - \frac{R_c}{8}\left[\frac{(u - u_c)^2}{R_c^2} + \frac{2\,(u - u_c)\sin\theta_q}{R_c}\right]^2 + \cdots \end{aligned} \tag{6.18}$$

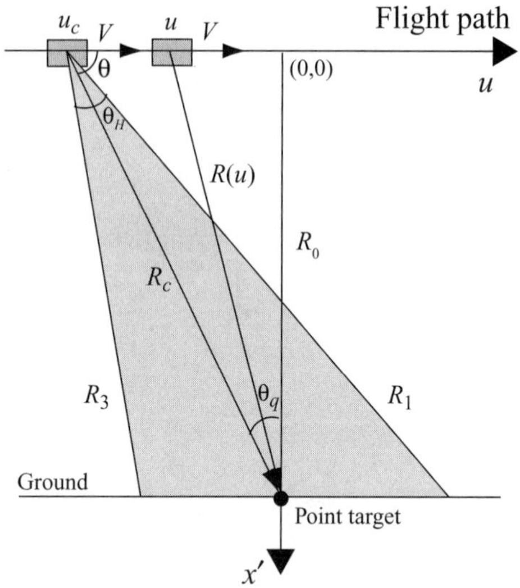

FIGURE 6.23 A simplified single-target squint SAR system.

Ignoring the higher-order items following the square bracket, Eq. (6.18) becomes

$$R(u) \approx R_c + (u - u_c)\sin\theta_q + \frac{(u - u_c)^2}{2R_c}. \tag{6.19}$$

The range migration ΔR is defined as the difference between $R(u)$ and R_c, that is, $\Delta R = R(u) - R_c$. The linear part of Eq. (6.19) is called *range walk*, and the quadratic part of Eq. (6.19) is called *range curvature*. Because $|u - u_c| \leq Ls/2$, the maximum range migration can be obtained from Eq. (6.19) as follows:

$$\begin{aligned} \Delta R_{\max} &= R_{\max}(u) - R_c \\ &= |\sin\theta_q|\frac{Ls}{2} + \frac{Ls^2}{8R_c} \\ &= \left(\frac{\lambda R_c}{2L}\right)\left(|\sin\theta_q| + \frac{\lambda}{4L}\right). \end{aligned} \tag{6.20}$$

Given a range resolution ΔR_r, in general, there will be no range migration correction required if $\Delta R_{\max}$ is less than $\Delta R_r/4$. Therefore, the criterion for determining whether range migration correction is required, is to check whether

$$\left(\frac{2\lambda R_c}{L}\right)\left(|\sin\theta_q| + \frac{\lambda}{4L}\right) < \Delta R_r. \tag{6.21}$$

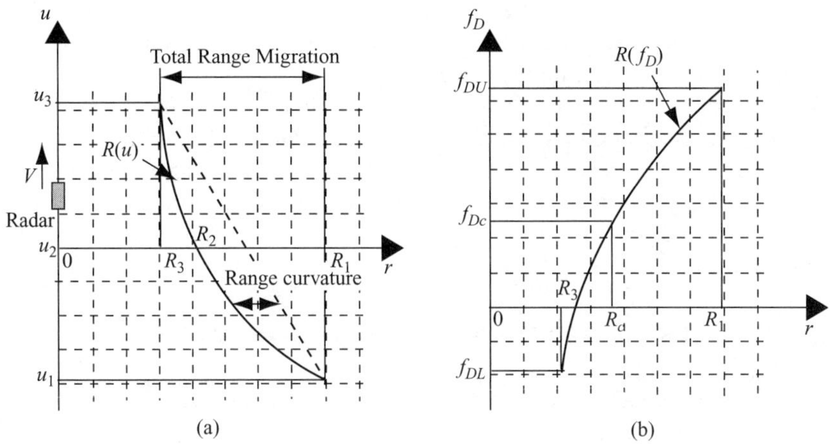

FIGURE 6.24 Single-target trajectory in squint SAR system.

For the broadside mode, $\theta_q = 0$ and $R_c = R_0$, this criterion becomes

$$\frac{R_0\lambda^2}{2L^2} < \Delta R_r. \tag{6.22}$$

Figure 6.24 is a graphical representation of a point target range migration under the squint SAR. Figure 6.24a shows slant range $R(u)$ plotted against radar position u. The two slant range points R_1 and R_3 are the start- and endpoints where the target is under the radar's illumination. R_2 is the point at which the center beam of the radar is on the target, which is normally represented as $R_2 = R_c$. The corresponding radar positions are u_1, u_2, and u_3. The difference between R_1 and R_3 is the total *range migration*. The distance between the curve $R(u)$ and the line linking R_1 and R_3 is the *range curvature*. The vertical dashed line represents the range cell lines; the horizontal dashed lines, the radar positions or azimuth cell lines. The intersection points of these two lines are the radar image data points. $R(u)$ represents a single-target trajectory seen by the radar through the synthetic aperture length. It migrates from range cell R_1 to range cell R_2, then stops at range cell R_3. The total range migration in this example is 5 range cells, which will cause distortion of the image if not corrected.

Figure 6.24b shows the corresponding range migration in the Doppler frequency domain, namely, the target trajectory $R(f_D)$ plotted against the slant range r. It illustrates the Doppler frequency migration through different range cells. The vertical dashed lines denote the range cell lines; the horizontal dashed lines, the Doppler frequency bin lines. $R(f_D)$ migrates from f_{DU} at range cell R_1 to f_{Dc} at range cell R_c, then stops at f_{DL} at range cell R_3. The total Doppler frequency migration in this example is 5 range cells.

In general, the range migration is relatively small for airborne-based platforms; yet for satellite-based platforms, the range migration is severe. The requirement for range migration correction depends on the choice of radar platforms, the methods used to process the radar image, and the waveform of radar (frequency bandwidth and beamwidth of the transmitted signal).

Imaging radar records the phase and polarization of the reflected signal. The radar image consists of pixels, which represent the radar backscatter from the ground targets. Darker areas represent low backscatter, while bright areas represent high backscatter. The radar images come with certain geometric distortions, such as slant range to ground distortion, image layover, and shadowing. This is discussed in the next section.

6.5 GEOMETRIC DISTORTIONS OF THE RADAR IMAGE

Because the image radar measures the distance to features in slant range rather than the true horizontal distance along the ground, this results in a varying image scale. Consider the example shown in Fig. 6.25, where the radar maps the three features with different height and shapes on the ground. The three features, described and labeled as $a,b,c,\ldots, j$, are mapped and appear as b', a', $c',\ldots, j'$, to the radar image plane. On the basis of Fig. 6.25, the geometric distortions of radar image are described below.

6.5.1 Layover

Layover occurs when the radar beam reaches the top of a tall feature (point b in Fig. 6.25) before it reaches the base (point *a* in Fig. 6.25). The return signal from the top of the feature will be received before the signal from the bottom will. As a result, the

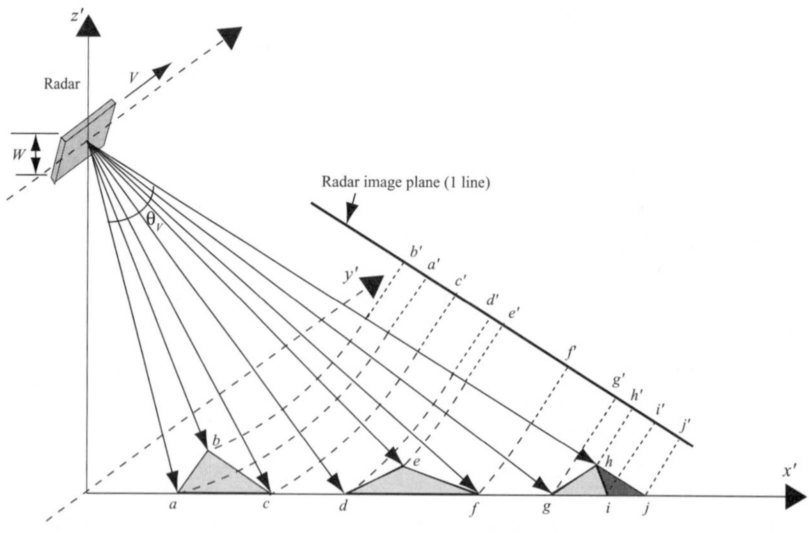

FIGURE 6.25 Geometric distortions of radar image.

top of the feature is displaced toward the radar from its true position on the ground, and "lays over" the base of the feature (*b*' to *a*').

6.5.2 Foreshortening

Foreshortening occurs when the radar beam reaches the base of a tall feature tilted toward the radar (e.g., a mountain) before it reaches the top. Because the radar measures distance in slant range, the slope (*d* to *e*) will appear compressed and the length of the slope will be represented incorrectly (d' to e').

6.5.3 Shadowing

The *shadowing* effect occurs when the radar beam cannot reach part of a tall feature (*h* to *i*). In the image plane, *h* to *i* will appear as dark. In addition, any target between *i* and *j* will also be shown as dark.

In addition to the distortions mentioned above other distortions can occur in the radar image, and they are described below.

6.5.4 Slant-to-Ground Range Distortion

The *slant range* is the distance seen by radar to the point target and is represented by R_s. The ground range distance is the horizontal distance along the ground corresponding to the point target measured in the slant range and is represented as R_g. The relationship between R_s and R_g is $R_g \sin\theta = R_s$, where θ is the radar incident angle. Features of the radar image displayed in slant range must be converted into ground range to reflect their real-world positions relative to one another.

6.5.5 Speckle

Speckle appears as a grainy "salt and pepper" texture in an image. This is caused by random interference from the multiple scattering returns that occur within each resolution cell and produce random bright and dark areas in the radar image. In order to clearly identify objects within an image, it may be necessary to clean up the image through speckle reduction. Speckle reduction can be achieved in either of two ways:

1. By *multilook processing*—multiple measurements of an object's (pixels) backscatter from different locations
2. By *spatial filtering*—calculating the average (often median) value of adjacent pixels to average out speckle

6.6 RADAR IMAGE RESOLUTION

The resolution of radar image consists of both range and angular resolutions. A resolution cell in the slant range is formed by ΔR_s by ΔR_a, while on the ground it is formed by ΔR_g by ΔR_a. Figure 6.26 displays the resolution cell in both slant range and on the ground.

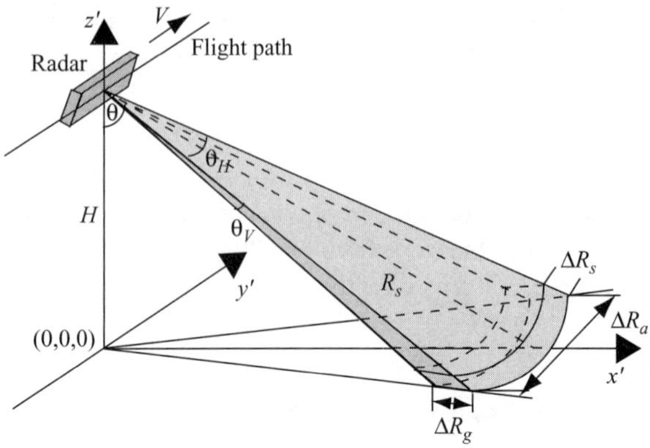

FIGURE 6.26 The resolution cell of a side-looking radar.

The term *resolution cell* refers to an area of the image, while a *pixel* (picture element) corresponds to the location of a digital sample in an image. Normally, at least 2 pixels per range resolution are needed. This is similarly true for angular resolution.

The range resolution of imaging radar is determined by the ability of the radar to distinguish between two point targets on the ground in the range direction. For real aperture radar (RAR), it is dictated by the time duration of the radar pulse T_p, and the angle of incidence θ, such that two targets on the ground can be distinguished only if they are separated by more than one pulsewidth.

For pulse-based RAR, the resolution in slant range depends on the transmitted pulse duration T_p or on the frequency bandwidth B of the pulse:

$$\begin{aligned}\Delta R_{sr} &= \frac{cT_p}{2} \\ &= \frac{c}{2B}.\end{aligned} \tag{6.23a}$$

The $\frac{1}{2}$ factor originates from the two-way travel of the radar signal. The corresponding ground range resolution is

$$\Delta R_{gr} = \frac{cT_p}{2\sin\theta} \tag{6.23b}$$

or

$$\Delta R_{gr} = \frac{c}{2B\sin\theta}. \tag{6.23c}$$

Along the flight track or azimuth direction, RAR resolution corresponds to the size of the antenna footprint on the ground. From Fig. 6.26, the footprint on the ground is $R_s\theta_H$, where R_s is the slant range from the radar to the center of the footprint and θ_H is the 3-dB beamwidth with $\theta_H = \lambda/L$. The RAR angular resolution in the azimuth direction is then

$$\begin{aligned} \Delta R_a &= R_s\theta_H \\ &= \frac{R_s\lambda}{L}. \end{aligned} \tag{6.24}$$

6.6.1 Example of Real Aperture Radar (RAR) Resolution: ERS-1/2-Imaging Radars

Major specifications of ERS-1 and ERS-2 radars are:

Pulse duration time T_p: 0.0371 ms
Average angle of incidence θ: 20°
Signal wavelength λ: 0.056 m
Mean range to a target on the earth R_s: 850 km
Radar moving speed V: 7 km/s
Antenna size $L \times W$: 10 × 1 m

Without considering the LFM signal for radar pulse, the ERS-1/2 radars yield the following results:

Resolution in slant range $\Delta R_{sr} = cT_p/2$
$= 5.56$ km
Resolution in ground range $\Delta R_{gr} = cT_p/(2\sin\theta)$
$= 16$ km
Resolution in azimuth $\Delta R_a = R_s\,\lambda/L$
$= 4.76$ km

Now consider linear frequency modulation. The ERS-1/2 satellite radars utilize the LFM signal with frequency bandwidth at 15.46 MHz and compressed pulse duration $T_p = 1/B$. The slant and ground range resolutions can be computed as

$$\begin{aligned} \Delta R_{sr} &= \frac{c}{2B} \\ &= 9.7\,\text{m} \\ \Delta R_{gr} &= \frac{c}{2B\sin\theta} \\ &= 28.4\,\text{m} \end{aligned} \tag{6.25}$$

This is much better than the pure pulse case of $\Delta R_{sr} = 5.56$ km and $\Delta R_{gr} = 16$ km.

Equation (6.24) shows that the azimuth resolution can be improved if the length of the antenna L increases. A synthetic aperture radar with a small physical antenna achieves a high resolution by synthesizing a long length of antenna. A long length of synthesized aperture corresponds to wider bandwidth of Doppler frequency, which in turn improves the azimuth resolution.

Computation of the Doppler frequency bandwidth requires computing the synthetic aperture length Ls and the upper and lower bounds of the Doppler frequency f_{DU} and f_{DL}.

The synthetic aperture length Ls equals the product of a 3-dB beamwidth multiplied by slant range R_s:

$$\begin{aligned} Ls &= R_s \theta_H \\ &= R_s \frac{\lambda}{L}. \end{aligned}$$

Consider the broadside SAR case. The upper bound of the Doppler frequency f_{DU} can be obtained as follows:

$$\begin{aligned} f_{DU} &= f_D \big|_{u=-Ls/2} \\ &= -\frac{2V}{\lambda}\frac{u}{R_s}\Big|_{u=-Ls/2} \\ &= \frac{V}{\lambda}\frac{Ls}{R_s} \\ &= \frac{V}{L}. \end{aligned}$$

Similarly, the lower bound of the Doppler frequency f_{DL} can be obtained as

$$\begin{aligned} f_{DL} &= -\frac{2V}{\lambda}\frac{Ls}{2R_s} \\ &= -\frac{V}{L}. \end{aligned}$$

The bandwidth of the Doppler frequency is

$$\begin{aligned} B_D &= f_{DU} - f_{DL} \\ &= \frac{2V}{L}. \end{aligned}$$

The along-track spatial resolution is the product of the synthesized LFM pulse duration T_a and the relative velocity of the flying vehicle; that is

$$\Delta R_a = V T_a, \tag{6.26a}$$

where $T_a = 1/B_D$. Therefore

$$\Delta R_a = \frac{V}{B_D} \\ = \frac{L}{2}. \tag{6.26b}$$

For an antenna length equal to 10 m, the azimuth resolution becomes 5 m, which is much smaller than the RAR case of 4.76 km.

The cross-range resolution for SAR is independent of the slant range R_s. Equation (6.26b) states that an arbitrarily high resolution can be obtained by using a shorter antenna. However, smaller radar antennas output less power, which reduce the target signal-to-noise power ratio. Therefore, a tradeoff between the antenna length and cross-range resolution must be made in designing the SAR antenna.

The criteria used to check whether range migration correction is required will now be applied on ERS-1/2 radars.

Given the wavelength $\lambda = 0.056$ m, the satellite–earth distance $R_0 = 850$ km, the antenna length $L = 10$ m, and—assuming that the squint angle $\theta_q = 0°$—the maximum range migration amount can be computed as follows, from Eq. (6.20):

$$\frac{R_0\lambda^2}{8L^2} = \frac{850 \times 10^3 \times 0.056^2}{8 \times 10^2} \\ = 3.333\text{ m}.$$

From Eq. (6.22), no range migration correction is required if

$$\frac{R_0\lambda^2}{8L^2} < \frac{\Delta R_{sr}}{4}.$$

However, the slant range resolution $\Delta R_{sr}/4 = 2.425$ m, from Eq. (6.25), is less than $(R_0\lambda^2/8L^2) = 3.333$ m. Therefore, range migration correction is required.

7

SYSTEM MODEL AND DATA ACQUISITION OF SAR IMAGE

The two-dimensional target locations (or function) of radar image can be obtained by processing the target-reflected signals in both the cross-range (azimuth) and range directions. Range migration correction may be required for some applications. The principles and basic characteristics of radar image processing have been covered in a previous chapter in terms of Doppler frequency. This chapter describes radar image processing in the wavenumber domain. Section 7.1 describes the range radar imaging in the wavenumber domain; Section 7.2 discusses the cross-range radar imaging. Both the broadside SAR and the squint SAR are covered. Section 7.3 reviews data acquisition and the frequency spectrum of radar images.

7.1 SYSTEM MODEL OF RANGE RADAR IMAGING

7.1.1 System Model

Figure 7.1 shows a radar system configuration for range processing. The radar beam is assumed to be narrow enough in the azimuth direction that all targets are located at $y = 0$ and randomly distributed along the range (or x axis) direction.

Figure 7.1a is a 3D display of the radar system. The radar, moving at speed V along the y axis, is located at $(0,0,H)$ and emits a pulse at $t = 0$. The ground area covered by the radar pulse in the range direction will reflect the pulse wave after a short delay. The shortest distance to the ground is R_N, and the longest distance is R_F. The echo delay time due to the target at R_N is $t_0 = (2R_N/c) = (2x_0/c \sin\theta)$. The echo delay time due to the target at R_F is $t_{N-1} = (2R_F/c) = (2x_{N-1}/c \sin\theta)$.

Digital Signal Processing Techniques and Applications in Radar Image Processing, by Bu-Chin Wang.

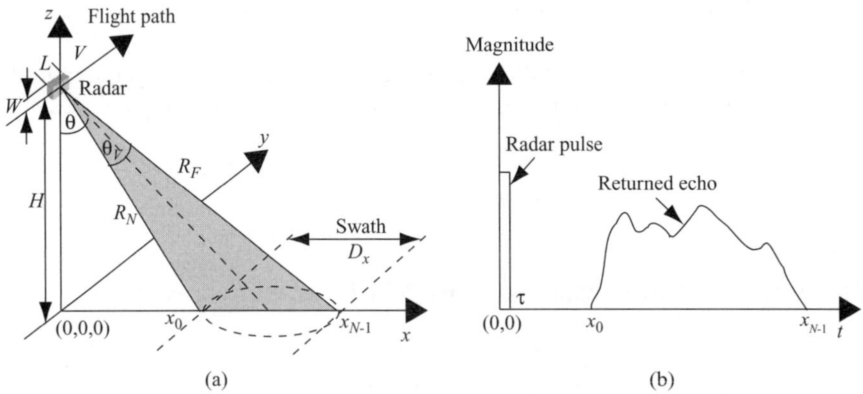

FIGURE 7.1 Geometry of a range imaging radar.

The corresponding ground ranges of t_0 and t_{N-1} are $x_0 = [(ct_0 \sin \theta)/2]$ and $x_{N-1} = [(ct_{N-1} \sin \theta)/2]$. The swath D_x, which ranges from x_0 to x_{N-1}, is related to the radar beamwidth as $D_x \approx \theta_V (R_N + R_F)/2 = \lambda (R_N + R_F)/(2W)$.

Figure 7.1b shows the relationship between radar pulse and returned echoes. Here the horizontal axis indicates the echo return time. The radar pulse emits at $t = 0$ and the echo returns at time t_0 through t_{N-1}. Since parameters t and x are linearly related by $t = 2x/c$, they are used interchangeably in the following discussion.

Consider a set of N discrete point targets, each located at x_n with a reflectivity σ_n, where $n = 0,1,2,\ldots, N-1$. The ideal target function, which identifies the location of each target along the x axis, is defined as follows:

$$f_0(x) = \sum_{n=0}^{N-1} \sigma_n \delta(x - x_n). \tag{7.1}$$

Figure 7.2 is an example of the ideal target function $f_0(x)$. The targets are randomly located along the x axis between x_0 and x_{N-1}. The magnitude of each target

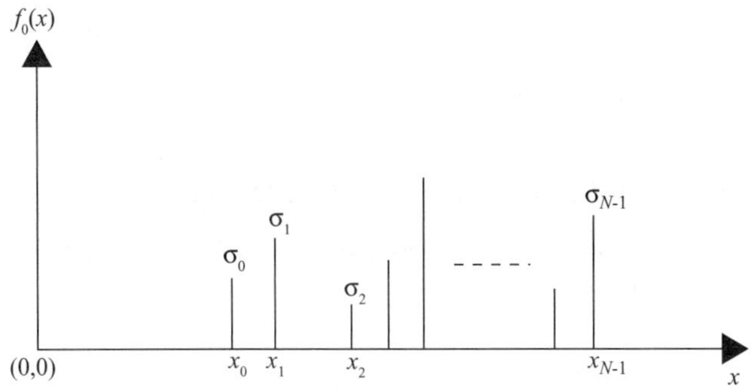

FIGURE 7.2 An ideal target function.

denotes reflectivity σ_n, and eventually reflects as a bright or dark dot on the radar image.

Let $p(t)$ be the radar signal illuminated on a one-dimensional target area in the range (x) direction as shown in Fig. 7.1. The returned echo signal $s(t)$ becomes

$$s(t) = \sum_{n=0}^{N-1} \sigma_n p(t - t_n). \tag{7.2}$$

The Fourier transform of Eq. (7.2), with $t_n = 2x_n/c$, can be expressed as

$$S(\omega) = \sum_{n=0}^{N-1} \sigma_n P(\omega)\exp\left(-j\omega\frac{2x_n}{c}\right), \tag{7.3}$$

where $P(\omega)$ is the Fourier transform of $p(t)$.

Since the delay time t_n and range x_n are linearly related and interchangeable, the Fourier transform of Eq. (7.1) can be expressed as

$$F_0(\omega) = \sum_{n=0}^{N-1} \sigma_n \exp\left(-j\omega\frac{2x_n}{c}\right). \tag{7.4}$$

From Eqs. (7.3) and (7.4), one obtains

$$S(\omega) = P(\omega)F_0(\omega). \tag{7.5}$$

The corresponding equation in time t or range x domain becomes

$$s(t) = p(t) * f_0(t)$$

or

$$s(x) = p(x) * f_0(x). \tag{7.6}$$

where the asterisk denotes convolution.

7.1.2 Reconstruction of Range Target Function

Mathematically, given the received signal spectrum $S(\omega)$, the ideal target function $f_0(x)$ can be derived from Eq. (7.5) as

$$f_0\left(\frac{ct}{2}\right) = \mathscr{F}^{-1}\left[\frac{S(\omega)}{P(\omega)}\right] = \sum_{n=0}^{N-1} \sigma_n \delta(t - t_n), \tag{7.7}$$

or

$$f_0(x) = \sum_{n=0}^{N-1} \sigma_n \delta(x - x_n).$$

However, for Eq. (7.7) to work properly, the function $P(\omega)$ must satisfy two conditions: (1) the bandwidth must be infinite, and (2) no zeros can exist in $P(\omega)$. Since $P(\omega)$ is the frequency spectrum of a radar pulse, its envelope has the sinc-like function. Therefore $P(\omega)$ cannot meet conditions 1 and 2 (above) to make Eq. (7.7) work correctly.

One practical method of finding the target function $f(x)$ for range imaging is to utilize the concept of a matched filter as discussed in Chapter 5. Let the matched filter function $P^*(\omega)$ be chosen as the complex conjugate of the transmitted pulse function $P(\omega)$, with the time-domain representation of $p^*(\omega)$ as $p^*(-t)$. The returned echo signal $s(t)$ can be considered as the output of a system (or target) function $f_0(t)$ with input $p(t)$. This output signal $s(t)$ is then applied to the matched filter to obtain the target function $f(t)$. The block diagram for solving the target function $f(t)$ is shown in Fig. 7.3.

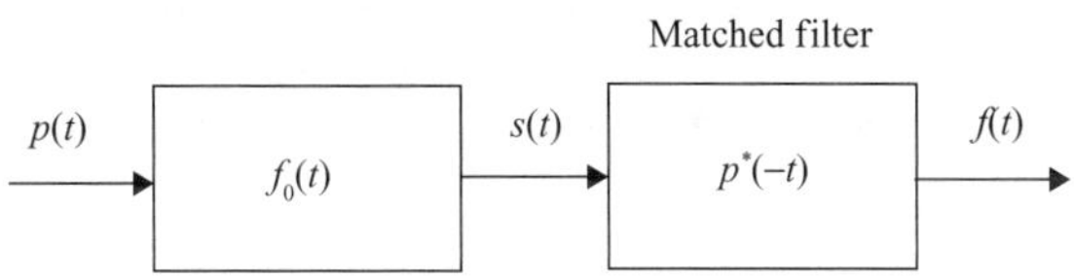

FIGURE 7.3 Matched filtering for range imaging.

The output of the matched filter can be expressed, in the frequency domain, as follows:

$$\begin{aligned} F(\omega) &= S(\omega)P^*(\omega) \\ &= F_0(\omega)P(\omega)P^*(\omega) \\ &= F_0(\omega)|P(\omega)|^2. \end{aligned} \tag{7.8}$$

Let psf(t) be a point spread function; that is

$$\text{psf}\,(t) = \mathfrak{F}^{-1}[|P(\omega)|^2]$$

or

$$\text{psf}\,(\omega) = |P(\omega)|^2, \tag{7.9}$$

where psf(ω) is the Fourier transform of psf(t); then

$$\begin{aligned} f(t) &= f_0(t) * \text{psf}\,(t) \\ &= \sum_{n=0}^{N-1} \sigma_n \,\text{psf}\left(t - \frac{2x_n}{c}\right). \end{aligned} \tag{7.10a}$$

Alternatively, the target function can be expressed in terms of range x as

$$f(x) = \sum_{n=0}^{N-1} \sigma_n \text{psf}(x - x_n). \tag{7.10b}$$

The point spread function psf (t) depends on the spectral shape of the transmitted radar signal $p(t)$. As an example, let $|P(\omega)| = 1$ for $\omega \in [-\omega_0, \omega_0]$; then psf($t$) becomes

$$\text{psf}(t) = \text{sinc}\left(\frac{\omega_0 t}{\pi}\right). \tag{7.11}$$

An example of a matched filter output, or reconstructed target function $f(x)$, is shown in Fig. 7.4. As can be seen, each target appears at a different location x_n with different magnitude, which corresponds to the reflection coefficient σ_n. When comparing Fig. 7.2 with Fig. 7.4, one can see that the nth target location changed from a point $\delta(x - x_n)$ in the ideal target function $f_0(x)$ to a segment occupied by psf$(x - x_n)$ in the real target function $f(x)$. The point spread function psf(t) becomes sharper, from Eq. (7.9), if the bandwidth of $p(t)$ increases. A sharper psf(x) improves the range resolution, and therefore the quality of the radar image.

The discussion above shows that range image processing detects the returned echo signal on the basis of a given transmitted signal. It depends only on the target distance or range, and is independent of the squint angle of the radar. Therefore, the prior discussion applies to both broadside SAR and squint SAR systems.

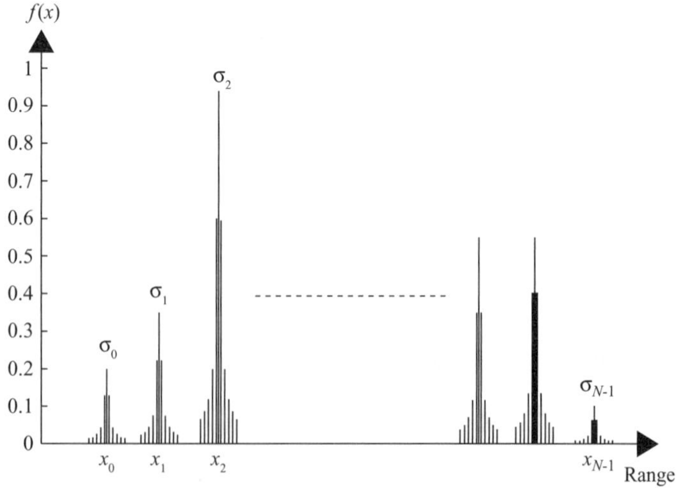

FIGURE 7.4 A reconstructed target function $f(x)$.

7.2 SYSTEM MODEL OF CROSS-RANGE RADAR IMAGING

This section discusses two cases of cross-range radar imaging. The broadside SAR is addressed first, followed by squint SAR.

7.2.1 Broadside Radar Case

7.2.1.1 System Model. The geometry of a typical radar system for cross-range image processing is shown in Fig. 7.5. Similar to but somewhat different from range image processing, the radar pulse is assumed to be narrow enough along the range direction so that all targets are located at the same range $x = X_c$ and randomly distributed along the y axis direction. It is assumed that all targets are located within the radar beamwidth.

Figure 7.5a shows a broadside SAR imaging system with M ground targets located along the cross-range (or y axis) direction. Each target has a reflection coefficient σ_m, where $m = 0,1,2,\ldots,M-1$. The imaging radar is located at $(0,u,H)$ and moves at speed V along the y axis. The radar illuminates the target area with beamwidth θ_H. The triangular shaded area shows a 2D plane where both radar and ground targets are in the same plane. Certainly the assumption that all targets have a range distance X_c is not practical in the real world. However, as will be discussed in Chapter 8, the radar image can be processed in range and cross-range directions independently if the range migration problems can be ignored. Therefore, the assumption that all targets in the cross-range direction are located at the same range X_c will not lose any generality in our future discussions.

Figure 7.5b is a simplified model based on the triangular shaded area of Fig. 7.5a. The x' axis is perpendicular to the flight path. The y direction is the cross-range direction, and the targets are all located within the beamwidth of the radar signal. All M targets are located at $x' = X_c'$, where $X_c' = (H^2 + X_c^2)^{1/2}$, and $y = y_m$ for $m = 0,1,\ldots,M-1$. We will use x and X_c instead of x' and X_c' in the following

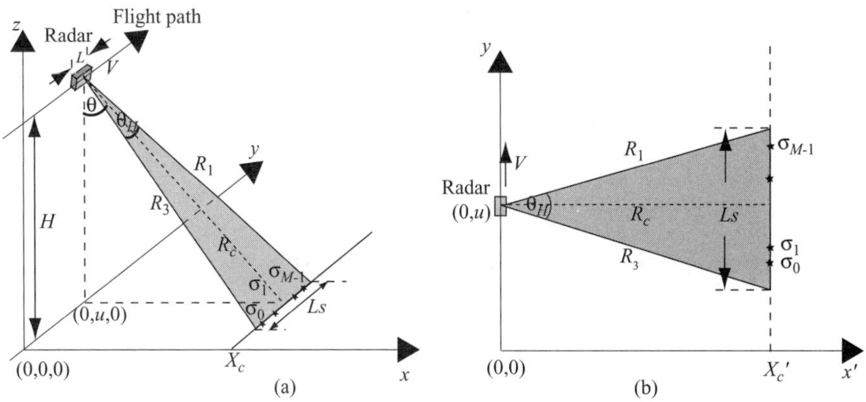

FIGURE 7.5 (a) A typical cross-range radar imaging system; (b) a simplified system.

discussion. With the radar located at $y = u$, the M targets are represented by their reflection coefficients as $\sigma_0, \sigma_1, \ldots, \sigma_{M-1}$.

Let $f_0(y)$ be an ideal target function in the cross-range domain, which identifies a group of M targets located along the y axis. Then $f_0(y)$ can be defined as follows, similar to the ideal target function described in range domain:

$$f_0(y) = \sum_{m=0}^{M-1} \sigma_m \delta(y - y_m). \tag{7.12}$$

Cross-range image processing is based on the phase history of returned signals from the targets, which are located along the y axis and illuminated under the radar beam. Consider a single target at distance R from the radar, which has a 3-dB beamwidth θ_H. The single-target phase history is computed during the time interval $T_1 = Ls/V$, where $Ls = R\theta_H$ is the synthetic aperture length and V is the velocity of the moving radar. For a group of M targets spread over a range of length D_y, the time interval for computing phase history becomes $T_a = (Ls + D_y)/V$.

Figure 7.6 shows the relationship between the radar beam and the targets. The M targets are assumed to be distributed between $-D_y/2$ and $D_y/2$. The radar moves between $-L_u$ and L_u along the y axis and continuously illuminate the targets, where $L_u = (Ls + D_y)/2$.

Also shown in Fig. 7.6 are the aspect angles of radar with respect to a particular target. The *aspect angle* is defined as the angle between the x axis and the line that connects the radar to the target. With radar located at u, the aspect angle with respect

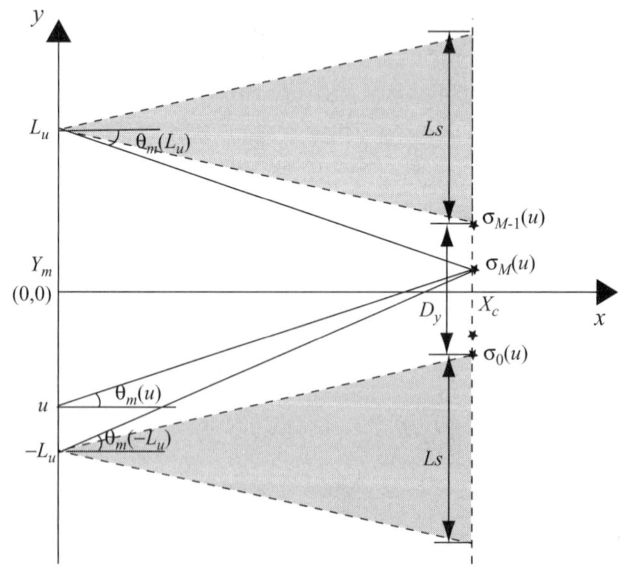

FIGURE 7.6 Relationship between radar beams and targets.

to the mth target is defined as

$$\theta_m(u) = \tan^{-1}\left(\frac{y_m - u}{x_m}\right), \tag{7.13}$$

where $x_m = X_c$ in our discussion.

Consider a system with radar located at $(0,u)$, where u is within the range $-L_u \le u \le L_u$, and the M targets are located at $\{(X_c, y)\}$ for $y = y_0, y_1, \ldots, y_{M-1}$. Let $p(t)$ be the transmitted signal and $R_m(u)$ be the slant range between the radar and the mth target:

$$R_m(u) = \sqrt{X_c^2 + (y_m - u)^2}.$$

The radar echo signal from M targets can then be expressed as

$$\begin{aligned} s(t,u) &= \sum_{m=0}^{M-1} \sigma_m p\left[t - \frac{2R_m(u)}{c}\right] \\ &= \sum_{m=0}^{M-1} \sigma_m p\left[t - \frac{2\sqrt{X_c^2 + (y_m - u)^2}}{c}\right]. \end{aligned} \tag{7.14}$$

For purposes of simplicity and to preserve generality, the radar signal $p(t)$ is assumed to be a single-frequency signal. Thus, $p(t) = \exp(j\omega t)$, and Eq. (7.14) becomes

$$s(t,u) = \exp(j\omega t) \sum_{m=0}^{M-1} \sigma_m \exp\left[-j2k\sqrt{X_c^2 + (y_m - u)^2}\right], \tag{7.15}$$

where $k = \omega/c = 2\pi/\lambda$. After a baseband conversion, Eq. (7.15) becomes

$$\begin{aligned} s_b(t,u) &= s(t,u)\exp(-j\omega t) \\ &= \sum_{m=0}^{M-1} \sigma_m \exp\left[-j2k\sqrt{X_c^2 + (y_m - u)^2}\right] \\ &= \sum_{m=0}^{M-1} s_{bm}(t,u), \end{aligned} \tag{7.16}$$

where

$$\begin{aligned} s_{bm}(t,u) &= \sigma_m \exp\left[-j2k\sqrt{X_c^2 + (y_m - u)^2}\right] \\ &= \sigma_m \exp\left[-j\frac{4\pi R_m(u)}{\lambda}\right]. \end{aligned} \tag{7.17}$$

To simplify the notation, $s(u)$ and $s_m(u)$ will be used to represent $s_b(t,u)$ and $s_{bm}(t,u)$ in the following discussion.

The instantaneous spatial frequency of $s_m(u)$ for u in the synthetic aperture region, $-L_u \leq u \leq L_u$, is

$$\begin{aligned} k_{um}(u) &= \frac{\partial}{\partial u}\left[-2k\sqrt{X_c^2 + (y_m - u)^2}\right] \\ &= \frac{2k(y_m - u)}{\sqrt{X_c^2 + (y_m - u)^2}} \\ &= 2k\frac{\sqrt{R_m^2(u) - X_c^2}}{R_m(u)} \\ &= 2k\sqrt{1 - \left(\frac{X_c}{R_m}\right)^2} \\ &= 2k\,\sin\theta_m(u), \end{aligned} \tag{7.18}$$

where $\theta_m(u)$ is the aspect angle of radar for the mth target when radar is located at $(0,u)$ and can be computed as follows, as defined in Eq. (7.13):

$$\theta_m(u) = \tan^{-1}\left(\frac{y_m - u}{X_c}\right).$$

Let an ideal reflector be located at $(X_c,0)$ with reflection coefficient $\sigma_r = 1$. Also let $s_{r0}(u)$ be a reference signal, which is the echo signal from this ideal reflector when radar is located at $(0,u)$:

$$s_{r0}(u) = \exp\left[-j2k\sqrt{X_c^2 + u^2}\right]. \tag{7.19}$$

From Eq. (7.12) with $y = u$, the ideal target function becomes

$$f_0(u) = \sum_{m=0}^{M-1} \sigma_m \delta\left(u - y_m\right). \tag{7.20}$$

The convolution of $s_{r0}(u)$ with $f_0(u)$ turns out to be

$$\begin{aligned} f_0(u) * s_{r0}(u) &= \int_{-\infty}^{\infty} f_0(\tau) s_{r0}(u - \tau)\, d\tau \\ &= \int_{-\infty}^{\infty} \sum_{m=0}^{M-1} \sigma_m \delta(\tau - y_m) \exp\left[-j2k\sqrt{X_c^2 + (u - \tau)^2}\right] d\tau \\ &= \sum_{m=0}^{M-1} \sigma_m \exp\left[-j2k\sqrt{X_c^2 + (u - y_m)^2}\right] \\ &= s(u), \end{aligned} \tag{7.21}$$

where $s(u)$ is the baseband signal shown in Eq. (7.16).

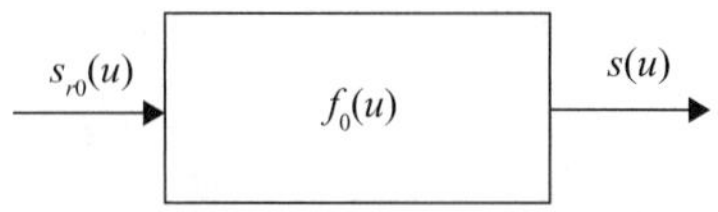

FIGURE 7.7 Relationship between received signal and reference signal.

Equation (7.21) states that the baseband echo signal $s(u)$ can be considered as the output function of a system, where $f_0(u)$ and $s_{r0}(u)$ are the system function and input function, respectively. Note that since the radar location u is a variable along the y axis, the parameters y and u are therefore interchangeable. Figure 7.7 shows the relationship between $s_{r0}(u)$, $f_0(u)$, and $s(u)$ for the cross-range target (image) processing, which is similar to that of the range target processing discussed in the previous section.

The purpose of cross-range target processing is to find the target location and its corresponding reflection coefficient. Equation (7.21) states that the target-reflected signals are formulated as the convolution of the ideal target function with the reference function.

The spatial frequency (or spatial Fourier) transform converts the signal from the spatial (y or u) domain to the spatial frequency (k_y or k_u) domain. In other words, given a signal $g(y)$, the spatial frequency transform of $g(y)$ is defined as

$$G\left(k_y\right) = \int_{-\infty}^{\infty} g(y)\exp\left(-jk_y y\right) dy. \tag{7.22}$$

The spatial frequency transform of $s_m(u)$, expressed in Eq. (7.17), can then be expressed as

$$\begin{aligned} S_m(k_u) &= \int_{-\infty}^{\infty} \sigma_m \exp\left[-j2k\sqrt{X_c^2 + (y_m - u)^2}\right] \exp(-jk_u u)du \\ &= \sigma_m \int_{-\infty}^{\infty} \exp\left[-j2k\sqrt{X_c^2 + (y_m - u)^2} - jk_u u\right]du. \end{aligned} \tag{7.23}$$

As discussed previously, the radar movement is limited from $u = -L_u$ to $u = L_u$ for a group of M targets. It is difficult to obtain a closed form of the integration shown in Eq. (7.23). Instead, the integration will be approximated by the "principle of stationary phase." This method ignores the integration regions at the far ends and is concerned only with some limited narrow regions, which are explained in the next section.

7.2.1.2 Principle of Stationary Phase.

Consider a general waveform

$$s(t) = a(t)\exp[j\phi(t)], \tag{7.24}$$

where $a(t)$ is a slow time-varying amplitude function relative to the phase function $\phi(t)$.

The Fourier transform of $s(t)$ is

$$S(\omega) = \int_{-\infty}^{\infty} a(t) \exp\{-j[\omega t - \phi(t)]\}\, dt. \tag{7.25}$$

Since $a(t)$ is a slow varying function relative to the phase function $\omega t - \phi(t)$, the contribution to the integral value from regions of fast fluctuation of $\omega t - \phi(t)$ will nearly cancel. The integral expressed above is therefore approximately equal to the integrand at the points where the phase function is nearly constant, or the instantaneous frequency of the phase function is zero. In other words, the time ranges around $t = t_s$ that satisfy

$$\frac{d[\omega t - \phi(t)]}{dt} = 0 \quad \text{or} \quad \omega = \frac{d\phi(t)}{dt} \tag{7.26}$$

will contribute to the integration in Eq. (7.25). Therefore, one can expand the integrand of Eq. (7.25) as a Taylor series around t_s:

$$\begin{aligned} a(t) &= a(t_s) + (t - t_s)a'(t_s) + \frac{(t - t_s)^2}{2} a''(t_s) + \cdots \\ \omega t - \phi(t) &= \omega t_s - \phi(t_s) + (t - t_s)\left[\omega - \phi'(t_s)\right] - \frac{(t - t_s)^2}{2} \phi''(t_s) + \cdots \\ &= \omega t_s - \phi(t_s) - \frac{(t - t_s)^2}{2} \phi''(t_s) + \cdots. \end{aligned} \tag{7.27}$$

Because $a(t)$ is a slow varying function, only the zeroth-order term of $a(t)$ will be retained. However, both the zeroth- and second-order terms of phase function $\omega t - \phi(t)$ will be used for integration. Assume that only one stationary point exists, Equation (7.25) then becomes

$$S(\omega) = a(t_s) \exp\{-j[\omega t_s - \phi(t_s)]\} \int_{-\infty}^{\infty} \exp\left[j \frac{(t - t_s)^2}{2} \frac{d^2\phi(t_s)}{dt^2}\right] dt. \tag{7.28}$$

Let

$$\frac{(t - t_s)^2}{2} \phi''(t_s) = \pm s^2, \tag{7.29}$$

where $\pm$ sign depends on the sign of $\phi''(t_s)$. Substituting Eq. (7.29) into Eq. (7.28), one obtains

$$S(\omega) = a(t_s) \exp\{-j[\omega t_s - \phi(t_s)]\} \sqrt{\frac{2}{|\phi''(t_s)|}} \int_{-\infty}^{\infty} \exp\left(\pm j s^2\right) ds. \tag{7.30}$$

Consider the following two cases of $\exp\left(\pm js^2\right)$;

1. Let $\exp(js^2) = \exp(-u^2)$; then

$$\int_{-\infty}^{\infty} \exp\left(js^2\right)ds = \frac{1}{\sqrt{-j}} \int_{-\infty}^{\infty} \exp\left(-u^2\right)du$$
$$= \sqrt{\pi}\exp\left(j\frac{\pi}{4}\right).$$

2. Let $\exp(-js^2) = \exp(-u^2)$; then

$$\int_{-\infty}^{\infty} \exp\left(-js^2\right)ds = \frac{1}{\sqrt{j}} \int_{-\infty}^{\infty} \exp\left(-u^2\right)du$$
$$= \sqrt{\pi}\exp\left(-j\frac{\pi}{4}\right).$$

Therefore

$$\int_{-\infty}^{\infty} \exp\left(\pm js^2\right)ds = \sqrt{\pi}\exp\left(\pm j\frac{\pi}{4}\right). \tag{7.31}$$

Equation (7.30) then becomes

$$S(\omega) = a\left(t_s\right)\sqrt{\frac{2\pi}{\left|\phi''\left(t_s\right)\right|}}\exp\left\{-j\left[\omega t_s - \phi\left(t_s\right) - \operatorname{sgn}\left[\phi''\left(t_s\right)\right]\frac{\pi}{4}\right]\right\}. \tag{7.32}$$

Example 7.1 Fourier Transform of Pulsed LFM Signal Consider a pulsed LFM signal, which is represented in Eq. (5.9) as follows:

$$p(t) = A_0 \operatorname{Rect}\left(\frac{t - T_p/2}{T_p}\right)\exp\left[j\left(2\pi f_c t + \pi\alpha t^2\right)\right].$$

Comparing it with Eq. (7.24), one obtains

$$a(t) = A_0 \operatorname{Rect}\left[\frac{t - T_p/2}{T_p}\right],$$
$$\phi(t) = 2\pi f_c t + \pi\alpha t^2.$$

From Eq. (7.26), one can compute the time parameter t_s as

$$\begin{aligned}\omega &= \left.\frac{d\phi}{dt}\right|_{t=t_s} \\ &= 2\pi f_c + 2\pi\alpha t_s\end{aligned}$$

or

$$t_s = \frac{f - f_c}{\alpha}$$

and

$$\begin{aligned}\left.\frac{d^2\phi}{dt^2}\right|_{t=t_s} &= 2\pi\alpha \\ a\left(t_s\right) &= A_0 \operatorname{Rect}\left(\frac{t_s - T_p/2}{T_p}\right) \\ &= A_0 \operatorname{Rect}\left\{\frac{\left[\left(f - f_c\right)/\alpha\right] - \left(T_p/2\right)}{T_p}\right\}.\end{aligned}$$

Therefore, the Fourier transform of $p(t)$ becomes

$$P(\omega) = \int_{-\infty}^{\infty} A_0 \operatorname{Rect}\left(\frac{t - T_p/2}{T_p}\right) \exp\left[-j\left(\omega t - \omega_c t - \pi\alpha t^2\right)\right] dt.$$

This equation can be approximated as follows, from Eq. (7.32):

$$\begin{aligned}P(\omega) &= A_0\sqrt{\frac{1}{\alpha}}\operatorname{Rect}\left(\frac{t_s - T_p/2}{T_p}\right)\exp\left[-j\left(\omega t_s - \omega_c t_s - \pi\alpha t_s^2 - \frac{\pi}{4}\right)\right] \\ &= A_0 \operatorname{Rect}\left(\frac{\omega - \omega_c}{2\pi\alpha T_p} - \frac{1}{2}\right)\exp\left(j\frac{\pi}{4}\right)\sqrt{\frac{1}{\alpha}}\exp\left[-j\frac{\left(\omega - \omega_c\right)^2}{4\pi\alpha}\right].\end{aligned}$$

If the constants of $\exp\left(j\pi/4\right)$ and $\alpha^{-1/2}$ are ignored, the Fourier transform of $p(t)$ becomes

$$P(f) = A_0 \operatorname{Rect}\left(\frac{f - f_c}{\alpha T_p} - 0.5\right)\exp\left[\frac{-j\pi\left(f - f_c\right)^2}{\alpha}\right].$$

This equation shown is an approximate form derived from the principle of stationary phase, which differs from that shown in Eq. (5.12). However, it is reasonably accurate as long as the time–bandwidth product of $p(t)$ is greater than 100, which normally is true in practical radar applications.

7.2.1.3 Spatial Fourier Transform of Cross-Range Target Response.

The computation of spatial Fourier transform of Eq. (7.23) will now be approximated by using the principle of stationary phase. For illustration purpose, we will use k_{um}, instead of k_u, to represent the spatial frequency corresponding to the mth target. Thus, to compute

$$S_m(k_{um}) = \sigma_m \int_{-\infty}^{\infty} \exp\left[-j2k\sqrt{X_c^2 + (y_m - u)^2} - jk_{um}u\right] du,$$

based on Eqs. (7.25) and (7.32), one can change t to u, ω to k_{um}. In addition, letting

$$\begin{aligned} a(t) &= \sigma_m, \\ \phi(t) &= -2k\sqrt{X_c^2 + (y_m - u)^2}, \end{aligned} \tag{7.33}$$

one can then obtain the corresponding u_s in k_u domain by using Eq. (7.26), that is

$$\frac{d\left[\omega t - \phi(t)\right]}{dt} = 0,$$

or

$$\frac{d\left[k_{um}u + 2k\sqrt{X_c^2 + (y_m - u)^2}\right]}{du} = 0.$$

Therefore

$$k_{um} = 2k\frac{y_m - u_s}{\sqrt{X_c^2 + (y_m - u_s)^2}}. \tag{7.34}$$

By solving Eq. (7.34) for u_s, which corresponds to t_s in the time domain, one obtains

$$u_s = y_m - \frac{k_{um}X_c}{\sqrt{4k^2 - k_{um}^2}}. \tag{7.35}$$

One can then compute the corresponding $\phi(u_s)$ and $(d^2\phi/du^2)\big|_{u=u_s}$ as follows:

$$\begin{aligned} \phi(u_s) &= -2k\sqrt{X_c^2 + (y_m - u)^2}\,\Big|_{u=u_s} \\ &= \frac{-4k^2X_c}{\sqrt{4k^2 - k_{um}^2}} \\ \left.\frac{d^2\phi}{du^2}\right|_{u=u_s} &= -\frac{4k^2 - k_{um}^2}{4k^2}\sqrt{4k^2 - k_{um}^2}. \end{aligned}$$

The symbol k_{um} is the spatial wavenumber related to the spatial frequency along the cross-range direction, which is quite small. The parameter $k = 2\pi/\lambda$ is the wavenumber related to the carrier frequency and is much larger than k_{um}. Therefore, for $k_{um} \ll k$, one obtains

$$\left.\frac{d^2\phi}{du^2}\right|_{u=u_s} \approx -\sqrt{4k^2 - k_{um}^2}$$
$$a(u_s) = \sigma_m. \tag{7.36}$$

The spatial frequency transform of $s_m(u)$, from Eq. (7.23), then becomes

$$S_m\left(k_u\right) = \sigma_m \frac{\sqrt{2\pi}\exp\left(-j\pi/4\right)}{\sqrt{4k^2 - k_{um}^2}} \exp\left(-j\sqrt{4k^2 - k_{um}^2}X_c - jk_{um}y_m\right). \tag{7.37a}$$

The term

$$\frac{\sqrt{2\pi}\exp\left(-j\pi/4\right)}{\sqrt{4k^2 - k_{um}^2}}$$

is a known and slowly-fluctuating amplitude function. It can be ignored in the cross-range imaging process. The spatial frequency spectrum of the echo signal due to the mth target located at (X_c, y_m) can then be expressed as

$$S_m(k_{um}) \approx \sigma_m \exp\left(-j\sqrt{4k^2 - k_{um}^2}X_c - jk_{um}y_m\right). \tag{7.37b}$$

The characteristics of the spatial frequency k_{um} are examined next.

Figure 7.8 shows the configuration of radar beams to be used for computing the spatial frequency band limitation. Figure 7.8a displays radar moving along the u- or y-axis direction with two radar beams at different locations, and one ground target

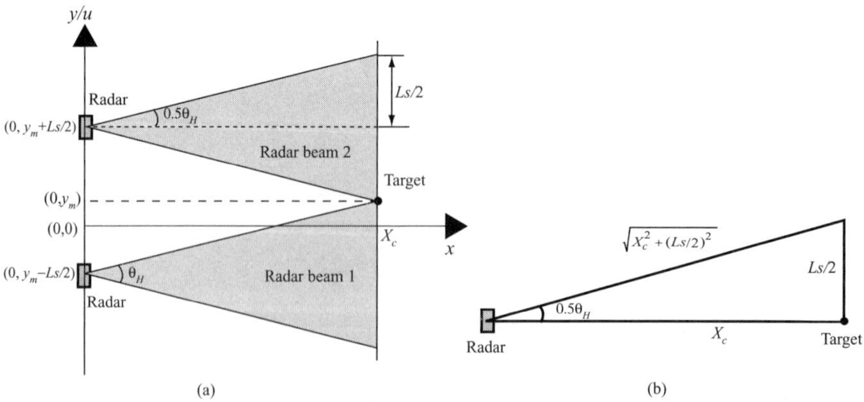

FIGURE 7.8 Computation of spatial frequency band limitation.

located at (X_c, y_m). Radar beam 1, originating from $(0, y_m - Ls/2)$ and shown as the shaded area with beamwidth θ_H, has its upper edge as the first beam illuminating the target. Beam 2, originating from $(0, y_m + Ls/2)$, has its lower edge as the last beam illuminating the same target. Figure 7.8b displays the geometric relationship between the beam angle θ_H, X_c, and Ls. Notice that Ls is the synthetic aperture length with beamwidth θ_H.

From Eq. (7.18), the spatial frequency k_{um} is expressed as

$$k_{um} = 2k \sin\theta_m(u), \tag{7.37c}$$

where $\theta_m(u)$ is the aspect angle of the mth target located at (X_c, y_m). For a group of M targets occupying an area with length D_y, the radar movement is within the range $[-L_u, L_u]$, where $L_u = (Ls + D_y)/2$. However, for any single target located at (X_c, y_m), the radar movement is within $[y_m - Ls/2,\ y_m + Ls/2]$. The spatial frequency k_{um} corresponding to a single target m is therefore band limited to

$$\begin{aligned} k_{um} &\in \left[2k \sin\theta_m\left(\frac{y_m + Ls}{2}\right), 2k \sin\theta_m\left(\frac{y_m - Ls}{2}\right)\right] \\ &= \left[-2k \sin\frac{\theta_H}{2},\ 2k \sin\frac{\theta_H}{2}\right]. \end{aligned} \tag{7.38a}$$

Refer to Fig. 7.8b, with $R = [X_c^2 + (Ls/2)^2]^{1/2}$ and $\theta_H = \lambda/L$ and $Ls = X_c\theta_H$, then $\sin(\theta_H/2) = Ls/(2R)$. Therefore

$$\begin{aligned} k_{um} &= \left[-\frac{kLs}{R}, \frac{kLs}{R}\right] \\ &= \left[\ \frac{kX_c\lambda}{RL}, \frac{kX_c\lambda}{RL}\right]. \end{aligned} \tag{7.38b}$$

The bandwidth B_{ku} and the center of the spatial frequency k_{uc} can be computed as follows:

$$B_{ku} = \frac{2kX_c\lambda}{LR} \tag{7.38c}$$

$$k_{uc} = 0. \tag{7.38d}$$

Equation (7.38c) states that the bandwidth of spatial frequency k_{um} is constant and independent of the target location. In other words, all targets occupying an area of length D_y in the cross-range direction will have the same spatial frequency spectrum, but with different phases. Therefore k_u, will be used instead of k_{um} for future discussion.

For a small beamwidth of θ_H, $\sin(\theta_H/2) \approx \theta_H/2$ or $R \approx X_c$. Equations (7.38a) and (7.38b) then become

$$k_u \approx [-k\theta_H, k\theta_H] = \left[-\frac{2\pi}{L}, \frac{2\pi}{L}\right]. \tag{7.39a}$$

The bandwidth B_{ku} and the center of the spatial wavenumber k_{uc} therefore become

$$B_{ku} \approx \frac{4\pi}{L} \tag{7.39b}$$

$$k_{uc} = 0. \tag{7.39c}$$

One can compute the changing rate of the spatial frequency k_u, from Eq. (7.34):

$$k'_u = \frac{\partial k_u}{\partial u} = \frac{-2kX_c^2}{\sqrt[3]{X_c^2 + (y_m - u)^2}}.$$

Normally $X_c \gg (y_m - u)$; therefore

$$k'_u \approx -\frac{2k}{X_c} = -\frac{4\pi}{\lambda X_c}. \tag{7.39d}$$

The slope of the spatial frequency is a constant, which implies that k_u varies linearly with the two ends of the k_u equal to $2\pi/L$ and $-2\pi/L$, respectively. From Eqs. (7.39a)–(7.39d), it is clear that the spatial frequency band limitation, bandwidth, center of band, and rate of change are all identical for all targets with the same slant range ($x = X_c$ in this case).

Since k_u in the spatial frequency domain corresponds to ω_D in the Doppler frequency domain, it is interesting to see that the parameters derived in the spatial frequency domain match those derived from the Doppler frequency discussed in Chapter 6. In other words, the spatial frequency k_u, the bandwidth B_{ku} and k_{uc} (the centroid of k_u) derived from Eqs. (7.37c), (7.39a), and (7.39b) match ω_D, B_D, and ω_{Dc}, respectively, derived in Eq. (6.8). This match comes from the conditions set by assigning $R_0 = X_c$, $r = R_m$, and $k_u = \omega_D/V$.

7.2.1.4 Reconstruction of Cross-Range Target Function. Once the spatial frequency spectrum of the mth target located at (X_c, y_m) is known, the spatial frequency spectrum for all target responses [$s_m(u)$, $m = 0, 1, \ldots, M-1$] can be obtained:

$$\begin{aligned} S(k_u) &= \sum_{m=0}^{M-1} \sigma_m \exp\left(-j\sqrt{4k^2 - k_u^2 X_c} - jk_u y_m\right) \\ &= \exp\left(-j\sqrt{4k^2 - k_u^2 X_c}\right) \sum_{m=0}^{M-1} \sigma_m \exp\left(-jk_u y_m\right). \end{aligned} \tag{7.40}$$

The spatial frequency spectrum of reference signal $s_{r0}(u)$, defined in Eq. (7.19), can be obtained by letting $m = 0$ and $y_m = 0$ in Eq. (7.40):

$$S_{r0}(k_u) = \exp\left(-j\sqrt{4k^2 - k_u^2 X_c}\right). \tag{7.41}$$

The spatial frequency transform of an ideal target function, defined in Eq. (7.20), is

$$F_0\left(k_u\right) = \sum_{m=0}^{M-1} \sigma_m \exp\left(-jk_u y_m\right). \tag{7.42}$$

From Eqs. (7.40)–(7.42), one obtains

$$S(k_u) = F_0(k_u) S_{r0}(k_u). \tag{7.43}$$

Again, the ideal target function $f_0(u)$ can be derived as the inverse spatial frequency transform of $S(k_u)/S_{r0}(k_u)$, where $S(k_u)$ and $S_{r0}(k_u)$ are the spatial frequency transforms of $s(u)$ and $s_{r0}(u)$, respectively. However, because of the bandwidth limitation, it cannot be solved in this way as discussed in the range imaging case, and a practical reconstruction method is again via the matched filter, which is defined as the complex conjugate of the reference function $S_{r0}(k_u)$. Multiplication of the spatial frequency function $S(k_u)$ by the matched filter function $S_{r0}^*\,(k_u)$, yields the following product:

$$\begin{aligned} F\left(k_u\right) &= S\left(k_u\right) S_{r0}^*\left(k_u\right) \\ &= \sum_{m=0}^{M-1} \sigma_m \exp\left(-jk_u y_m\right). \end{aligned} \tag{7.44}$$

Since the spatial frequency transform of each target is band-limited, one can rewrite Eq. (7.37b) as

$$S_m\left(k_u\right) = \sigma_m I_m\left(k_u\right) \exp\left(-j\sqrt{4k^2 - k_u^2 X_c} - jk_u y_m\right), \tag{7.45}$$

where

$$I_m(k_u) = 1 \quad \text{for} \quad k_u \in \left[-\frac{2\pi}{L}, \frac{2\pi}{L}\right], \\ = 0 \quad \text{otherwise.} \tag{7.46}$$

Equation (7.44) then becomes

$$F(k_u) = \sum_{m=0}^{M-1} \sigma_m I_m(k_u) \exp(-jk_u y_m). \tag{7.47}$$

Assuming a small beamwidth, the inverse spatial frequency transform of $I_m(k_u)$ is

$$\begin{aligned} i_m(u) &= \frac{1}{2\pi}\int_{-2\pi/L}^{2\pi/L} I_m(k_u)\exp(jk_u u)dk_u \\ &= \frac{1}{j2\pi u}\left[\exp\left(\frac{ju2\pi}{L}\right) - \exp\left(\frac{-ju2\pi}{L}\right)\right] \\ &\approx \frac{2}{L}\text{sinc}\left(\frac{2u}{L}\right). \end{aligned} \tag{7.48}$$

Since u and y are interchangeable, the inverse spatial frequency transform of $F(k_y)$ becomes

$$\begin{aligned} f(y) &= \sum_{m=0}^{M-1} \sigma_m i_m(y - y_m) \\ &= \sum_{m=0}^{M-1} \frac{2}{L}\sigma_m \,\text{sinc}\left(\frac{2(y - y_m)}{L}\right). \end{aligned} \tag{7.49}$$

Let $\text{psf}_m(y) = i_m(y)$ be difined as the point spread function of the mth target, similar to the point spread function used in range imaging shown in Eq. (7.11), the reconstructed cross-range target function becomes

$$f(y) = \sum_{m=0}^{M-1} \sigma_m \text{psf}_m(y - y_m). \tag{7.50}$$

Comparing Eqs. (7.49) and (7.50) with Eqs. (7.10) and (7.11) in Section 7.1.2, one obtains the following observations:

- In both cases, the target functions are a group of sinc functions in the range x and cross-range y directions, respectively.

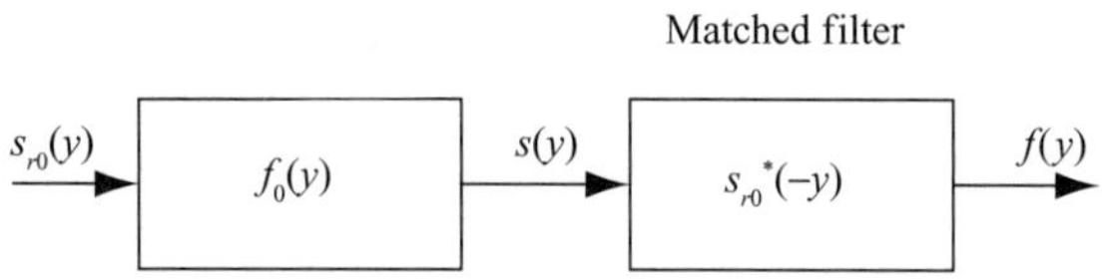

FIGURE 7.9 Matched filtering for cross-range imaging.

- The function $i_m(y)$ in the cross-range y direction can be considered as the point spread function $\mathrm{psf}_m(y)$, while the point spread function $\mathrm{psf}(x)$ in the range x direction is computed as $\mathfrak{F}^{-1}\{|P(\omega)|^2\}$.
- In both cases, a reference function is needed for reconstruction purposes. A matched filter defined as the complex conjugate of the time-reversed transmitted signal is chosen as the reference function in range image processing. For the case of cross-range image processing, a matched filter is chosen as the complex conjugate of the reversed slow time of the echo signal from an ideal reflector located at $(X_c,0)$ (or at the center of the cross-range targets).

On the basis of the discussion above, the cross-range image processing under the broadside case can be represented in a system block diagram as shown in Fig. 7.9.

7.2.2 Squint Radar Case

7.2.2.1 System Model. Figure 7.10 displays the geometry of a squint-mode radar system for cross-range imaging. Similar to the broadside mode, all targets are located at $x = X_c$ and spread along the y axis with all targets located within the radar beamwidth. Each target is represented by its reflection coefficient σ_m, $m = 1,2,\ldots,M$. The radar is located at $(0,0,H)$ and moves at speed V along the y axis. However, in contrast to the broadside-mode radar, the center beam of the squint mode radar aims at the target area with a squint angle θ_q, which is the angle between the range axis and the center beam of the radar.

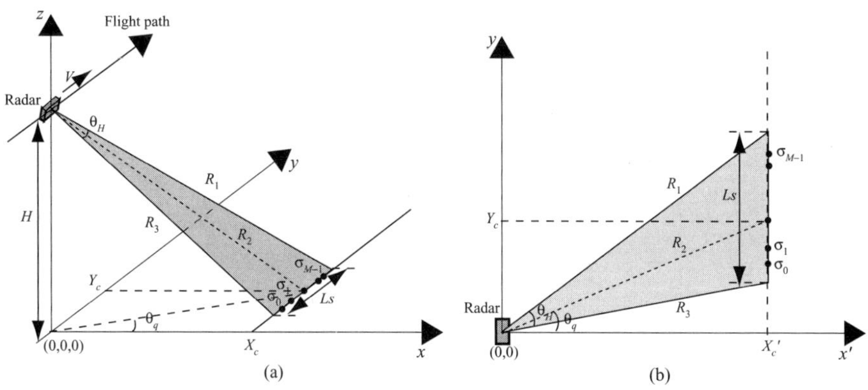

FIGURE 7.10 A squint mode cross-range imaging system.

Figure 7.10a shows the configuration of a cross-range imaging system, while Fig. 7.10b shows a simplified 2D model of Fig. 7.10a; Fig. 7.10b can also be considered as the overhead view of Fig. 7.10a. The center of the target area along the y axis is at (X_c', Y_c) in Fig. 7.10b, with R_2 as the distance from the radar to the center of the target area. The distance R_2 equals $(X_c^2 + Y_c^2 + H^2)^{1/2}$ in Fig. 7.10a and equals $(X_c'^2 + Y_c^2)^{1/2}$ in FIg. 7.10b, where $X_c' = (X_c^2 + H^2)^{1/2}$. To simplify the discussion, X_c will be used instead of X_c' in the following discussion. R_1 and R_3 are the shortest and the longest distances from the radar to the target, respectively. The squint angle can be computed as $\theta_q = \tan^{-1}(Y_c/X_c)$.

Figure 7.11 displays the relationship between the target area and radar beam. All targets are located at $x = X_c$, and spread along the y axis with the center of the target area at $y = Y_c$. The length of targets spreading around Y_c is D_y. The radar moves along the y axis from $y = -L_u$ to $y = L_u$, where it continuously illuminates the targets located inside the region of D_y. The length $2L_u = 2Ls + D_y$, where Ls is the synthetic aperture length. Here the length D_y is assumed to be less than Ls.

Consider the case when the radar is located at $y = u$, $-L_u \leq u \leq L_u$. Let the targets be located at $\{(X_c, y_m)\}$, where $m = 0,1,\ldots, M-1$, and

$$y_m \in \left[Y_c - \frac{D_y}{2}, Y_c + \frac{D_y}{2}\right].$$

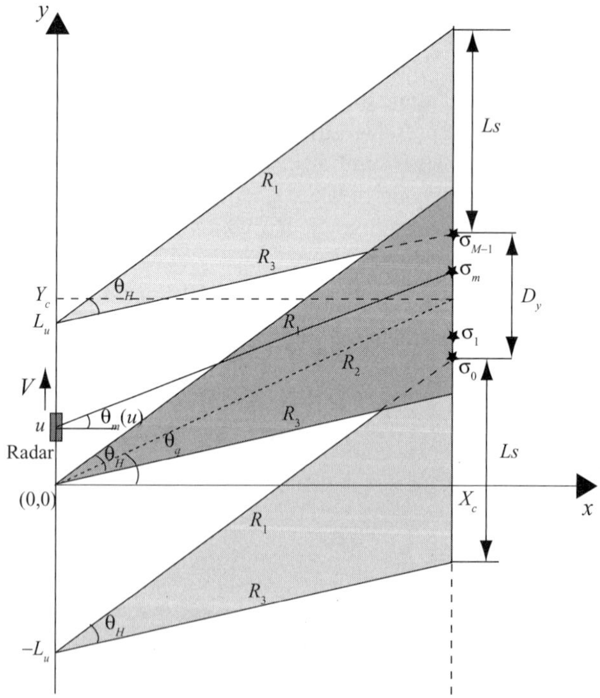

FIGURE 7.11 Relationship between targets and squint radar beam.

The received echo signal due to the target located at (X_c, y_m) is

$$s_m(t, u) = \sigma_m p\left[t - \frac{2R_m(u)}{c}\right].$$

The received echo signal due to a group of M targets can then be expressed as

$$s(t, u) = \sum_{m=0}^{M-1} \sigma_m p\left[t - \frac{2R_m(u)}{c}\right],$$

where

$$R_m(u) = \sqrt{X_c^2 + (y_m - u)^2}.$$

The pulsed LFM signal is normally used for radar image processing. Again, we will assume that the transmitted signal is a single frequency; that is, $p(t) = \exp(j\omega t)$. The baseband received signal becomes

$$\begin{aligned} s_b(t, u) &= s(t, u)\exp(-j\omega t) \\ &= \sum_{m=0}^{M-1} \sigma_m \exp\left[-j2k\sqrt{X_c^2 + (y_m - u)^2}\right], \end{aligned} \tag{7.51}$$

which is identical to Eq. (7.16) for the broadside case.

By defining an ideal reflector with reflection coefficient $\sigma_r = 1$, located at (X_c, Y_c), one can express the reference signal $s_{r0}(u)$ received by a radar located at $y = u$ as

$$s_{r0}(u) = \exp\left(-j2k\sqrt{X_c^2 + (Y_c - u)^2}\right). \tag{7.52}$$

Notice that the value of $y_m \in [Y_c - (D_y/2), Y_c + (D_y/2)]$, where Y_c is nonzero in the squint radar case. Similar to the broadside radar case, the ideal target function $f_0(u)$ for squint radar is also defined with $Y_c = 0$. Therefore, by defining $y'_m = y_m - Y_c$, one can see that all targets are located at $y'_m \in \left[-D_y/2, D_y/2\right]$ and the ideal target function becomes

$$\begin{aligned} f_0(u) &= \sum_{m=0}^{M-1} \sigma_m \delta\left(u - y'_m\right) \\ &= \sum_{m=0}^{M-1} \sigma_m \delta\left(u - y_m + Y_c\right). \end{aligned} \tag{7.53}$$

The convolution of $s_{r0}(u)$ with $f_0(u)$ turns out to be

$$\begin{aligned}
f_0(u)^* s_{r0}(u) &= \int_{-\infty}^{\infty} f_0(\tau) s_{r0}(u-\tau) d\tau \\
&= \int_{-\infty}^{\infty} \sum_{m=0}^{M-1} \sigma_m \delta(\tau - y_m + Y_c) \exp\left[-j2k\sqrt{X_c^2 + (Y_c + \tau - u)^2}\right] d\tau \\
&= \sum_{m=0}^{M-1} \sigma_m \exp\left[-j2k\sqrt{X_c^2 + (y_m - u)^2}\right] \\
&= s(u).
\end{aligned} \tag{7.54}$$

By comparing Eqs. (7.51) and (7.54), one can see that the baseband signal $s_b(t,u)$ is identical to $s(u)$, which is the output of a system with input signal $s_{r0}(u)$ and system function $f_0(u)$. Therefore, the same system block diagram used for broadside-mode radar, as shown in Fig. 7.7, can also be applied to squint-mode radar.

7.2.2.2 Spatial Fourier Transform of Cross-Range Target Response. From Eq. (7.23), the spatial Fourier transform of $s_m(u)$ can be expressed as

$$\begin{aligned}
S_m(k_u) &= \int_{-\infty}^{\infty} \sigma_m \exp\left[-j2k\sqrt{X_c^2 + (y_m - u)^2}\right] \exp(-jk_u u)\, du \\
&= \sigma_m \int_{-\infty}^{\infty} \exp\left[-j2k\sqrt{X_c^2 + (y_m - u)^2} - jk_u u\right] du.
\end{aligned}$$

Based on Eqs. (7.33) and (7.34), and following the same steps as used in the broadside case, one obtains

$$a(t) = \sigma_m$$

$$\phi(t) = -2k\sqrt{X_c^2 + (y_m - u)^2}$$

and

$$\begin{aligned}
k_u &= 2k \frac{y_m - u}{\sqrt{X_c^2 + (y_m - u)^2}} \\
&= 2k \frac{y_m - u}{R_m(u)}
\end{aligned}$$

or

$$k_{\mathrm{u}} = 2k\ \sin\theta_m(u),$$

where $\theta_m(u)$ is the aspect angle of the target at (X_c, y_m). The spatial Fourier transform $S_m(k_u)$ can then be approximated as follows:

$$S_m(k_u) = \exp\left(-j\sqrt{4k^2 - k_u^2}X_c\right)\exp\left(-jk_u y_m\right). \tag{7.55}$$

Given a group of targets that occupies a length D_y and centers on $y = Y_c$, the radar movement must be within $[-L_u, L_u]$ with $L_u = Ls + D_y/2$ for the targets to be under radar beam illumination. However, for a single target located at (X_c, y_m), the radar movement is in the range $[-Ls/2+y_m-Y_c, Ls/2+y_m-Y_c]$. The spatial frequency k_{um} for the mth target is therefore band-limited to

$$k_{um} \in \left[2k\ \sin\theta_m\left(y_m - Y_c + \frac{Ls}{2}\right),\ 2k\ \sin\theta_m\left(y_m - Y_c - \frac{Ls}{2}\right)\right]. \tag{7.56a}$$

Figure 7.12 shows the configuration of radar beams used for computing the spatial frequency band limitation based on a single target located at (X_c,y_m), where the radar moves upward along the y axis. Radar beam 1, shown as the shaded area, has its upper edge as the first beam illuminating the target. Beam 2, also shown as the shaded area, has its lower edge as the last beam illuminating the same target. The squint angle θ_q is the angle between the x axis and the line from the origin (0,0) to the center of targets (X_c,Y_c). The squint angle can be computed as $\theta_q = \tan^{-1}(Y_c/X_c)$. The corresponding

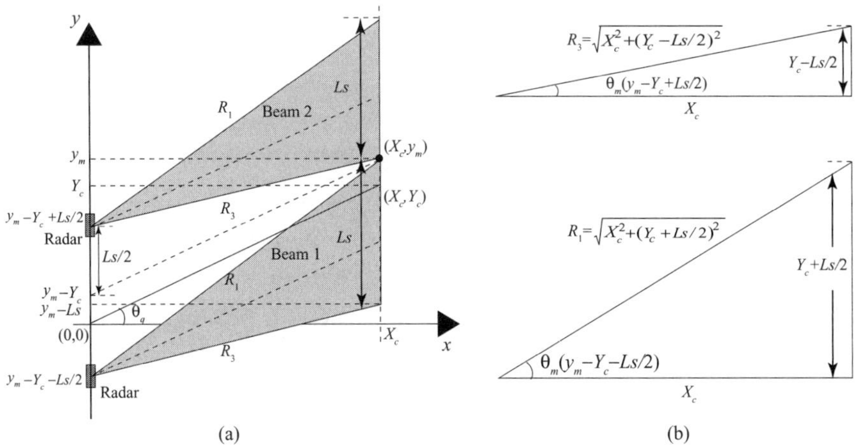

FIGURE 7.12 Computation of spatial frequency band limitation for squint radar.

$\sin\theta_m(y_m - Y_c + Ls/2)$ and $\sin\theta_m(y_m - Y_c - Ls/2)$ can be computed as

$$\sin\theta_m\left(y_m - Y_c + \frac{Ls}{2}\right) = \frac{Y_c - Ls/2}{\sqrt{X_c^2 + (Y_c - Ls/2)^2}}$$
$$\sin\theta_m\left(y_m - Y_c - \frac{Ls}{2}\right) = \frac{Y_c + Ls/2}{\sqrt{X_c^2 + (Y_c + Ls/2)^2}}.$$

For a radar with small beamwidth and small squint angle, $X_c \gg Ls$ and $X_c \gg Y_c$; therefore

$$\sin\theta_m\left(y_m - Y_c + \frac{Ls}{2}\right) \approx \frac{Y_c}{X_c} - \frac{L_s/2}{X_c} \approx \theta_q - \frac{\theta_H}{2},$$
$$\sin\theta_m\left(y_m - Y_c - \frac{Ls}{2}\right) \approx \frac{Y_c}{X_c} + \frac{L_s/2}{X_c} \approx \theta_q + \frac{\theta_H}{2}.$$

The spatial frequency k_{um} is then band-limited as

$$\begin{aligned} k_{um} &\in \left[2k\left(\theta_q - \frac{\theta_H}{2}\right), 2k\left(\theta_q + \frac{\theta_H}{2}\right)\right] \\ &= \left[-\frac{2\pi}{L} + 2k\theta_q, \frac{2\pi}{L} + 2k\theta_q\right], \end{aligned} \tag{7.56b}$$

which is independent of the location of the mth target. The bandwidth and the center of the spatial frequency k_{um} or simply k_u can then be computed as follows:

$$\begin{aligned} B_{ku} &= \frac{4\pi}{L}, \\ k_{uc} &= 2k\theta_q. \end{aligned} \tag{7.56c}$$

The change rate of k_u is similar to Eq. (7.39d), and can be expressed as

$$k_u' \approx -\frac{2k}{X_c} = -\frac{4\pi}{\lambda X_c}. \tag{7.56d}$$

Therefore, all targets at the same range distance ($x = X_c$ in this case) will share the same band limitation, bandwidth, center of band, and rate of spatial frequency change as shown in Eqs. (7.56a)–(7.56d). The spatial frequency bandwidth B_{ku} and the slope of the spatial frequency k_u' are identical to those in the broadside case. However, the center of the spatial frequency $k_{uc} = 2k\theta_q$ is different from that in the broadside case.

The spatial frequency spectra corresponding to all received signals $\{s_m(u),\ m = 0,1,\ldots, M-1\}$ can therefore be expressed as follows, from Eq. (7.55):

$$S(k_u) = \exp\left(-j\sqrt{4k^2 - k_u^2}X_c\right)\sum_{m=0}^{M-1}\exp(-jk_u y_m). \tag{7.57}$$

7.2.2.3 Reconstruction of Cross-Range Target Function. The reference signal received from an ideal reflector is shown in Eq. (7.52) as

$$s_{r0}(u) = \exp\left(-j2k\sqrt{X_c^2 + (Y_c - u)^2}\right).$$

The corresponding spatial Fourier transform can be obtained as

$$S_{r0}(k_u) = \exp\left(-j\sqrt{4k^2 - k_u^2}X_c - jk_u Y_c\right). \tag{7.58}$$

The ideal target function for a group of M targets can be expressed as follows, from Eq. (7.53):

$$\begin{aligned} f_0(u) &= \sum_{m=0}^{M-1}\sigma_m\delta(u - y'_m) \\ &= \sum_{m=0}^{M-1}\sigma_m\delta(u - y_m + Y_c). \end{aligned}$$

The spatial frequency transform of the ideal target function is

$$F_0(k_u) = \exp(jk_u Y_c)\sum_{m=0}^{M-1}\sigma_m\exp(-jk_u y_m). \tag{7.59}$$

From Eqs. (7.57)–(7.59), one obtains

$$S(k_u) = F_0(k_u)S_{r0}(k_u),$$

which is identical to Eq. (7.43) derived in the broadside case.

The matched filter, defined as the complex conjugate of the reference function $S_{r0}(k_u)$, is again used to reconstruct the target function. The output of the matched filter is

$$\begin{aligned} F(k_u) &= S(k_u)S_{r0}^*(k_u) \\ &= \exp(jk_uY_c)\sum_{m=0}^{M-1}\sigma_m \exp(-jk_u y_m). \end{aligned} \tag{7.60a}$$

By comparing Eq. (7.60a) with Eq. (7.44), one can see that the only difference is the phase function $\exp(jk_uY_c)$. The phase factor $\exp(jk_uY_c)$ of $F(k_u)$ implies that in the spatial domain the target function $f(u)$ is shifted downward along the u or y axis by Y_c. This is because of the ideal target function $f(u)$ in the squint case was defined with $Y_c = 0$, which in turn caused the phase factor $\exp(jk_uY_c)$ to appear in the spatial frequency function $F(k_u)$. Since the true squint target area is centered on $y = Y_c$, therefore, the correct target function in the squint case is to remove the phase factor $\exp(jk_uY_c)$ from Eq. (7.60a). Thus, the true target function in the squint radar is obtained by taking the inverse spatial frequency transform of $F'(k_u)$, which is defined as follows:

$$\begin{aligned} F'(k_u) &= F(k_u)\exp(-jk_uY_c) \\ &= \sum_{m=0}^{M-1}\sigma_m \exp(-jk_u y_m). \end{aligned} \tag{7.60b}$$

The spatial frequency transform of each target is band-limited as shown in Eq. (7.56b); therefore one can rewrite Eq. (7.60b) as

$$F'(k_u) = \sum_{m=0}^{M-1}\sigma_m I_m(k_u)\exp(-jk_u y_m), \tag{7.61a}$$

where

$$\begin{aligned} I_m(k_u) &= 1 \quad \text{for} \quad k_u \in= \left[-\frac{2\pi}{L} + 2k\theta_q, \frac{2\pi}{L} + 2k\theta_q\right] \\ &= 0 \quad \text{otherwise.} \end{aligned} \tag{7.61b}$$

The band limitation of $I_m(k_u)$ is the same as that shown in Eq. (7.56b). Notice that $I_m(k_u)$ is a bandpass signal with the center of the spatial frequency $k_{uc} = 2k\theta_q$ and bandwidth $B_{ku} = 4\pi/L$. Let $I_{bm}(k_u)$ be the baseband signal of $I_m(k_u)$:

$$\begin{aligned} I_{bm}(k_u) &= 1 \quad \text{for} \quad k_u \in \left[-\frac{2\pi}{L}, \frac{2\pi}{L}\right] \\ &= 0 \quad \text{otherwise.} \end{aligned} \tag{7.62}$$

Then, $I_{bm}(k_u) = I_m(k_u - 2k\theta_q)$, and the inverse spatial frequency transform of $I_{bm}(k_u)$ can be computed as follows:

$$i_{bm}(u) = \frac{1}{2\pi}\int_{-2\pi/L}^{2\pi/L} I_{bm}(k_u)\exp(jk_u u)\,dk_u$$

$$= \frac{1}{j2\pi u}\left[\exp\left(ju\frac{2\pi}{L}\right) - \exp\left(-ju\frac{2\pi}{L}\right)\right]$$

$$= \frac{2}{L}\,\text{sinc}\left(\frac{2u}{L}\right).$$

Therefore, the inverse spatial frequency transform of $I_m(k_u)$ becomes

$$i_m(u) = i_{bm}(u)\exp(j2k\theta_q u).$$

The final reconstructed cross-range target function can be obtained as

$$f'(y) = \sum_{m=0}^{M-1}\frac{2}{L}\sigma_m\exp(j2k\theta_q y)\,\text{sinc}\left(\frac{2(y-y_m)}{L}\right), \quad (7.63)$$

which is similar to Eq. (7.49), except for an extra phase factor $\exp(j2k\theta_q y)$.

From the preceding discussion, one can see that the techniques used for cross-range image processing are quite similar for both broadside radar and squint radar. However, there are two major differences: (1) the reference function is defined differently—the broadside radar deals with an ideal reflector located at $(X_c,0)$, and the squint radar deals with an ideal reflector located at (X_c,Y_c); and (2) the nonzero squint angle θ_q causes an extra phase factor $\exp(j2k\theta_q y)$ in the squint radar case. Overall, the system block diagram of Fig. 7.9 works for both broadside and squint radar cases.

7.3 DATA ACQUISITION, SAMPLING, AND POWER SPECTRUM OF RADAR IMAGE

The 2D radar imaging data is formed by digitizing the received radar signal at sampling frequency f_s, which satisfies the Nyquist requirement. This signal in general is a pulse or linear FM signal with carrier frequency ranges from tens of megahertz to tens of gigahertz, and its bandwidth could be a few megahertz to hundreds of megahertz. The 2D radar imaging data are arranged in a matrix-like array of complex numbers. Each row of data corresponds to one reflected radar pulse, while each column of data contains information on the same target from successive reflected radar

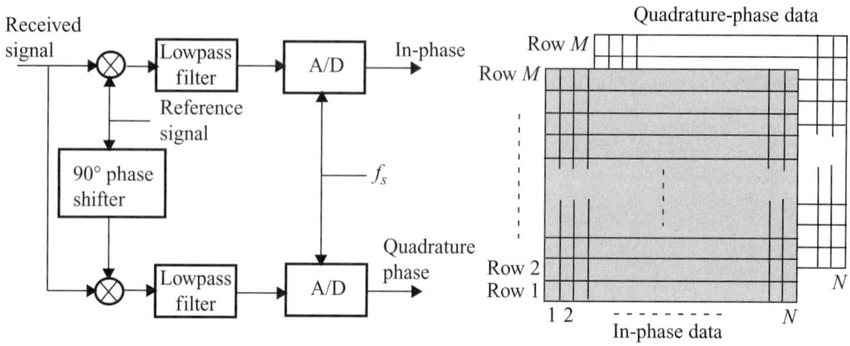

FIGURE 7.13 I–Q radar signal generation.

pulses at a constant time interval. Each column of data serves as the along-track direction imaging data, and is equivalently sampled by the pulse repetition frequency f_{PRF}.

The received radar signal is a downconverted intermediate signal. It is first passed through an in-phase–quadrature–phase splitter to produce the in-phase and quadrature-phase (or I–Q) components. These two signals are then lowpass-filtered and digitized to render a complex number pair that serves as one row element of the 2D radar image data. Figure 7.13 displays the generation of I–Q signals from a received radar signal, where the reference signal is dependent on the transmitted signal.

The sampling frequency f_s in the range direction must satisfy the Nyquist requirement; that is, $f_s \geq B$, where B is the bandwidth of the transmit signal. Assuming $f_s = B$, the range sample spacing is therefore

$$\Delta x = \frac{c}{B}.$$

Along the azimuth direction, the radar moves with speed V and transmits a pulse at the time interval $\mathrm{PRI} = 1/f_{\mathrm{PRF}}$. The sampling frequency is therefore equal to f_{PRF}. In the spatial frequency domain, the wavenumber k_u or k_y equals ω_D/V in the Doppler frequency domain. Given the bandwidth of $B_{ku} = 4\pi/L$, the corresponding Doppler frequency bandwidth $B_{\mathrm{Dop}} = 2V/L$. The along track sampling frequency f_{PRF} must satisfy the relationship $f_{\mathrm{PRF}} \geq B_{\mathrm{Dop}}$. Assuming $f_{\mathrm{PRF}} = B_{\mathrm{Dop}}$, the azimuth sample spacing can then be computed as follows:

$$\begin{aligned}\Delta u &= \frac{V}{f_{\mathrm{PRF}}} \\ &= \frac{L}{2}.\end{aligned}$$

7.3.1 Digitized Doppler Frequency Power Spectrum

7.3.1.1 Broadside SAR. For broadside SAR, the Doppler centroid $f_{Dc} = 0$, and the bandwidth of Doppler frequency is

$$B_{\text{DoP}} = \frac{2V}{L}.$$

Since each point of the azimuth data corresponds to the received signal along the flight path at a fixed-range cell, the Doppler frequency power spectrum is discrete and equivalently sampled at the rate of f_{PRF}. The observed Doppler frequency power spectrum is therefore centered on the origin with replicas appearing at multiples of the sampling frequency f_{PRF}. To prevent the Doppler frequency spectrum from aliasing to each other, the pulse repetition frequency f_{PRF} must be chosen to be greater than B_D. As an example Fig. 7.14 displays the Doppler frequency spectra of the broadside SAR image, where the true or absolute Doppler spectrum is shown inside the dotted box. The maximum Doppler frequencies appear at two sides of the absolute Doppler spectrum but with opposite signs. Their values are $-V/L$ and V/L, respectively.

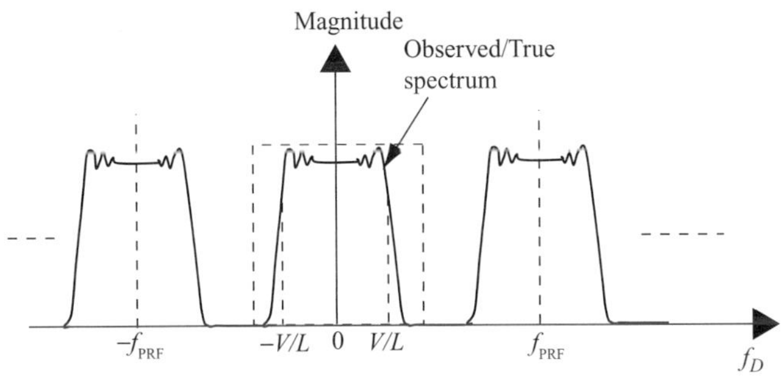

FIGURE 7.14 Doppler frequency spectra of a broadside SAR.

7.3.1.2 Squint SAR. For a nonzero squint angle θ_q, the Doppler frequencies corresponding to the two ends of the 3-dB radar beamwidth, namely, f_{DU} and f_{DL}, can be computed as follows, from Eq. (6.16d) or (6.16e):

$$f_{DU} = \frac{2V}{\lambda} \sin(\theta_q + 0.5\,\theta_H),$$

$$f_{DL} = \frac{2V}{\lambda} \sin(\theta_q - 0.5\,\theta_H).$$

The Doppler frequency bandwidth is then

$$B_D = f_{DU} - f_{DL}.$$

The Doppler frequency centroid for a nonzero squint angle is as follows, from Eq. (6.16c):

$$f_{Dc} = \frac{2V}{\lambda} \sin \theta_q.$$

The Doppler frequency centroid computed above is the true or absolute centroid frequency, with f_{DU} and f_{DL} located at two sides of f_{Dc}. Because of the built-in sampling frequency f_{PRF}, the absolute Doppler frequency power spectrum is then repeated at the multiples of f_{PRF}. Let f_{Dc} be the absolute Doppler centroid and f_{DC}' be the observed Doppler centroid. The relationship between the absolute Doppler centroid and the observed Doppler centroid is given by

$$f_{Dc} = f_{Dc}' + M_{\text{amb}} f_{\text{PRF}},$$

where M_{amb} is the Doppler ambiguity and is the smallest integer defined by

$$M_{\text{amb}} = \left[\frac{f_{Dc}}{f_{\text{PRF}}} \right].$$

As an example, Fig. 7.15 displays both the true (absolute) Doppler frequency spectrum and the digitized and observed spectra of a squint SAR with $M_{\text{amb}} = 2$.

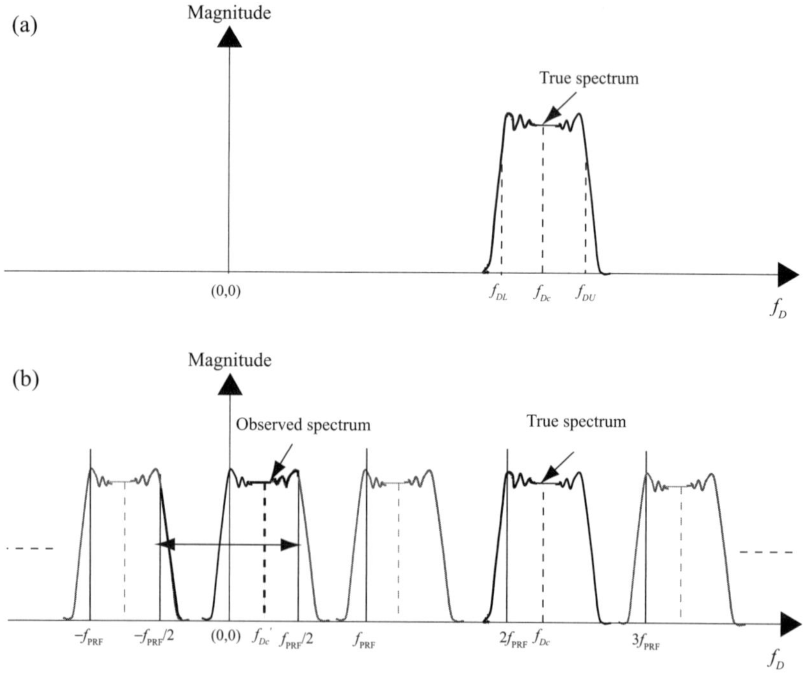

FIGURE 7.15 Doppler frequency spectra of a squint SAR.

Figure 7.15a shows the true frequency spectrum, where the Doppler frequency centroid f_{Dc} is nonzero with f_{DU} and f_{DL} located at two sides. Figure 7.15b shows the digitized and observed Doppler frequency spectra. The observed Doppler frequency centroid is located away from the origin and labeled as $f_{DC}{}'$. The true Doppler frequency centroid f_{Dc} differs from the observed Doppler frequency centroid f_{Dci} by 2 times the sampling frequency f_{PRF}. The observed spectrum is no longer symmetric around the origin because the Doppler frequency centroid f_{Dc} is not a multiple integer of f_{PRF}.

Both the true and the observed Doppler frequencies serve as important parameters for radar image reconstruction, which is discussed later.

8

RANGE–DOPPLER PROCESSING ON SAR IMAGES

In Chapters 6 and 7, we discussed the basics of radar images and provided a system model for analyzing and processing them. The $M \times N$ array of 2D radar image raw data is formed by M bursts of radar pulses with N samples each. The radar signal itself is a pulsed LFM waveform, which can be used to measure the target's range or identify its location through the matched filtering. The phase variation for various pulsed LFM waveforms in the azimuth direction leads to the concept that another LFM type of signal exists in the azimuth direction. This signal is not directly generated from the radar, but derived indirectly from the many LFM pulses sent and received by the radar within the synthetic aperture length. This azimuth-direction-based LFM signal plays a key role in fine-tuning the target function in the azimuth direction, and greatly improves the radar image quality.

Also mentioned previously is that radar image processing can be independently processed in the range and azimuth direction if range migration can be neglected. However, as we described in Section 6.6 (of Chapter 6), the range migration could become severe for radar with large synthetic aperture length (or wide antenna beamwidth), or when the radar beam has a nonzero squint angle toward the target. Therefore, the range cell migration correction is needed to decouple the azimuth process from that of the range.

The SAR image processing therefore can be summarized as three major tasks: (1) range compression, (2) range cell migration correction, and (3) azimuth compression. Figure 8.1 shows the block diagram for SAR image processing. Here the raw data refer to the demodulated in-phase–quadrature-phase (I–Q) baseband data.

Digital Signal Processing Techniques and Applications in Radar Image Processing, by Bu-Chin Wang.

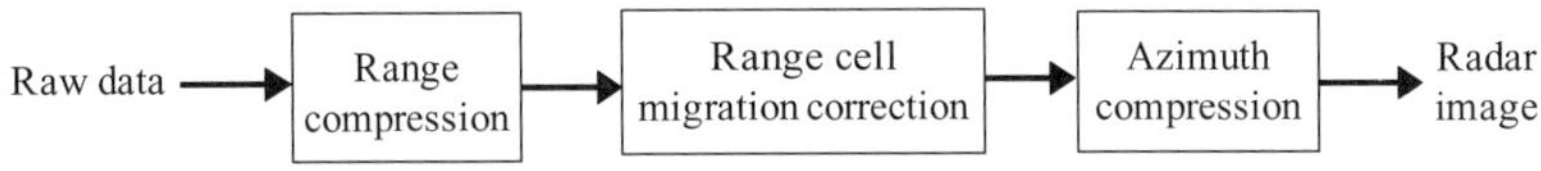

FIGURE 8.1 Major tasks of SAR radar image processing.

The raw data, after range compression, provide valuable information on the targets' locations in the range direction but do present two problems: (1) the range-compressed signal is spread in the azimuth direction, and (2) the signal migrates to other range cells. The purpose of SAR processing, ideally, is to convert the raw data into a single pixel in the final processed image. Several techniques are available to deal with the three tasks listed in the preceding paragraph. Each technique has its advantages in either computation efficiency or high-quality imaging, and has its own beamwidth limitation and/or bandwidth restriction. In this chapter and Chapter 9, two SAR image processing techniques will be presented: the range–Doppler and the Stolt interpolation. The former technique processes the SAR image based on Doppler frequency shifts, while the latter is based on the wavefront reconstruction theory. The range–Doppler algorithm, developed in the early 1980s, has been the most commonly used algorithm for processing SAR data images. It is computationally efficient and is an accurate approximation for processing radar images. The Stolt interpolation technique was applied to SAR image processing in the early 1990s. It is computation-intensive, but provides some advantages over the range–Doppler method. The range–Doppler algorithm is described in this chapter, and the Stolt interpolation technique is covered in Chapter 9.

We start the SAR image data generation in Section 8.1, where the principles of radar image formation and the method for synthesizing radar image data are reviewed. Section 8.2 provides a description and examples of how broadside SAR data are synthesized. Both single target and multiple targets are covered. Section 8.3 provides the same discussion and examples as in Section 8.2, but for the squint SAR case. The algorithm of range–Doppler processing on an SAR image is reviewed in Section 8.4, and the simulation results for both broadside and squint SAR are described in Section 8.5.

We will focus only on airborne-based radar image data processing; therefore, the estimation of Doppler centroid and Doppler frequency ambiguity, which is critical for satellite-based squint SAR systems, is not addressed here.

8.1 SAR IMAGE DATA GENERATION

The pulsed LFM radar transmitted waveform can be written as follows, from Eq. (5.11)

$$p(t) = A_0 \operatorname{Rect}\left(\frac{t - T_p/2}{T_p}\right) \exp\left[j2\pi f_c t + j\pi\alpha t^2\right], \tag{8.1}$$

where A_0 is the amplitude of the transmitter signal and $\operatorname{Rect}(t/T_p)$ is a rectangular gate function with T_p as the pulse duration time. The symbol f_c is the carrier frequency, and α is the LFM pulse chirp rate. Equation (8.1) states that the signal $p(t)$ starts at $t = 0$ and ends at $t = T_p$; therefore, for simplicity, we will use $a(t)$ to substitute the gate function and let $A_0 = 1$ in the following discussion.

The echo from K reflectors with reflection coefficient σ_k at ranges R_k, for $k = 1, 2, \ldots, K$ can be represented by a sum of delayed signals as

$$\begin{aligned} s(t) &= \sum_{k=0}^{K} \sigma_k a\left(\frac{t - 2R_k}{c}\right) \exp\left[j2\pi f_c\left(\frac{t - 2R_k}{c}\right) + j\pi\alpha\left(\frac{t - 2R_k}{c}\right)^2\right] \\ &= \sum_{k=0}^{K} \sigma_k a(t - \tau_k) \exp\left[j2\pi f_c (t - \tau_k) + j\pi\alpha (t - \tau_k)^2\right], \end{aligned} \tag{8.2}$$

where $\tau_k = 2R_k/c$ is the echo delay time due to the kth target. For a target located at (x_k, y_k) and with the radar position at (0,0) and assuming the radar height $H = 0$, the delay time can be computed as follows:

$$\tau_k = \frac{2R_k}{c} = \frac{2\sqrt{x_k^2 + y_k^2}}{c}. \tag{8.3a}$$

If the delay time τ_k were known, the range R_k could be computed as

$$R_k = \frac{c\tau_k}{2}. \tag{8.3b}$$

The baseband signal $s_b(t)$ can be obtained by removing the carrier frequency followed by a lowpass filter through the guadrature demodulation process. The demodulated baseband signal can be represented as:

$$s_b(t) = \sum_{k=0}^{K} \sigma_k \left|a(t - \tau_k)\right|^2 \exp\left[-j2\pi f_c \tau_k + j\pi\alpha (t - \tau_k)^2\right]. \tag{8.4a}$$

Letting n be the time index of the received and digitized LFM signal, one can then express $s_b(t_n)$ as

$$s_b(t_n) = \sum_{k=0}^{K} \sigma_k \left|a(t_n - \tau_k)\right|^2 \exp\left[-j2\pi f_c \tau_k + j\pi\alpha (t_n - \tau_k)^2\right], \tag{8.4b}$$

where $n = 0, 1, 2, \ldots, N - 1$. Here N is the total number of received range samples, which is a function of the pulse duration time, sampling frequency, and number of targets.

Derivation of Eq. (8.4b) is based on one short duration pulse transmitted from one radar position. If the radar moves along a direction perpendicular to or at some angle from the radar beam, and emits signal at the rate of f_{PRF}, then a 2D received data array can be generated.

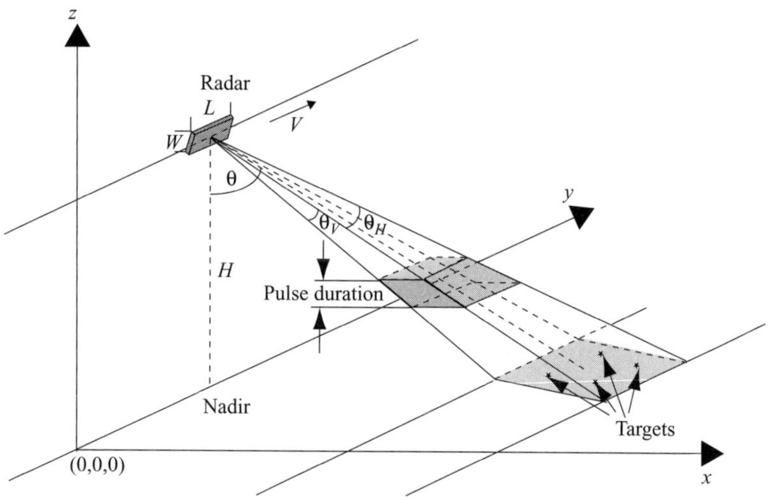

FIGURE 8.2 System model of radar image generation.

Let u_i be the moving radar position along the flight path or y axis, with $i = 1,2,\ldots,$ M. Here M is the total number of processed azimuth lines, which must be equal to or greater than the number of azimuth lines (Naz) within the synthetic aperture length Ls.

Equation (8.4b) can then be modified as

$$r_b(u_i, t_n) = \sum_{k=0}^{K} \sigma_k \left| a\left(t_n - \tau_{u_ik}\right)\right|^2 \exp\left[-j2\pi f_c \tau_{u_ik} + j\pi\alpha(t_n - \tau_{u_ik})^2\right], \quad (8.4c)$$

where $\tau_{u_ik} = (2R_{u_ik}/c)$ is the received signal delay time from the kth target with the radar located at position u_i.

A system model used to generate a broadside SAR image is shown in Fig. 8.2, where the radar is at height H above the ground (or x–y plane) and moves at speed V along the y-axis direction. The symbol θ_H is the radar beamwidth along the y-axis direction, while θ_V is the radar beamwidth along the x-axis direction. The radar center beam is always perpendicular to the flight path in the broadside SAR system. The squint SAR system is similar to that shown in Fig. 8.2, except that the radar center beam is not perpendicular to the flight path, but has a squint angle θ_q.

The following assumptions are made to simplify the system model of Fig. 8.2:

- Ignore the background objects, so that the received signal is noise-free.
- The target is an ideal reflector and stationary; therefore, motion compensation is not required.
- Ignore the effect of the inclination angle. The slant range is considered equal to the ground range.

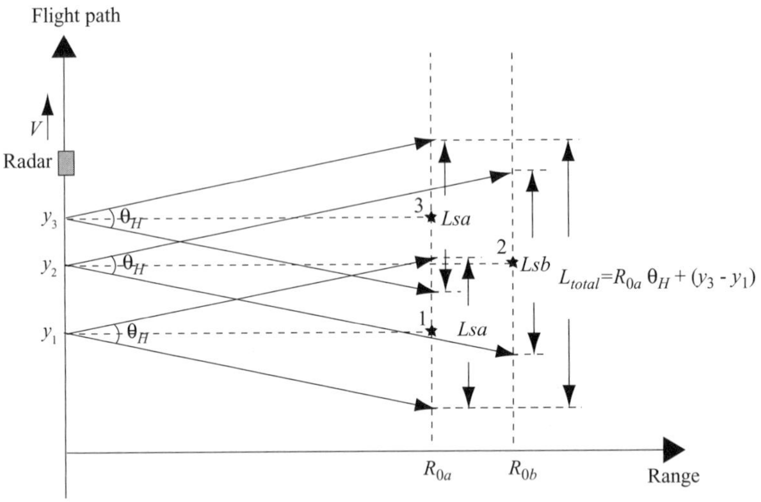

FIGURE 8.3 A simplified broadside SAR system for radar image generation.

A simplified 2D version of Fig. 8.2 is shown in Fig. 8.3, where the radar height $H = 0$ and the y axis and flight path are merged.

Three targets are shown in Fig. 8.3, and they are located at (R_{0a}, y_1), (R_{0b}, y_2), and (R_{0a}, y_3), respectively. The vertical axis represents the radar moving direction, while the horizontal axis corresponds to the range, which is perpendicular to the flight path. The 3-dB radar beamwidth is represented as θ_H. The three radar positions y_1, y_2, and y_3 are the positions when the corresponding targets are illuminated under the center beam of the radar. The synthetic aperture length for the targets at range R_{0a} is $Lsa = R_{0a}\theta_H$, and for the target at range R_{0b} this length is $Lsb = R_{0b}\theta_H$. The total synthetic aperture length of the three targets is $L_{\text{total}} = Lsa + (y_3 - y_1)$.

Figure 8.4 is a simplified 2D version of Fig. 8.2 for a squint SAR system. Here a squint angle θ_q is shown as the angle between the radar center beam to the target and the range axis. The three radar positions with the center beam on the corresponding targets are u_1, u_2, and u_3. The three targets are located at (R_{0a}, y_1), (R_{0b}, y_2), and (R_{0a}, y_3), respectively. The synthetic aperture length for the targets located at range R_{0a} and R_{0b} can be computed as follows:

$$Lsa = R_{0a}[\tan(\theta_q + 0.5\theta_H) - \tan(\theta_q - 0.5\theta_H)]$$
$$Lsb = R_{0b}[\tan(\theta_q + 0.5\theta_H) - \tan(\theta_q - 0.5\theta_H)].$$

The total synthetic aperture length in this case is

$$L_{\text{total}} = Lsa + (y_3 - y_1)$$

Figure 8.3 will now be used to synthesize the broadside SAR image data. The single target case will first be described, followed by a case with three targets.

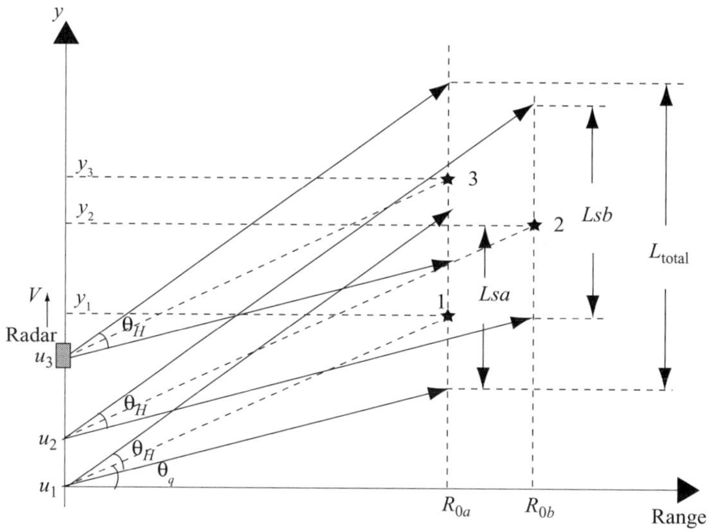

FIGURE 8.4 A simplified squint SAR system for radar image generation.

8.2 SYNTHESIS OF A BROADSIDE SAR IMAGE DATA ARRAY

8.2.1 Single-Target Case

The radar image generated from a single-target response can be modeled as shown in Fig. 8.5 with a target located at (R_0, y_2) and $y_2 = u_2$. The three radar positions u_1, u_2, and u_3, are the positions where the radar beam is starting, in the center, and ending to illuminate the target, respectively. The symbols R_1 and R_3 denote the slant ranges when the radar is at locations $(0, u_1)$ and $(0, u_3)$, respectively. The slant range R_2 equals R_0, which is the shortest distance between target and radar.

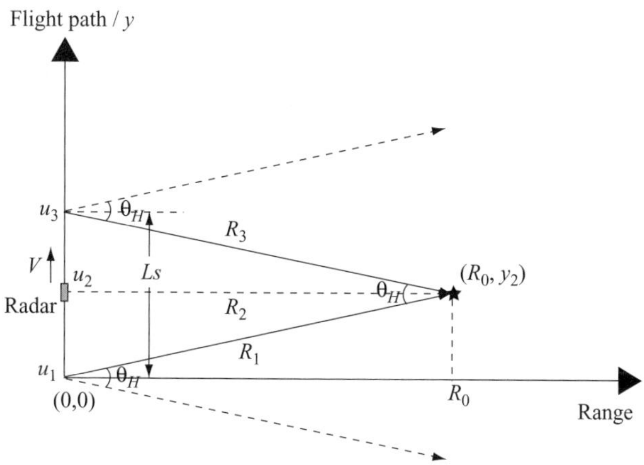

FIGURE 8.5 Single-target broadside SAR system for radar image generation.

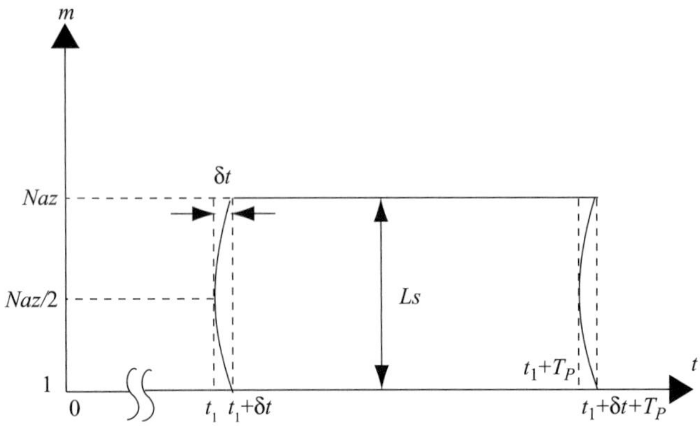

FIGURE 8.6 Received signal array from Fig. 8.5.

Figure 8.6 displays the model of a received radar signal array. The vertical axis m represents the azimuth samples. The horizontal axis t represents the delay time of the radar pulse returned from the target. Each azimuth sample (or line) corresponds to the radar position u as shown in Fig. 8.5. There are Naz number of azimuth lines within the synthetic aperture length Ls, and Naz = integer of (Ls/A_s), where A_s is the sample spacing between the azimuth lines. The azimuth location $m = 1$ is the position where the radar beam starts to illuminate the target, and $m = Naz$ is the last position when the radar beam ceases to illuminate the target. At radar position $m = Naz/2$, the target is under the center beam of the radar. This is also the position when the target has the shortest distance from the radar. The return time of the radar pulse starts at t_1 and ends at $t_1 + T_p + \delta t$. Here $t_1 = 2R_0/c = 2R_2/c$ is the closest target reflection time after the radar beam is transmitted. T_p is the time duration of the radar transmitting signal, and $\delta t = 2(R_1 - R_2)/c = 2(R_3 - R_2)/c$ is the echo time difference between the center radar beam and the two edges of the 3-dB radar beamwidth.

The curved lines inside the two vertical dotted lines, which occur at both $t = t_1$ and $t = t_1 + T_p$, correspond to the echo time for different slant ranges when the radar moves within the synthetic aperture length Ls. It is also the target trajectory seen by radar during the radar movement within Ls. For most airborne-based broadside SARs, the time difference δt is negligible (within $^1/_4$ range resolution time limit); therefore, a straight line will be used instead. However, the phase differences among various slant ranges must be preserved.

The time sample n and the slant range sample n' are defined as follows:

$$n = \frac{t}{\Delta t} = t f_s$$
$$n' = \frac{R}{R_s} = \frac{R f_s}{c}.$$

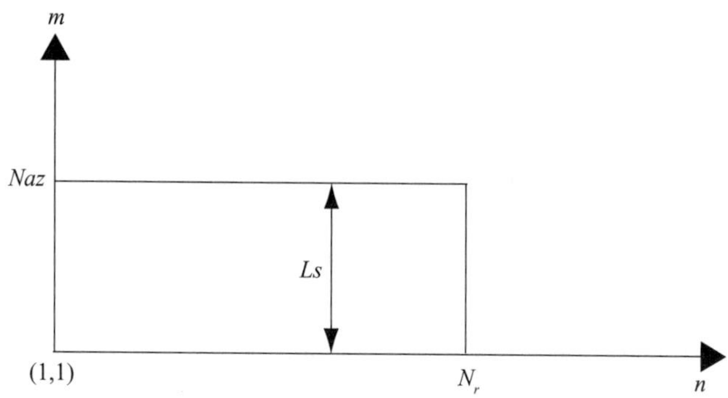

FIGURE 8.7 A simplified and digitized received signal array from Fig. 8.6.

In general, time t and slant range R are related by $t = R/c$; therefore $n = n'$. However, when time t refers to the target echo time, the relationship of the time and slant range becomes $t = 2R/c$. Therefore the relation between the time sample and the slant range sample becomes $n = 2n'$. In the equation displayed above, f_s is the sampling frequency and R_s is the slant range sample spacing.

For purposes of illustration, the time sample will be used for discussions on SAR image processing in this chapter, and the slant range sample will be used in Chapter 9.

Figure 8.7 shows a simplified and digitized plot of Fig. 8.6. The horizontal time axis is replaced with time sample n. For simplicity, the received time sample $n = 1$ starts from $t = t_1$ and ends at $n = N_r =$ integer of $(T_p f_s)$, where T_p is the pulse duration time.

The following airborne-based SAR radar parameters are used for synthesizing the received signal array. The same set of parameters, with minor differences, will be used for both the single target and multiple target cases, and serves both the broadside and squint SAR systems:

Radar beam carrier frequency $f_c = 1 \times 10^{10}$ Hz
Speed of wave $c = 3 \times 10^8$ m/s
Radar beam carrier wavelength $\lambda = c/f_c = 3$ cm
Symmetric LFM time duration $T_p = 6.033$ μs
Range FM rate $\alpha = 4 \times 10^{12}$ Hz/s
Closest range of target $R_0 = 7500$ m
Range sampling frequency $f_s = 30$ MHz
Range bandwidth $= \alpha T_p = 24.13$ MHz
Radar moving speed $V = 200$ m
Azimuth FM rate $= -2V^2/(R_0\lambda) = -355.56$ Hz/s
Pulse repetition frequency $f_{\mathrm{PRF}} = 500$ Hz
Antenna length $L = 1$ m

On the basis of these parameters, the following data can be computed:

Slant range sample spacing $R_s = c/f_s = 10$ m
Radar 3-dB beamwidth $\theta_H = \lambda/L = 0.03$ radian
Azimuth sample spacing $A_s = V/f_{\text{PRF}} = 0.4$ m
Synthetic aperture length $Ls = R_0\,\theta_H = 225$ m
Doppler frequency bandwidth $B_{\text{Dop}} = 2V/L = 400$ Hz
Number of azimuth samples = integer of $(Ls/A_s) = 563$
Number of time samples within the LFM pulse $N_r = (f_s T_p) = 181$
Number of time samples $N_{\text{tot}} = N_r = 181$
The slant range $R_1 = R_3 = R_0/\cos(0.5\theta_H) = 7500.8$ m

The slant range difference, namely, $R_1 - R_0 = 7500.8 - 7500 = 0.8$ m, is less than $^1/_4$ of the range resolution, which is $(c/2B)/4 = 1.56$ m in this setup. Therefore, range migration correction is not needed in this broadside SAR case.

Let both the reflection coefficient and the amplitude of the transmitted signal be equal to one. The baseband signal returned from a single target can then be expressed as follows, from Eq. (8.4c):

$$r_b(u_i, t_n) = \exp\lfloor -j2\pi f_c \tau_{u_i} + j\pi\alpha(t_n - \tau_{u_i})^2 \rfloor. \tag{8.5a}$$

For a target located at (R_0, y_1) with the radar position at $(0, u_i)$, the echo delay time τ_{u_i} can be computed as

$$\tau_{u_i} = \frac{2\sqrt{(u_i - y_1)^2 + R_0^2}}{c} \tag{8.5b}$$

Figure 8.8 shows the baseband signal waveform of LFM radar with the parameters and assumptions listed above. Figure 8.8a displays the real part of the signal and Fig. 8.8b displays the imaginary part of the signal.

The received signal array based on Eq. (8.5) is displayed in Fig. 8.9; Fig. 8.9a shows the real part of the received signal array, while Fig. 8.9b displays the imaginary part of the signal array. The vertical axis (ordinate) shows the number of azimuth lines (or cross-range samples), where one out of every 30 azimuth lines was sequentially selected throughout the synthetic aperture length. The azimuth lines run from line 1 to Naz, with Naz = integer of $(Ls/A_s) = 563$, where A_s is the azimuth sample spacing. The horizontal axis (abscissa) shows the time samples, with sampling rate at 30 MHz. The total number of time samples N_r equals the sampling frequency times the pulse duration, that is, $f_s T_p = 181$ time samples for each azimuth line. Although there is no range migration, the phase changes due to the slant range difference among various azimuth lines are obvious.

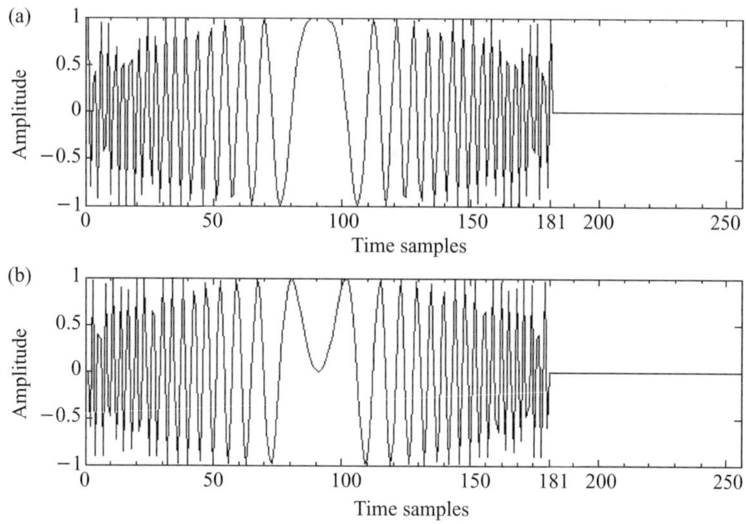

FIGURE 8.8 Waveforms of the real and imaginary parts of a baseband symmetric LFM signal.

8.2.2 Multiple-Target Case

A 2D broadside SAR signal array based on Fig. 8.3 is depicted in Fig. 8.10, with three targets located at (R_{0a},y_1), (R_{0b},y_2), and (R_{0a},y_3), respectively. Here the vertical axis (ordinate) represents the azimuth lines (or cross-range samples), while the horizontal axis (abscissa) shows time. Again, the curved lines occur at the beginning and end of the pulse duration time, which can be neglected as described before. *Lsa* and *Lsb* are the synthetic aperture lengths corresponding to range R_{0a} and R_{0b}, while <*r*1>,

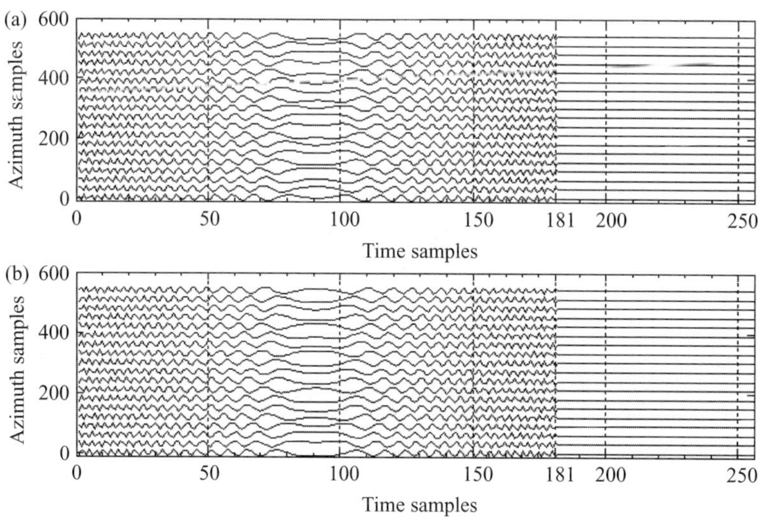

FIGURE 8.9 Waveforms of received baseband signal from Fig. 8.5.

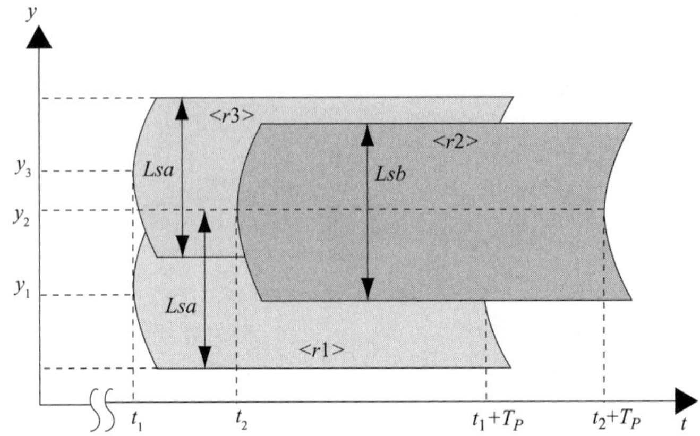

FIGURE 8.10 Received signal arrays from Fig. 8.3.

<r2>, and *<r3>* are the received signal arrays corresponding to targets 1, 2, and 3, respectively. The time variables t_1 and t_2 are the target echo time, with $t_1 = 2R_{0a}/c$ and $t_2 = 2R_{0b}/c$. T_p is the transmitted LFM pulse duration time.

A simplified version of Fig. 8.10 is shown in Fig. 8.11, where the time axis is digitized to become a time sample axis. *Naza* and *Nazb* are the number of azimuth lines corresponding to *Lsa* and *Lsb*, respectively. Namely, *Naza* = integer of (Lsa/A_s) and *Nazb* = integer of (Lsb/A_s) with A_s as the azimuth sample spacing. D_1 is the sample number of overlapped segment in azimuth direction between signal arrays *<r1>* and *<r3>*; that is, D_1 = integer of $((Lsa - (y_3 - y_1))/A_s)$. D_2 is the azimuth line difference between the two starting lines of *<r1>* and *<r2>*; that is D_2 = integer of $((y_2 - y_1) - (Lsb2 - Lsa2))/A_s$. N_{Rb} is the range sample difference between R_{0a} and R_{0b}, which is defined later.

The size of the received signal array is $M \times N$, with $M = 2 \times Naza - D_1$ and N = integer of $(f_sT_p) + 2N_{Rb}$.

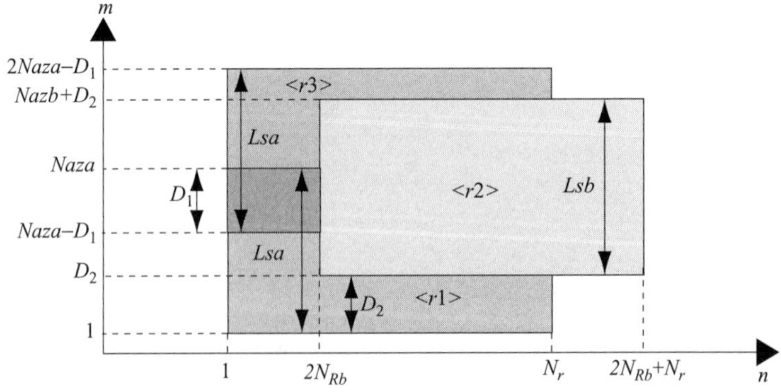

FIGURE 8.11 A simplified and digitized signal array from Fig. 8.10.

Using the same radar parameters used in the single target broadside case, the additional data of target locations and associated data are listed below:

Target 1 location: (R_{0a}, y_1)
Target 2 location: (R_{0b}, y_2)
Target 3 location: (R_{0a}, y_3)
Synthetic aperture length corresponding to R_{0a}:

$$Lsa = R_{0a}\,\theta_H$$

Synthetic aperture length corresponding to R_{0b}:

$$Lsb = R_{0b}\,\theta_H$$

$Naza =$ integer of (Lsa/A_s)
$Nazb =$ integer of (Lsb/A_s)
Number of range sample difference between R_{0a} and R_{0b}

$$N_{Rb} = \text{integer of}\left(\frac{R_{0b} - R_{0a}}{R_s}\right).$$

Again, let both the reflection coefficient and the amplitude of the transmitted signal be equal to one; the baseband signal returned from the three targets can be expressed as follows, from Eq. (8.4c):

$$\begin{aligned} r_b(u_i, t_n) = {} & \exp\lfloor -j2\pi f_0 \tau_{u_i 1} + j\pi\alpha(t_n - \tau_{u_i 1})^2 \rfloor \\ & + \exp\lfloor -j2\pi f_0 \tau_{u_i 2} + j\pi\alpha\left(t_n - \tau_{u_i 2}\right)^2 \rfloor \\ & + \exp\lfloor -j2\pi f_0 \tau_{u_i 3} + j\pi\alpha(t_n - \tau_{u_i 3})^2 \rfloor. \end{aligned} \tag{8.6a}$$

Given the three targets located at (R_{0a}, y_1), (R_{0b}, y_2), and (R_{0a}, y_3), and the radar position $(0, u_i)$, the time delay amounts $\tau_{u_{i1}}$, $\tau_{u_{i2}}$, and $\tau_{u_{i3}}$ can be computed as follows:

$$\begin{aligned} \tau_{u_i 1} &= 2 \cdot \frac{\sqrt{(u_i - y_1)^2 + R_{0a}^2}}{c}, \\ \tau_{u_i 2} &= 2 \cdot \frac{\sqrt{(u_i - y_2)^2 + R_{0b}^2}}{c}, \\ \tau_{u_i 3} &= 2 \cdot \frac{\sqrt{(u_i - y_3)^2 + R_{0c}^2}}{c}. \end{aligned} \tag{8.6b}$$

The three targets locations are listed below.

$y_1 = 0$ m
$y_2 = 100$ m
$y_3 = 150$ m
$R_{0a} = 7500$ m
$R_{0b} = 7650$ m

Accordingly, the following data can be computed.

$Lsa = 225$
$Lsb = 229.5$
$Naza = Lsa/A_s = 563$
$Nazb = Lsb/A_s = 573$
$D_1 = \text{integer of } [(Lsa - (y_3 - y_1))/A_s] = 187$
$D_2 = \text{integer of } \{[(y_2 - y_1) - (Lsb2 - Lsa2)]/A_s\} = 244$
$N_{Rb} = \text{integer of } [(R_{0b} - R_{0a})/R_s] = 15$
Number of azimuth lines $M_{\text{tot}} = 2 \times Naza - D_1 = 939$
Number of time samples $N_{\text{tot}} = c \times T_p + 2N_{Rb} = 211.$

The received signal arrays corresponding to targets 1–3, in terms of time samples, are shown in Fig. 8.12: Figs. 8.12a and 8.12b show the real and imaginary parts of the signal array <r1>, Figs. 8.12c and 8.12d show the real and imaginary

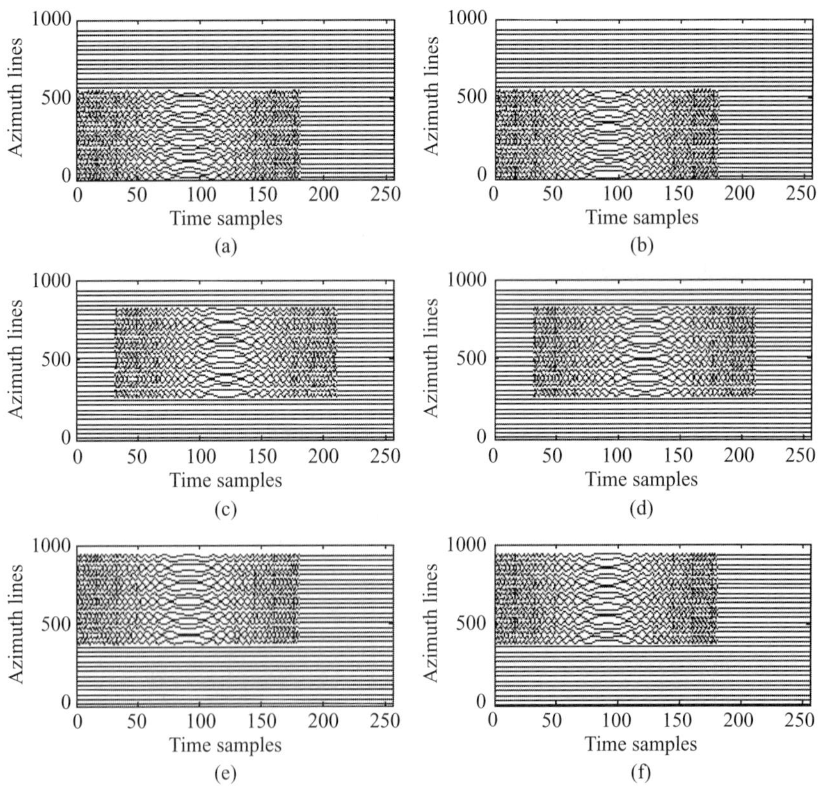

FIGURE 8.12 Waveforms of the individual received signal from Fig. 8.10.

parts of the signal array <$r2$>, and Figs. 8.12e and 8.12f show the real and imaginary parts of the signal array <$r3$>. The vertical axis shows the number of azimuth lines, where one out of every 30 azimuth lines was sequentially selected throughout the synthetic aperture length. The horizontal axis shows the time samples, with a 30 MHz sampling rate. The size of signal arrays <$r1$> is 1–563 (1 to *Naza*) in the azimuth direction and 1–181 (1 to N_r) in the time sample direction. The size of the signal array <$r2$> is 244–817 (D_2 to $D_2 + Nazb$) in the azimuth direction and 30–211 (30 to N_{tot}) in the time sample direction. The size of signal array <$r3$> is 377–939 ($Naza - D_1 + 1$ to $2Naza - D_1$) in azimuth direction and 1–181 in the time sample direction.

The overall received signal array, which is the sum of the three signal arrays <$r1$>, <$r2$> and <$r3$>, is displayed in Fig. 8.13. The real part is shown in Fig. 8.13a; the imaginary part, in Fig. 8.13b. Again, the vertical axis (ordinate) shows the number of azimuth lines, where one out of every 30 azimuth lines was sequentially selected throughout the total synthetic aperture length. The horizontal axis (abscissa) shows the time samples. The size of the overall received signal array is from rows 1–939 ($2Naza - D_1$) in the azimuth direction, and from column 1 to column 211 in the time sample direction.

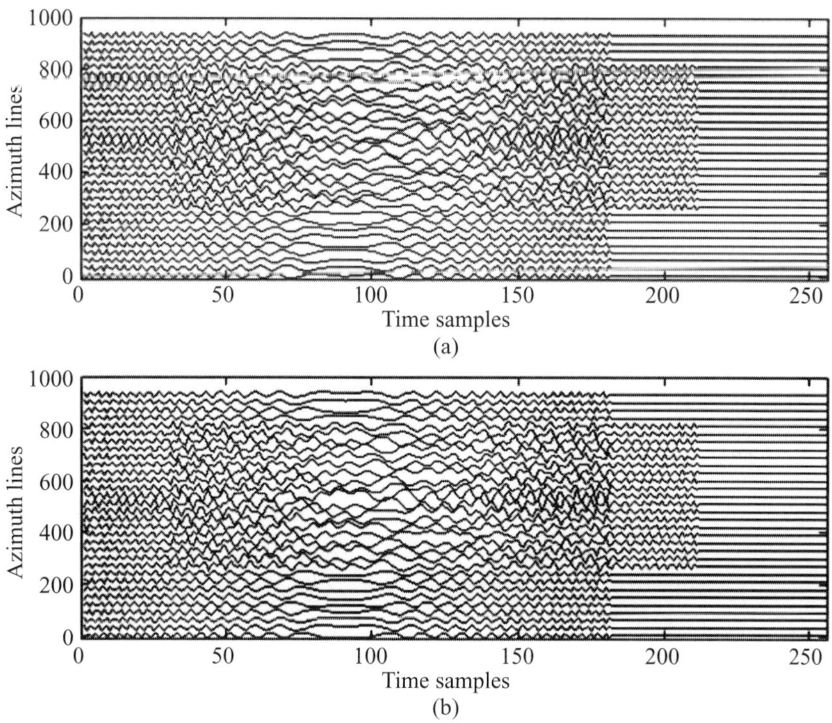

FIGURE 8.13 Waveforms of the received signals from Fig. 8.10.

8.3 SYNTHESIS OF A SQUINT SAR IMAGE DATA ARRAY

8.3.1 Single-Target Case

The squint radar image generated from a single target response can be modeled as shown in Fig. 8.14, with the target located at (R_0,0). The three radar positions, u_1, u_2, and u_3, are the locations where the radar beam is beginning, in the center, and ceasing to illuminate the target. The symbols R_1 and R_3 denote the slant ranges when the radar beam is at locations (0,u_1) and (0,u_3), respectively. The symbol R_2 denotes the slant range when the radar beam is at location (0,u_2). The symbol θ_q is the squint angle between the range axis and the radar center beam when the radar is at position (0,u_2).

Figure 8.15 displays the time–azimuth relationship of the received signal array. The vertical axis shows the positions of the radar, while the horizontal axis shows the return time of the echo signal. The received signal array appears as a curved parallelogram. The longest time for echo to arrive occurs at radar position u_1, with $t_1 = 2R_1/c$; and the shortest time to arrive occurs at radar position u_3, with $t_3 = 2R_3/c$. At radar position u_2, the echo arrival time is in between t_1 and t_3.

Figure 8.16 shows a simplified plot of Fig. 8.15. The time is replaced with time samples, and the radar position u is replaced with azimuth sample number. As can be seen, the slant range difference causes significant range migration. The maximum range migration amount is $\Delta N = (R_1 - R_3)/R_s$.

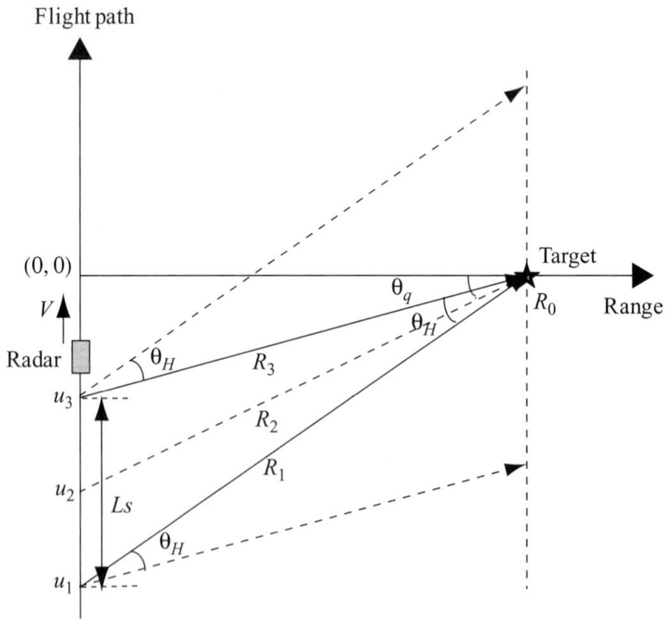

FIGURE 8.14 System model of a single-target squint SAR.

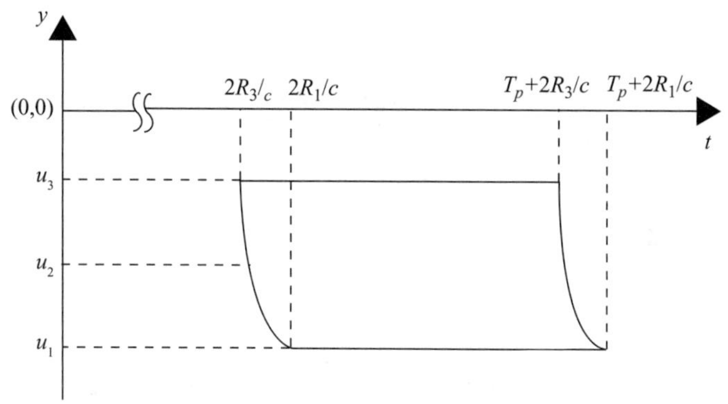

FIGURE 8.15 A received signal array from Fig. 8.14.

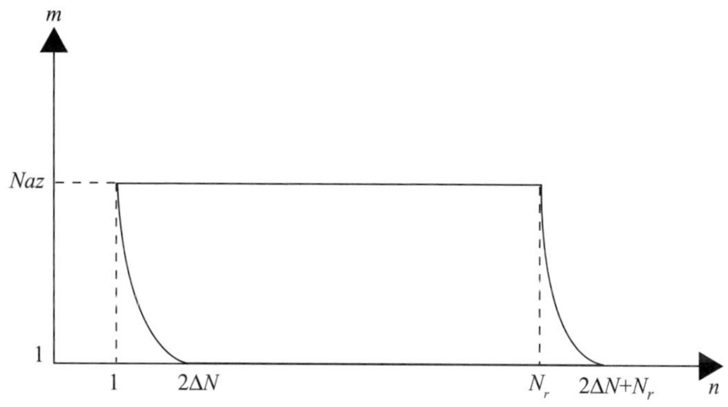

FIGURE 8.16 A digitized signal array from Fig. 8.15.

The same radar parameters used in the single-target broadside case, together with the squint angle $\theta_q = 6^\circ$, are used to compute the following data:

$$\begin{aligned}
\text{Synthetic aperture length } Ls &= R_0[\tan(\theta_q + 0.5\,\theta_H) - \tan(\theta_q - 0.5\,\theta_H)] \\
&= 227.5 \text{ m} \\
\text{Longest slant range } R_1 &= R_0/\cos(\theta_q + 0.5\theta_H) \\
&= 7554.1 \text{ m} \\
\text{Shortest slant range } R_3 &= R_0/\cos(\theta_q - 0.5\theta_H) \\
&= 7530.3 \text{ m}
\end{aligned}$$

The maximum range migration amount, namely, $(R_1 - R_3) = 7554.1 - 7530.3 = 23.8$ m, is larger than $^1/_4$ of the sample spacing, range resolution, which is $(c/2B)/4 = 1.56$ m) in this setup. Therefore, range migration occurs, and range migration correction is needed in this squint SAR case.

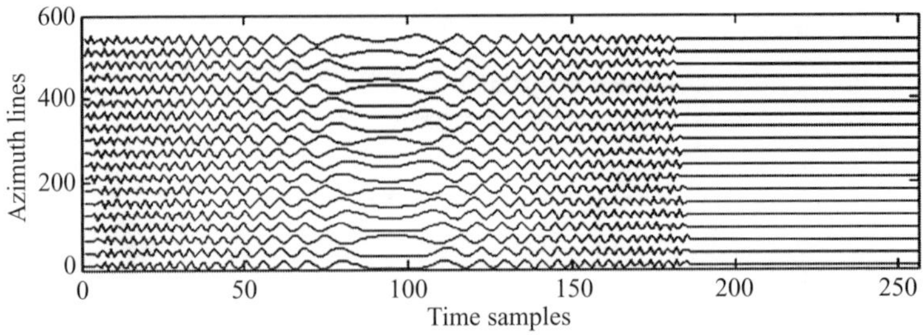

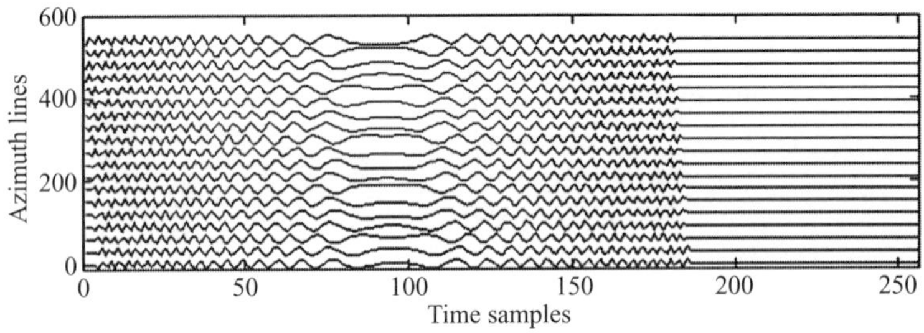

FIGURE 8.17 Waveforms of a received baseband signal from Fig. 8.14.

With the same baseband transmitting signal shown in Fig. 8.8, the received baseband signal array can be derived, from Eq. (8.5), and is shown in Fig. 8.17: Fig. 8.17a shows the real part of the received signal array; Fig. 8.17b, the imaginary part of the signal array. The vertical axis shows the number of azimuth lines, where one out of every 30 azimuth lines was sequentially selected throughout the synthetic aperture length. The azimuth line runs from line 1 to line Naz, with Naz = integer of $(Ls/A_s) = 569$. The horizontal axis represents the time samples. The total number of time samples $N_r = 181$ for each azimuth line.

As can be seen in Fig. 8.17, the range migration exists and range cell migration correction is therefore needed to process this type of SAR signal.

8.3.2 Multiple-Target Case

A three-target squint SAR system based on Fig. 8.4 is shown in Fig. 8.18, where the three targets are located at (R_{0a},y_1), (R_{0b},y_2), and (R_{0a},y_3), respectively. The radar position of u_1 and u_3 corresponds to the place where the radar beam starts and ends its illumination of the target areas. Position u_2 is where the radar center beam illuminates the target 2. The vertical axis represents the azimuth lines; the horizontal axis, the range. Lsa and Lsb are the synthetic aperture lengths corresponding to the targets located at range R_{0a} and R_{0b}. R_{1a} and R_{3a} are the longest and shortest slant ranges from the radar to the targets located at range R_{0a}, while R_{1b} and R_{3b} are the

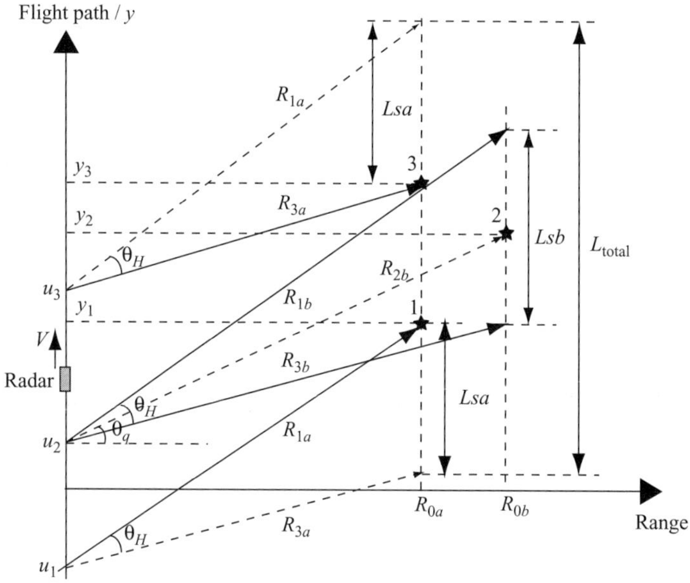

FIGURE 8.18 System model of a three-target squint SAR.

longest and shortest slant ranges to the target located at range R_{0b}. The squint angle is represented as θ_q.

From Fig. 8.18, the following parameters can be derived:

$$Naza = \frac{Lsa}{A_s},$$

$$Nazb = \frac{Lsb}{A_s},$$

$$R_{1a} = \frac{R_{0a}}{\cos(\theta_q + 0.5\theta_H)},$$

$$R_{3a} = \frac{R_{0a}}{\cos(\theta_q - 0.5\theta_H)},$$

$$R_{1b} = \frac{R_{0b}}{\cos(\theta_q + 0.5\theta_H)},$$

$$R_{3b} = \frac{R_{0b}}{\cos(\theta_q - 0.5\theta_H)},$$

$$N_{Rb} = \text{integer of}\left(\frac{R_{0b} - R_{0a}}{R_s}\right),$$

$$\Delta r = R_{0b} - R_{0a}.$$

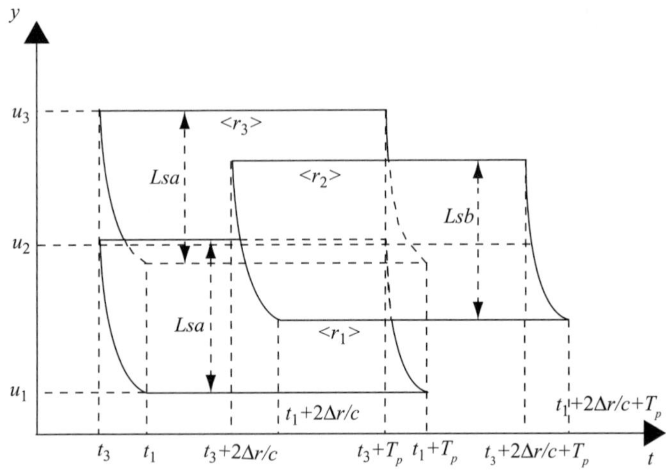

FIGURE 8.19 The received signal arrays from Fig. 8.18.

In addition, the echo arrival time can be computed as follows:

$$t_1 = \frac{2R_1}{c},$$

$$t_3 = \frac{2R_3}{c}.$$

On the basis of the three-target squint SAR system and parameters, the time–azimuth relationship of the received signal array can be plotted as shown in Fig. 8.19.

Figure 8.20 shows the simplified and digitized signal array based on Fig. 8.19 with abscissa (horizontal axis) changes from time to discrete time sample and ordinate (vertical axis) changes from radar position to azimuth line. The data array size is $M \times N$, with $M = 2\,Naza - D_1$ and $N = \Delta N_b + N_r + \Delta N_r$.

The parameters ΔN_a, ΔN_b, $\Delta \mathrm{N}_r$, D_1, and D_2 are defined below:

$$\begin{aligned}
\Delta N_a &= \text{integer of } [2(R_{1a} - R_{3a})/R_s] \\
\Delta N_b &= \text{integer of } [2(R_{1b} - R_{3b})/R_s] \\
D_1 &= \text{integer of } [Lsa - (y_3 - y_1)] \\
D_2 &= \text{integer of } [y_2 - R_{0b} \tan \theta_q] \\
\Delta N_r &= \text{integer of } [2\Delta r/R_s]
\end{aligned}$$

By applying all data and parameters used in multiple targets for the broadside SAR system, with the additional squint angle $\theta_q = 6°$, the following parameters can

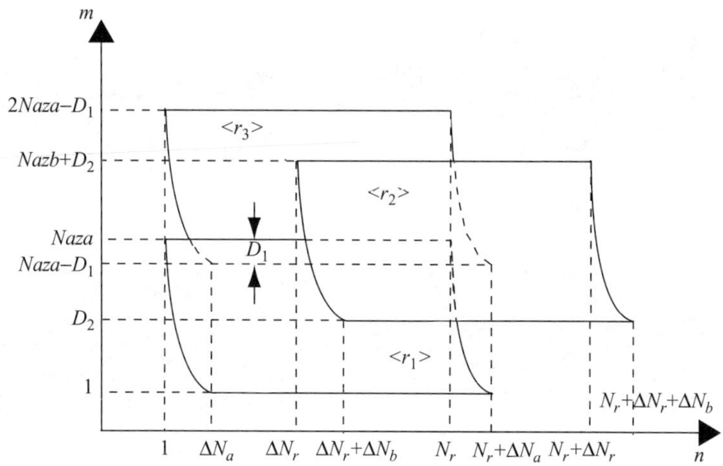

FIGURE 8.20 The digitized signal arrays from Fig. 8.19.

be computed:

$Lsa = 227.5$

$Lsb = 232$

$Naza = Lsa/A_s = 569$

$Nazb = Lsb/A_s = 581$

$R_{1a} = R_{0a}/\cos(\theta_q + 0.5\theta_H) = 7554.1$

$R_{3a} = R_{0a}/\cos(\theta_q - 0.5\theta_H) = 7530.3$

$R_{1b} = R_{0b}/\cos(\theta_q + 0.5\theta_H) = 7705.2$

$R_{3b} = R_{0b}/\cos(\theta_q - 0.5\theta_H) = 7680.9$

$D_1 =$ integer of $((Lsa - (y_3 - y_1))/A_s) - 193$

$D_2 =$ integer of $[y_2 - R_{0b}\tan\theta_q - (Nazb - Naza)/2] = 204$

$\Delta N_a =$ integer of $[2(R_{1a} - R_{3a})/R_s] = 5$

$\Delta N_b =$ integer of $[2(R_{1b} - R_{3b})/R_s] = 5$

$N_r =$ integer of $(f_s T_p) = 181$

$\Delta N_r =$ integer of $[2\Delta r/R_s] = 30$

Number of azimuth lines $M = 2Naza - D_1 = 945$

Number of time samples $N = \Delta N_b + N_r + \Delta N_r = 216$

Using the data presented above, the received signal array waveforms corresponding to targets 1–3 can be synthesized and are shown in Fig. 8.21: Figs. 8.21a and 8.21b show the real and imaginary parts of signal array <*r*1>; Figs. 8.21c and 8.21d, the real and imaginary parts of signal array <*r*2>; Figs. 8.21e and 8.21f, the real and imaginary parts of signal array <*r*3>. The dimensions of signal arrays <*r*1> and <*r*3> are the same, namely, from row 1 to row 569 (*Naza*) and from

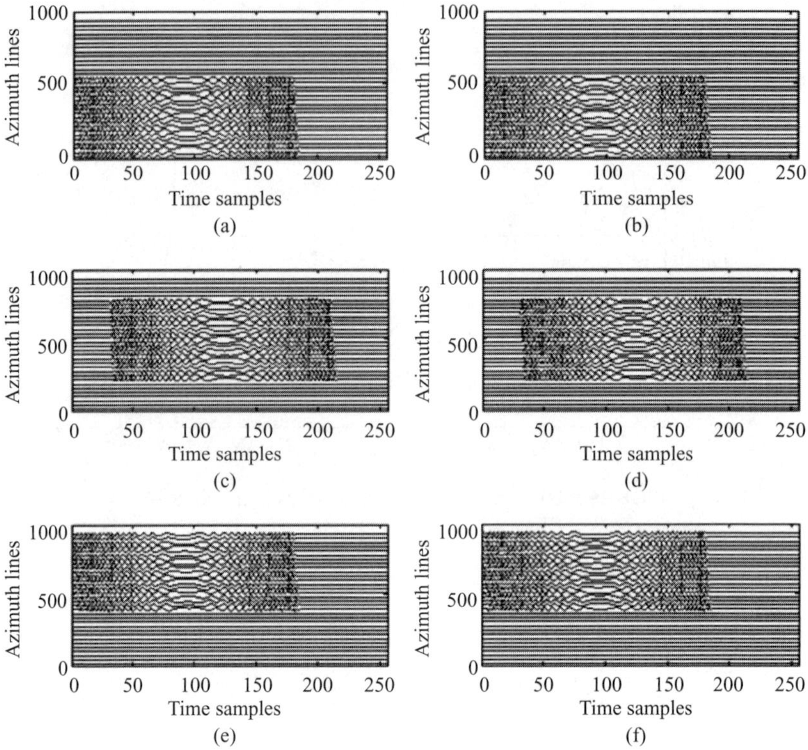

FIGURE 8.21 Waveforms of the individual received signal from Fig. 8.18.

column 1 to column 181 (N_r). Signal array <$r2$> ranges from row 244 to row 817 ($D_2 + Nazb$) and from column 30 [$2\Delta r/R_s$] to column 216.

The waveform of received signal array, which is the sum of three targets, is shown in Fig. 8.22. The real and imaginary parts of the waveform are shown in Figs. 8.22a and 8.22b, respectively.

8.4 RANGE–DOPPLER PROCESSING OF SAR DATA

The synthesis of 2D SAR data discussed in the previous section shows that the received data from a point target have a waveform with size $M \times N$ in both time and azimuth directions. The spread in time is caused by the time duration of the transmitted LFM pulse. The spread in azimuth is because the target is under the radar illumination, which causes the radar to receive the echo signal for the period during which the radar moves through the whole path of synthetic aperture length.

The three major functions of SAR data processing—namely, range compression, range cell migration correction, and azimuth compression—can be implemented by a typical range–Doppler processor shown as a block diagram in Fig. 8.23.

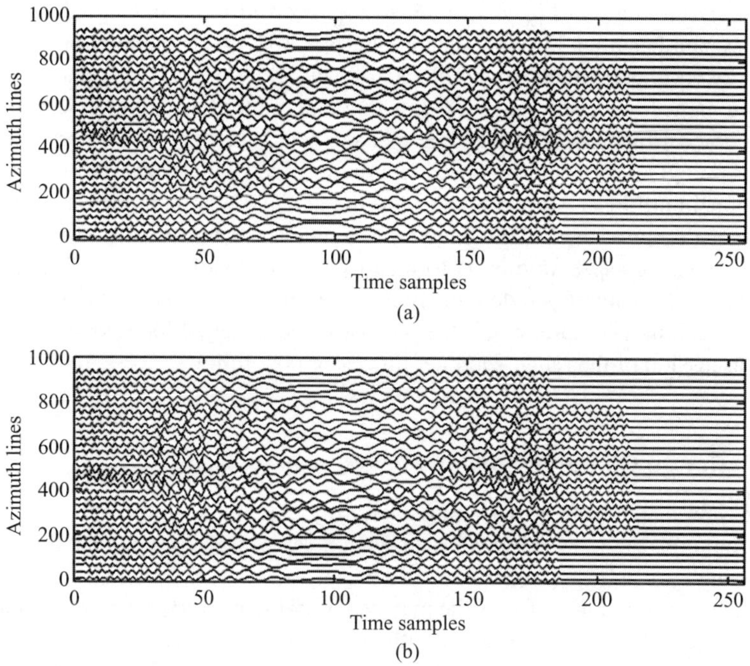

FIGURE 8.22 Waveforms of the received signals from Fig. 8.18.

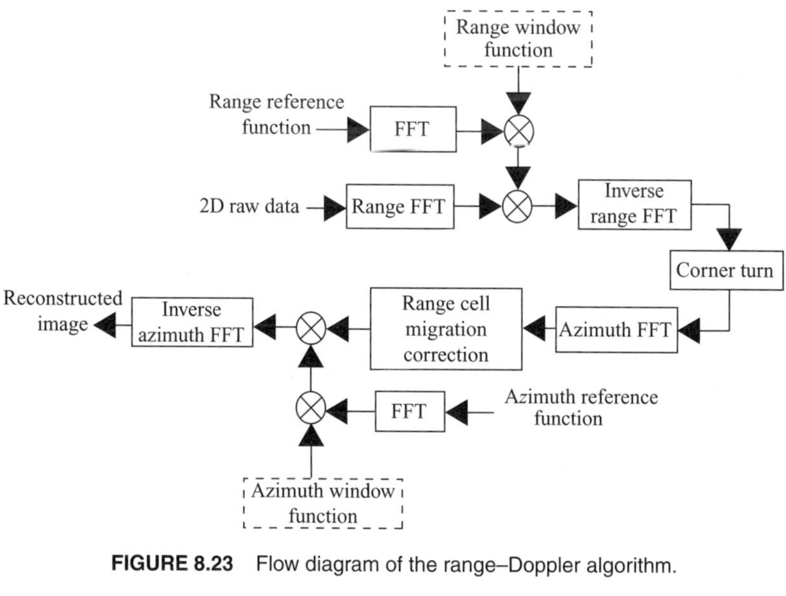

FIGURE 8.23 Flow diagram of the range–Doppler algorithm.

The 2D raw data in Fig. 8.23 refer to the received baseband signal, which is a complex number after in-phase–quadrature-phase (I–Q) demodulation. The *range compression* is performed by the blocks of range FFT (on 2D raw data), FFT (on range reference function), range window function, multiplier, and inverse range FFT. The "corner turn" operation is a data management scheme, that converts the range compressed 2D data from row-based to column-based data for the subsequent operation of azimuth FFT. After azimuth FFT, the 2D data will be in the range–Doppler frequency domain. *range cell migration correction*, if required, is performed at this stage. *Azimuth compression* is performed by the blocks of FFT (on azimuth reference function), azimuth window function, multiplier, and inverse azimuth FFT. The output of azimuth compression is the reconstructed image of focused targets.

A discussion of the range–Doppler processor now follows.

8.4.1 Range Compression

Figure 8.24 displays an $M \times N$ raw data array together with the mth row of data. Each row consists of N digitized data, which serve as 1D range signal for image processing.

Range compression performs the matched filtering on every row of range samples. The matched filter $h(t)$ is designed with $h(t) = s(-t)^*$, where $s(t)$ equals the transmitted signal, which has been explored in detail in Chapter 5 and is not repeated here. The matched filter serves as a reference function and is operated on every row of data during the range compression process. A window function, such as the Hanning window, is normally applied to the range reference function in order to reduce the sidelobe effect. After matched filtering on every row of 2D data, which is implemented in the frequency domain using the FFT technique, the inverse FFT is then applied to obtain the 2D time-domain range compressed signal again.

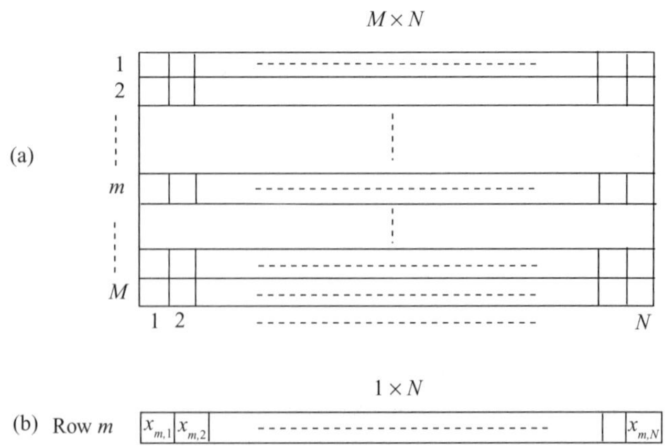

FIGURE 8.24 (a) An $M \times N$ 2D data array; (b) mth row of 2D data array.

8.4.2 Corner Turn

Range compression processes the $M \times N$ data array on a row basis, namely, $1 \times N$, and the output remains the same quantitatively. Appropriate zero padding on the data array and the selection of FFT size is needed to reduce the edge effect of circular convolution. After range compression, the corner turn is required to shuffle the row-based data to become column-based data, which is needed for the azimuth FFT operation.

Figure 8.25 demonstrates the "corner turn" operation. The nth column data, which are shown in Fig. 8.25a as an $M \times 1$ array, are rotated horizontally and appears as a $1 \times M$ row data array in Fig. 8.25b.

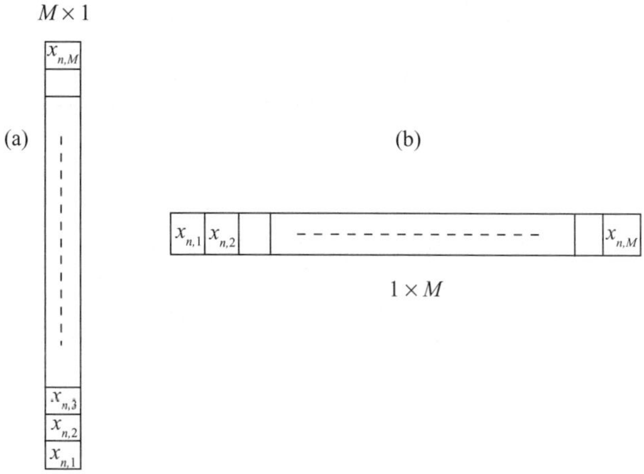

FIGURE 8.25 Operation of a corner turn.

The fast, modular, and multiple memory banks addressing schemes and mass memory storage capability due to the rapid evolution of digital signal processor (DSP) and integrated circuit (IC) technology, have made the corner turn much easier and faster to operate on column-based data.

8.4.3 Range Cell Migration Correction

Range cell migration correction operates on the range-compressed data in the time or range domain. The amount of range migration can be corrected either in range–Doppler frequency or range–spatial (or slow time) domain. However, as discussed in Chapter 6 and shown in Fig. 6.17, the multiple equal slant range targets and their trajectories appear randomly along the azimuth direction in the range–spatial domain, but show up as one only in the range–Doppler frequency domain. Therefore, only the range–Doppler frequency domain is discussed here.

8.4.3.1 Computation of Range Migration Amount. The range (which is different from slant range) axis is the line passing the target and perpendicular to the radar flight path. Let R_0 be the closest distance between radar and target and r be the slant range between the radar and the target. The slant range difference can be computed as follows:

$$\begin{aligned}\Delta R &= r - R_0 \\ &= \frac{R_0}{\cos\phi} - R_0.\end{aligned} \tag{8.7a}$$

Here ϕ is the angle between the range axis and the line connecting the radar to the target. Equation (8.7a) shows the amount of range migration required for correction with respect to R_0. From Eq. (6.3a), one obtains

$$f_D = \frac{2V}{\lambda}\sin\phi.$$

Therefore

$$\Delta R = R_0\left(1 - \left(\frac{\lambda f_D}{2V}\right)^2\right)^{-(1/2)} - R_0. \tag{8.7b}$$

Let N be the sample length of FFT and f_{PRF} be the pulse repetition frequency along the azimuth direction; then the kth bin of Doppler frequency can be expressed as

$$f_{Dk} = k\frac{f_{\mathrm{PRF}}}{N},$$

and

$$\Delta R_k = R_0\left(1 - \left(\frac{\lambda k f_{\mathrm{PRF}}}{2VN}\right)^2\right)^{-(1/2)} - R_0.$$

For a high-frequency LFM signal, $\lambda k f_{\mathrm{PRF}} \ll 2VN$; therefore

$$\Delta R_k \approx \frac{R_0 \lambda^2 k^2 f_{\mathrm{PRF}}^2}{8V^2N^2}. \tag{8.7c}$$

The number of range cells required for correction becomes

$$\begin{aligned}\Delta N_{Rk} &= \frac{2\Delta R_k}{R_s} \\ &\approx \frac{R_0 \lambda^2 k^2 f_{\mathrm{PRF}}^2}{4V^2N^2R_s}\end{aligned} \tag{8.7d}$$

Equation (8.7d) is then computed for $k = 1, 2 \ldots N$ with respect to the target-dependent closest distance R_0. It works well for broadside SAR systems, where $f_{Dc} = 0$. Modifications of Eqs. (8.7) are required for the squint SAR system to make the range cell migration correction with respect to the slant range distance R_3. Notice that R_3 is the shortest slant range between the radar and target, which corresponds to the Doppler frequency f_{DL}. Therefore

$$\begin{aligned}\Delta R &= r - R_3 \\ &= \frac{R_0}{\cos\phi} - \frac{R_0}{\cos\left(\theta_q - 0.5\theta_H\right)}\end{aligned} \tag{8.8a}$$

and

$$\begin{aligned}\Delta N_{Rk} &= \frac{2\Delta R_k}{R_s} \\ &\approx \frac{R_0\lambda^2}{4V^2R_s}\left(f_{DL} + \frac{kf_{\mathrm{PRF}}}{N}\right)^2 - \frac{R_0\lambda^2 f_{DL}^2}{4V^2R_s}\end{aligned} \tag{8.8b}$$

After range migration correction, the range differences among slant ranges are corrected and aligned along either with the reference range R_0 or slant range R_3. If R_0 is chosen, as is true for the broadside SAR system, the range cell migration correction is equivalently converts the slant range (not perpendicular to the flight path) to the range (perpendicular to the flight path). Notice that f_{DL} in Eq. (8.8b) refers to the true, not folded, low band edge of Doppler frequency.

Since the arrival time of the target-reflected signal varies while the radar moves through the whole synthetic aperture length, the peak of the range-compressed pulse may not be sampled accurately by the A/D converter with sampling frequency f_s. In other words, the peak of the range-compressed pulse may appear as a fraction of the sample away from the the sample digitized by the A/D converter, and may become unavailable for image processing.

Two major steps are used to correct the range cell migration: (1) fractional range sample interpolation and (2) range sample shift. These two steps can be performed in one process; however, for illustration purposes, they will be described separately.

Figure 8.26 shows an example of a 2D data array before range migration correction. This 2D data array resembles the range-compressed radar signal from a single-target squint SAR case. As can be seen, the range-compressed pulsed data along the Doppler frequency direction do not line up but spread along the range direction, and the magnitude of the pulse peaks varies because the data are not correctly sampled at the peak points. The correction process first computes the range migration amount on all row data corresponding to each Doppler frequency bin. The fractional part of the range migration amount is used to interpolate the digitized row samples. The integer part of the range migration amount is then used to shift the row data with respect to a reference sample. Further examples dealing with the squint SAR system are given later.

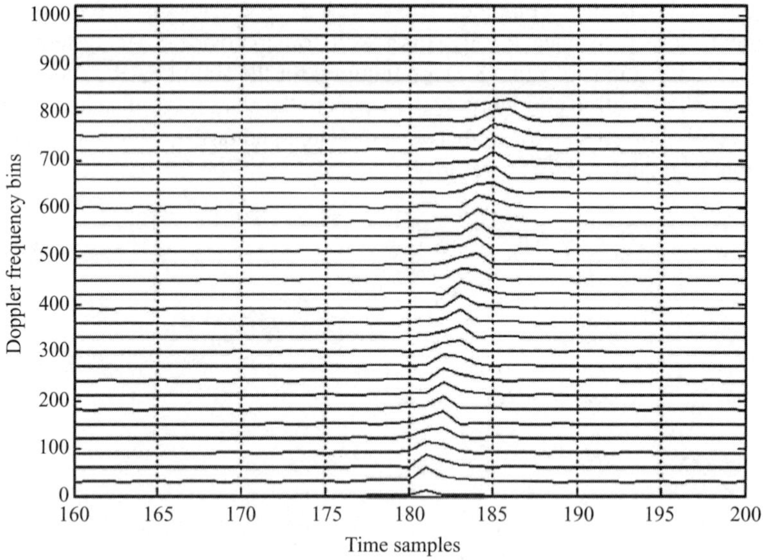

FIGURE 8.26 A range-compressed signal array in range–Doppler frequency domain.

8.4.3.2 Fractional Range Sample Interpolation. The fractional part of the range sample migration amount, which is computed from Eq. (8.8b) in this example, is used to interpolate the digitized row samples. The set of 16 sinc filters with 8-tap coefficients, as discussed in Section 1.5 of Chapter 1, serves as a good example of this interpolation process. Each computed fractional sample is first quantized with a 16-step quantizer. On the basis of this quantized amount, a corresponding sinc filter is then chosen. The entire row of the data is then convolved with this sinc filter, and the output of convolution is the interpolated signal with the peak positioned appropriately. This process continues for every row of the data array along the Doppler frequency axis. Notice that for multiple targets, multiple pulses will appear in the row-based data. However, in the range–Doppler frequency domain, they all share the same amount of range migration. Therefore, one sinc filter works for all the fractional sample interpolation processes.

Figure 8.27 displays the 2D data array after the fractional range sample interpolation process. By comparing the plots between Figs. 8.26 and 8.27, one can see that the peaks of the compressed and interpolated pulses in Fig. 8.27 appear to be aligned along time samples 181–185.

8.4.3.3 Range Sample Shift. The integer part of ΔN_{Rk}, shown in Eq. (8.8b), serves as the base for the range sample shift. In this example, range sample 181 is chosen as the reference sample R_3. The peaks previously aligned on different range samples are shifted left, based on integer part of ΔN_{Rk}, to align with sample 181. Figure 8.28 displays the compressed pulse array after the sample shift.

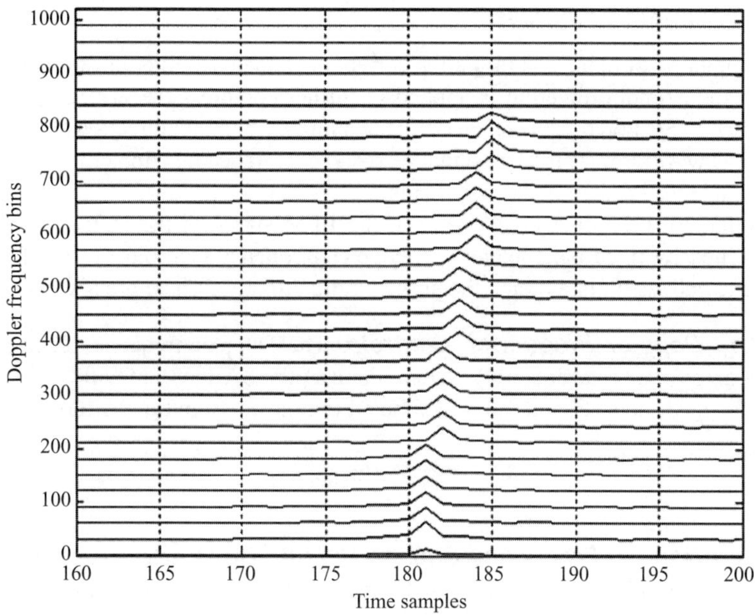

FIGURE 8.27 A range-compressed signal array after fractional interpolation.

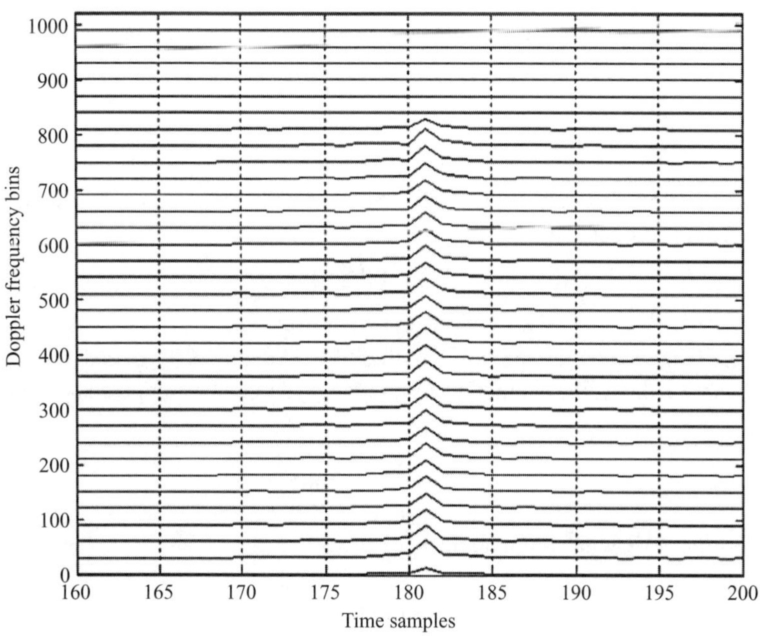

FIGURE 8.28 A range-compressed signal array after sample shift.

8.4.4 Azimuth Compression

On completion of range cell migration correction, a matched filter or reference function in the azimuth direction is applied to further compress the 2D data array. As discussed in Section 6.4, the received signal along the azimuth direction also appears as a LFM waveform. This LFM signal serves as a reference function or matched filter in the same way as in range compression.

Similar to the ordinary LFM signal, the LFM signal in the azimuth direction is specified by three parameters: the Doppler frequency centroid f_{Dc}, the Doppler frequency-changing rate β, and the pulse duration time T_a. These are discussed next.

8.4.4.1 Doppler Frequency Centroid. The Doppler frequency centroid f_{Dc} equals zero for a broadside SAR system. For a squint SAR system with given squint angle θ_q, f_{Dc} is nonzero and can be computed as follows, from Eq. (6.13a) or (6.16c):

$$f_{Dc} = \frac{2V}{\lambda}\sin\theta_q.$$

8.4.4.2 Doppler Frequency Change Rate β. The Doppler frequency changing rate β in terms of slow time s can be computed as, from Eq. (6.16a):

$$\beta = \dot{f}_D = -\frac{2V^2}{\lambda R_0}\cos^3\theta_u.$$

Since only small radar beamwidth and small squint angle are considered, the value of $\cos^3\theta_u \approx 1$. Therefore

$$\beta \approx -\frac{2V^2}{\lambda R_0}.$$

8.4.4.3 Pulse Duration Time T_a. The pulse duration time of azimuth LFM signal can be computed as

$$T_a = \frac{Ls}{V},$$

where Ls is the synthetic aperture length and V is the velocity of radar. Given the radar beamwidth θ_H, the synthetic aperture length Ls can be computed as

1. Broadside SAR:

$$Ls = 2R_0 \tan(0.5\theta_H)$$

2. Squint SAR with squint angle θ_q:

$$Ls = R_0 \tan(\theta_q + 0.5\theta_H) - R_0 \tan(\theta_q - 0.5\theta_H).$$

Since R_0 is dependent on target location, the synthetic aperture length Ls and consequently the pulse duration time T_a are range-dependent. Therefore, the azimuth matched filter is also range-dependent, and the azimuth compression must be performed with a different matched filter, which is very time-consuming. In practical applications, only a limited number of azimuth matched filters are used, with each filter shared by a small segment of range samples.

8.5 SIMULATION RESULTS

The radar image data synthesized in Section 8.2 will now be processed by the range–Doppler algorithm. The broadside SAR system with one and three targets is described first, followed by the squint SAR system with one and three targets.

8.5.1 Broadside SAR with Single Target

The parameters and data array described in Section 8.2.1 for a single-target broadside SAR will be adopted for target reconstruction. Since the baseband signal of an LFM radar is represented as

$$p_b(t) = \exp(j\pi\alpha t^2),$$

where $\alpha = 4 \times 10^{12}$ Hz/s and $T_p = 6 \times 10^{-6}$ s, the range matched filter is chosen as

$$\begin{aligned} h_r(t) &= p_b^*(-t) \\ &= \exp(-j\pi\alpha t^2). \end{aligned}$$

The azimuth matched filter can be designed as

$$h_a(s) = \exp(-j\pi\beta s^2).$$

The parameters Ls, T_a and β can be computed as follows:

$$\begin{aligned} Ls &= 2R_0 \tan(0.5\theta_H) \\ &= 225\,\text{m} \\ T_a &= \frac{Ls}{V} \\ &= 1.125\,\text{s} \\ \beta &= -\frac{2V^2}{\lambda R_0} \\ &= -355.56\,\text{Hz/s} \end{aligned}$$

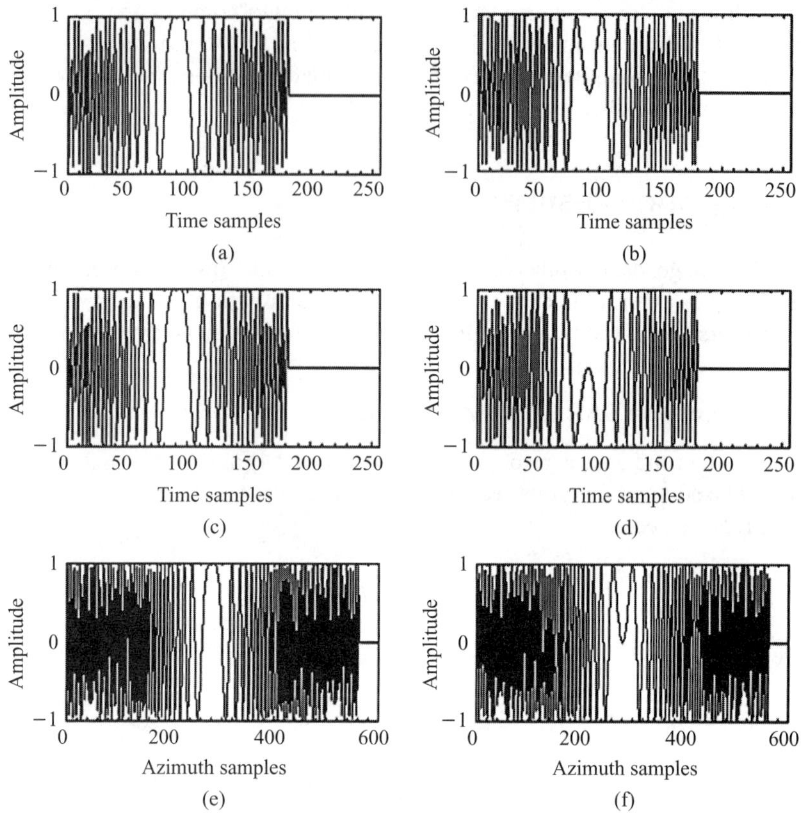

FIGURE 8.29 Waveforms of transmitter baseband signal, range reference function, and azimuth reference function.

Figure 8.29 displays the waveforms of baseband signal $p_b(t)$ together with the range reference function $h_r(t)$ and the azimuth reference function $h_a(s)$. The real part of the signal is shown in Figs. 8.29a, 8.29c, and 8.29e; the imaginary part, in Figs. 8.29b, 8.29d, and 8.29f. The $p_b(t)$, $h_r(t)$ and $h_a(s)$ are shown in Figs. 8.29a and 8.29b, 8.29c and 8.29d, and 8.29e and 8.29f, respectively. The sampling frequency is 30 MHz for range (or time) signal and 500 Hz for azimuth signal. Notice that no windowing function is used in generating the reference functions.

As can be seen, there are 181 samples in the range reference function, which corresponds to the sample numbers in the 6-μs pulse duration time. The real part of the range reference function $h_r(t)$ is identical to that of the transmitting baseband signal $p_b(t)$, while the imaginary part of $h_r(t)$ is negative to that of $p_b(t)$. The sample number of azimuth reference function is $Naz = Ls/A_s = 563$. The azimuth pulse duration time is $T_a = 1.125$ s.

Figure 8.30 shows the frequency spectrum of both the range and azimuth reference functions. The spectrum of the range reference function is based on a 256-point

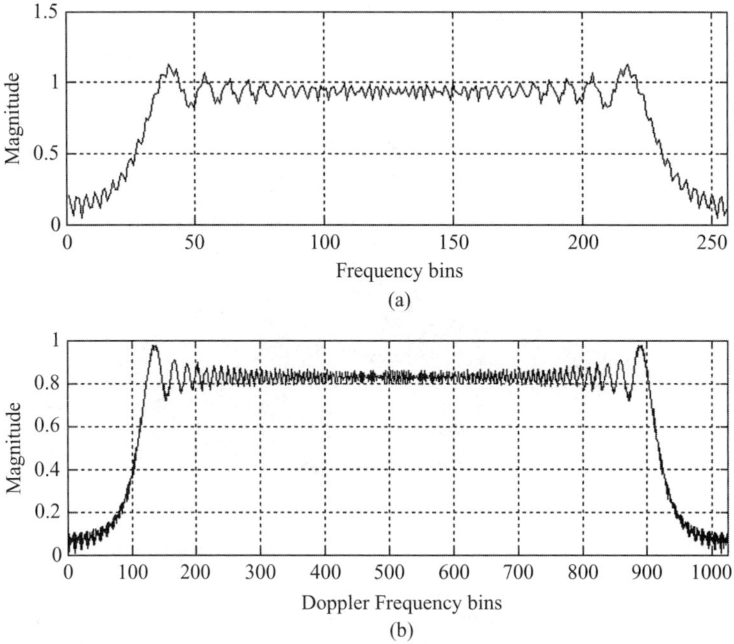

FIGURE 8.30 Frequency spectra of range and azimuth matched filters.

DFT with origin at bin 128 and sampling frequency of 30 MHz. The Doppler frequency spectrum of azimuth reference function is based on a 1024-point DFT with origin at bin 512 and sampling frequency of 500 Hz.

The synthesized single-target received data array shown in Fig. 8.9 will be used for range compression. The operation of range compression transforms every row of the time sample data into the frequency domain using a 256-point DFT. The frequency-domain data are then multiplied by the 256-point Fourier-transformed range reference function. The result is then inverse-Fourier-transformed back to the time-domain. The time-domain range-compressed data are shown in two displays, the 3D view in Fig. 8.31 and the 2D view in Fig. 8.32.

Figure 8.31 shows the magnitude of range-compressed data in the spatiotemporal domain. All compressed pulses are aligned along time sample 181 with equal magnitude and span through the whole synthetic aperture length Ls. The number 181 is the sample length of transmitter pulse.

Figure 8.32 is another view of the range-compressed data array. The horizontal axis (abscissa) shows the time samples; and the vertical axis (ordinate), the azimuth lines, together with magnitudes of the compressed pulses for each azimuth line. There are 563 rows along the vertical axis, and only 1 out of every 30 rows is displayed here. As can be seen, all compressed pulses align at time sample 181.

From Fig. 8.32, one can see that the range migration problem does not occur. Therefore, no range cell migration correction is required in this case.

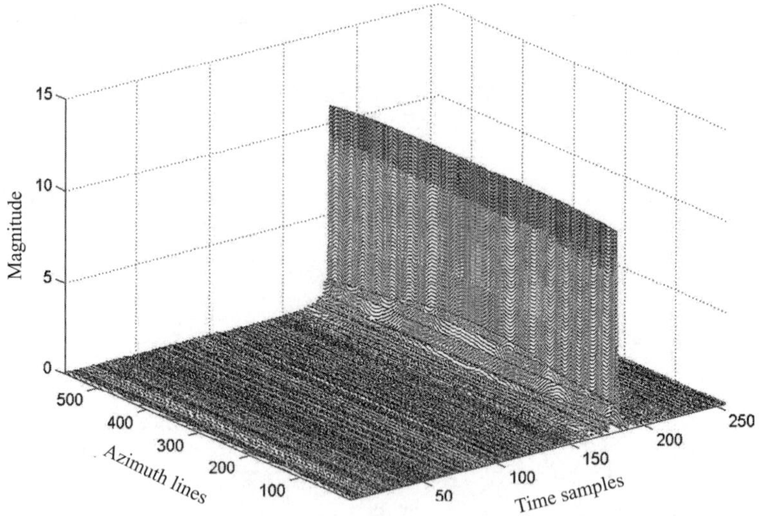

FIGURE 8.31 3D view of a range-compressed signal array based on Fig. 8.5.

The number of azimuth samples within the synthetic aperture length is $Naz = Ls/A_s = 563$. Transformation of the range-compressed signal array into Doppler frequency domain is performed by applying a 1024-point DFT on every column of the range-compressed signal array. Since $Naz = 563$, zero padding is required for the 1024-point DFT. The magnitude of the center-shifted Doppler frequency spectrum, with origin at bin 512, is shown as a 3D view in Fig. 8.33. The two perpendicular axes

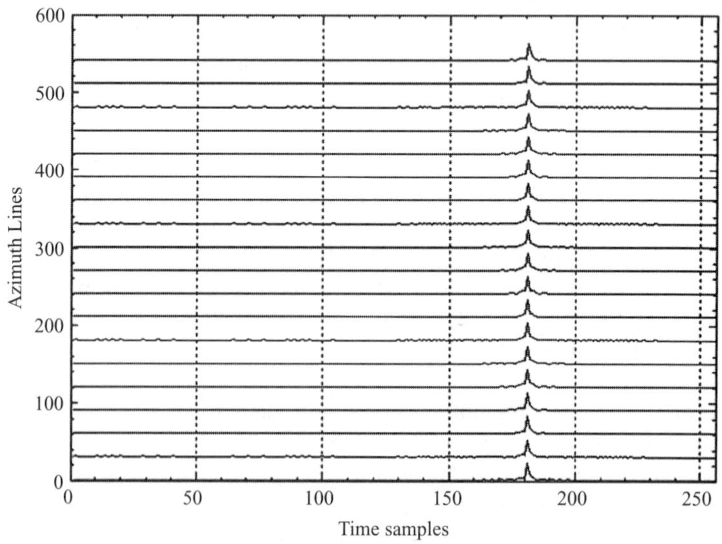

FIGURE 8.32 2D view of a range-compressed signal array based on Fig. 8.31.

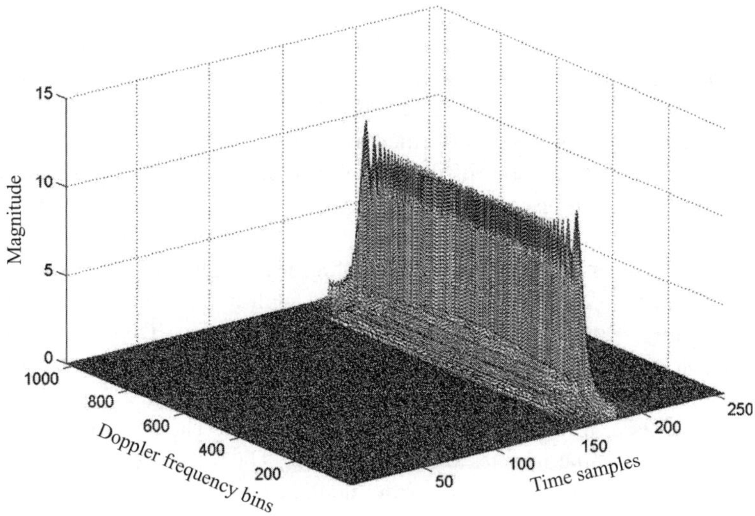

FIGURE 8.33 3D view of a range–Doppler frequency spectrum based on Fig. 8.31.

of the base plane are the time samples and the Doppler frequency bins, respectively. As can be seen, the Doppler frequency spectrum aligns along time sample 181, and the center of the Doppler frequency spectrum is at bin 512 with a bandwidth of ~800 frequency bins. Given the azimuth sampling frequency $f_{\mathrm{PRF}} = 500$ Hz, the Doppler frequency bandwidth can be computed as $500 \times 800/1024 \approx 400$ Hz.

Another way to view the display of the center-shifted Doppler frequency spectrum is shown in Fig. 8.34. Here the horizontal axis shows time samples; the vertical

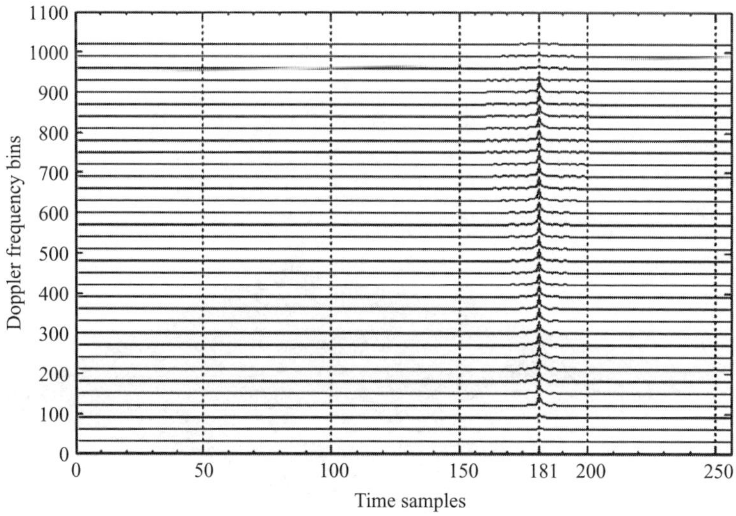

FIGURE 8.34 2D view of a range–Doppler frequency spectrum based on Fig. 8.33.

axis, Doppler frequency bins. There are 1024 rows along the vertical axis, and only 1 out of every 30 rows is displayed here. Similar to that of Fig. 8.32, the Doppler frequency spectrum aligns at time sample 181, and the center of the Doppler frequency spectrum is at bin 512 with ~400 Hz bandwidth.

The azimuth compression is performed by applying the range-dependent azimuth matched filter to every column of the Doppler frequency data array. Only one azimuth matched filter is used in this single-target example. The azimuth matched filter $h_{az}(s)$ is first transformed into the Doppler frequency domain, using a 1024-point DFT, to become $H_{az}(\omega_D)$. Without using any windowing functions, the azimuth matched filter $H_{az}(\omega_D)$ is then multiplied with every column of the data array in the range–Doppler frequency domain. A 1024-point IDFT is then applied on every column of the azimuth-compressed data array to obtain the reconstructed target.

Figure 8.35 is a 3D view of the reconstructed target. The two axes of the base plane correspond to time samples and azimuth lines. The reconstructed target appears as an impulse-like signal with very narrow sidelobes around the peak. The impulse-like target appears at time sample 181 and azimuth line 563. These two numbers correspond to the sample length of range pulse duration and azimuth pulse duration, respectively.

Figure 8.36 shows the cross sectional view of Fig. 8.35: Fig. 8.36a shows the display in terms of the azimuth lines with time sample 181; Fig. 8.36b, in terms of the time samples with azimuth line 563. The magnitude of the pulse is 181, which matches the sample length of the range pulse duration.

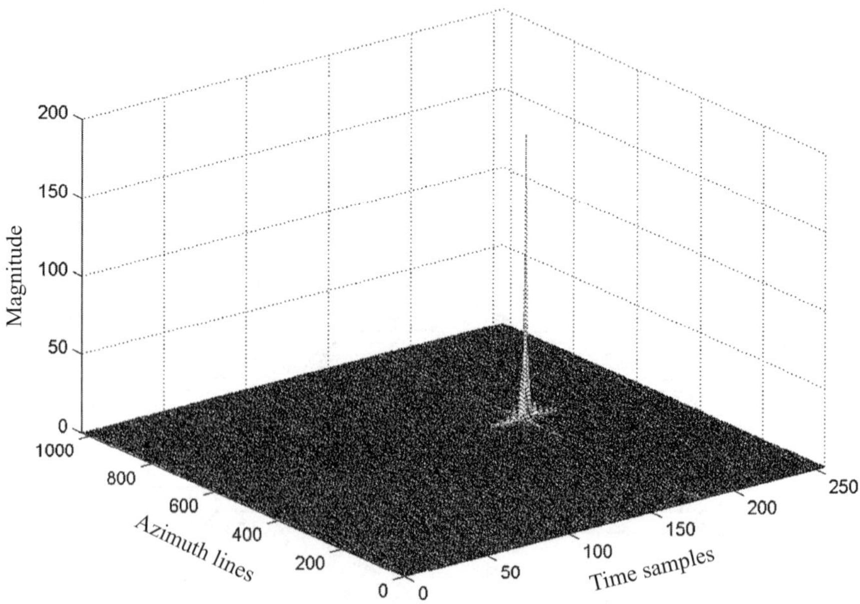

FIGURE 8.35 3D view of a reconstructed single-target function based on Fig. 8.33.

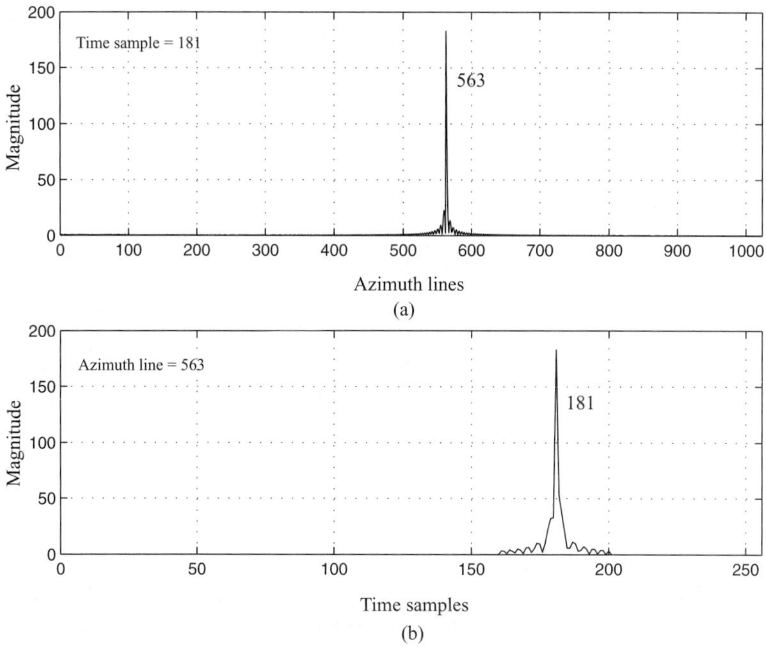

FIGURE 8.36 Cross-sectional view of a reconstructed single-target function based on Fig. 8.35.

8.5.2 Broadside SAR with Multiple Targets

The system model for broadside SAR with three targets shown in Fig. 8.3, together with the received data array shown in Fig. 8.13, will be used to reconstruct the targets.

Both the LFM transmitter signal and the matched filter for range compression will remain the same as in the single-target case:

$$\begin{aligned} h_r(t) &= p_b^*(-t) \\ &= \exp\left(-j\pi\alpha t^2\right). \end{aligned}$$

The azimuth matched filter $h_{az}(s)$ and the Doppler frequency change rate β are range-dependent. There are three targets located at two different ranges in this example: therefore, two sets of $h_{az}(s)$ and β must be designed:

$$\begin{aligned} h_{az1}(s) &= \exp\left(-j\pi\beta_1 s^2\right), \\ h_{az2}(s) &= \exp\left(-j\pi\beta_2 s^2\right), \\ \beta_1 &\approx -\frac{2V^2}{\lambda R_{0a}} \\ &= -355.56\,\text{Hz/s}, \\ \beta_2 &\approx -\frac{2V^2}{\lambda R_{0b}} \\ &= -348.58\,\text{Hz/s}. \end{aligned}$$

The azimuth pulse duration times T_{a1} and T_{a2}, corresponding to $h_{az1}(s)$ and $h_{az2}(s)$, respectively, are computed as follows:

$$T_{a1} = \frac{Lsa}{V} = \frac{2R_{0a}\tan(0.5\theta_H)}{V} = 1.125\ \text{s},$$

$$T_{a2} = \frac{Lsb}{V} = \frac{2R_{0b}\tan(0.5\theta_H)}{V} = 1.148\ \text{s}.$$

The synthesized three-target data array, shown in Fig. 8.13, is then range-compressed by first transforming every row of range data into the frequency domain using a 256-point DFT. The frequency-domain data are then multiplied by the 256-point Fourier transformed range reference function. The result is then inverse-Fourier-transformed back to the time domain.

Figure 8.37 is a 3D view of the range-compressed data array. There are two columns of a thin wall-like array. The first array has two magnitudes and lines up along time sample 181; the second array has one magnitude and aligns along time sample 211. The first array corresponds to two targets located at the same range

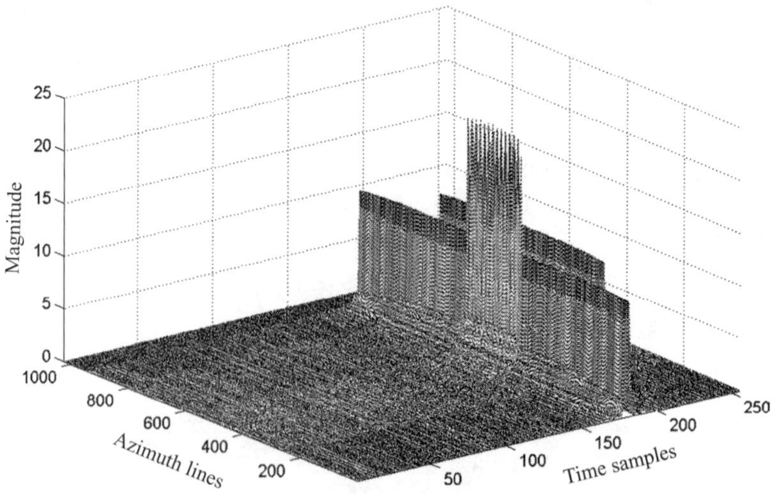

FIGURE 8.37 3D view of a range-compressed signal array based on Fig. 8.3.

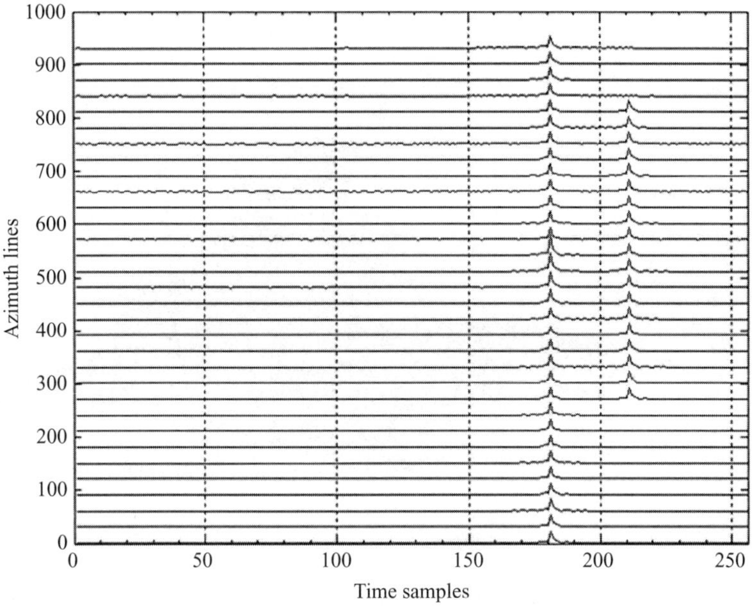

FIGURE 8.38 2D view of a range-compressed signal array based on Fig. 8.37.

$R_{0a} = 7500$ m. The larger magnitude portion of the first array occurs at the overlapped region between the two synthetic aperture lengths $Lsa1$ and $Lsa2$. The second array corresponds to the target located at range $R_{0b} = 7650$ m.

Figure 8.38 shows a different view of the same range-compressed data array. The horizontal axis represents time samples; the vertical axis, azimuth lines, together with magnitude of the compressed pulse for each azimuth line. There are $2Naza - D_1 = 939$ rows along the vertical axis, and only 1 out of every 30 rows is displayed here. As can be seen, there are two sets of column arrays: one extending through the whole length of azimuth samples (939) and aligning at time sample 181, the other extending with shorter azimuth samples (563) and lining up at time sample 211. Again, the range migration problem does not occur here, and range cell migration correction is not required in this case.

The next step in reconstructing the targets is transformation of the data array from the range–azimuth domain into the range–Doppler frequency domain. The total synthetic aperture length along the azimuth direction due to targets 1 and 3 is $2Naza - D_1 = 949$, the sample length of the azimuth matched filter is $Naza = 563$, and the output sample length of the azimuth matched filter is $(949 + 563 - 1) = 1511$. Therefore, a 2048-point DFT with zero padding on azimuth samples is adopted. By applying 2048-point DFT to every column of the range-compressed data array, the magnitude of the center-shifted Doppler frequency spectrum, with origin at bin 1024, is displayed in Fig. 8.39. The two perpendicular axes of the base plane are the time samples and the Doppler frequency bins, respectively. As expected, two columns of the Doppler frequency spectrum array exist; one lines

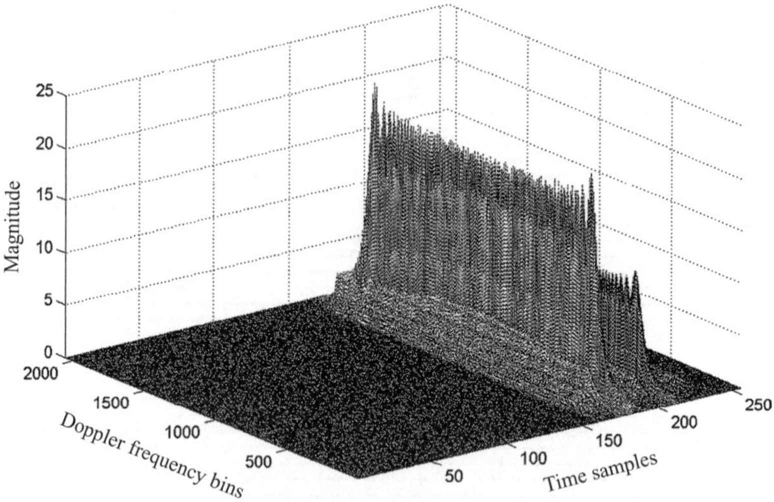

FIGURE 8.39 3D view of a range–Doppler frequency spectrum based on Fig. 8.37.

up along time sample 181, and the other aligns at time sample 211. The left one has larger magnitude than the right one, due to the existence of multiple targets. Both columns of the Doppler frequency spectrum array are centered on frequency bin 1024 with a bandwidth of $\sim$1600 frequency bins. Since the sampling frequency $f_s = f_{\text{PRF}} = 500$ Hz, the Doppler frequency bandwidth can be computed as $500 \cdot (1600/2048) \approx 391$ Hz.

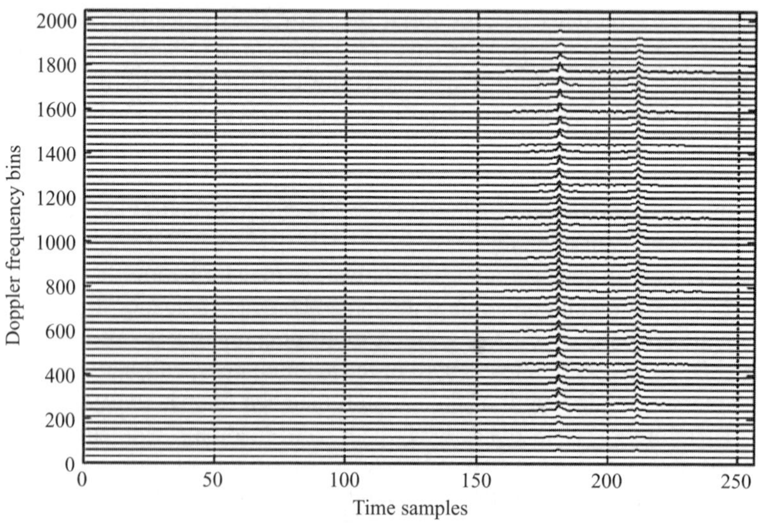

FIGURE 8.40 2D view of a range–Doppler frequency spectrum based on Fig. 8.39.

Figure 8.40 shows another view of the center-shifted Doppler frequency spectrum. Here the horizontal axis represents time samples; and the vertical axis, Doppler frequency bins. Similar to that of Fig. 8.38, both columns of the Doppler frequency spectrum array align at two time samples: one at 181 and the other at 211. Both arrays are centered at Doppler frequency bin 1024. The magnitude of the left column array seems to have a smaller value at some part of the frequency bins. This is due to the phase cancellation of two targets located at the same range of R_{0a}. Actually, both spectrum arrays have about the same bandwidth, but the left one has larger magnitude than the right one because of the existence of two targets.

The azimuth matched filter is then applied on every column of the data array in the Doppler frequency domain to perform the azimuth compression. Because of the existence of two targets, two azimuth matched filters, $h_{az1}(s)$ and $h_{az2}(s)$, are used. They are first transformed into the Doppler frequency domain, using a 2048-point DFT with zero padding. One is applied to the region around time sample 181, and then multiplied by every column of the data array in that region. The other is applied to the region around time sample 211, and then multiplied by every column of the data array in that region. A 2048-point IDFT is then applied on every column of the whole data array, and the reconstructed targets appear.

Figure 8.41 is a 3D view of the reconstructed targets. The x and y axes denote time samples and azimuth lines, respectively. The z axis represents magnitude. There are three impulse-like targets with about the same magnitude; two of them are located at (181, 563) and (181, 939), respectively, while the third one is located at (211, 818).

Figure 8.42 displays a cross-sectional view of Fig. 8.41 at two different time samples. Figure 8.42a shows the magnitude of the target distributed along the azimuth

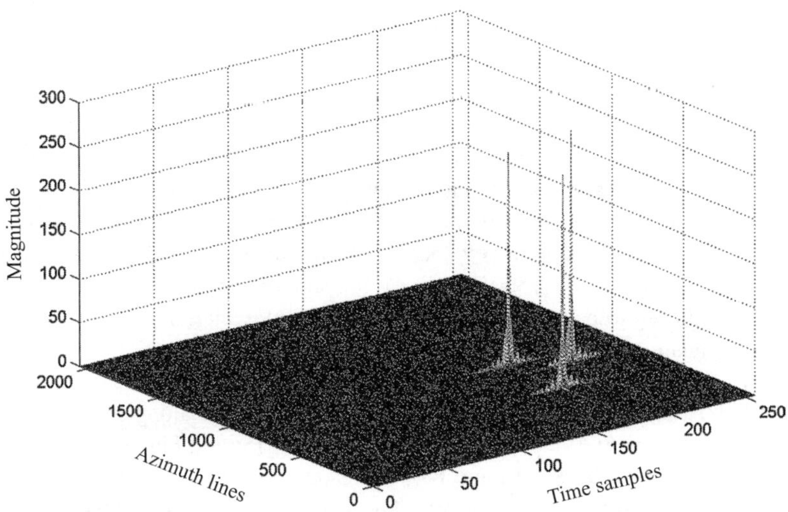

FIGURE 8.41 3D view of a reconstructed target function based on Fig. 8.39.

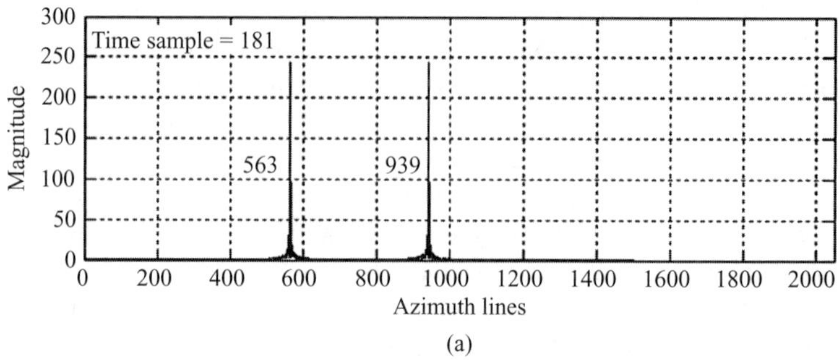

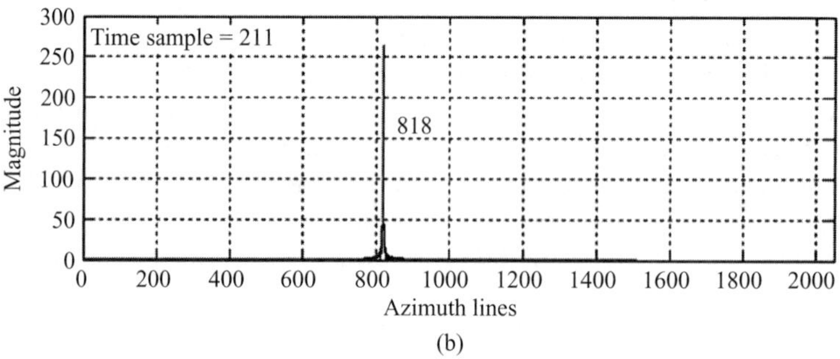

FIGURE 8.42 Cross-sectional view of Fig. 8.41 at range samples 181 and 211.

axis at time sample 181; Fig. 8.42b shows the magnitude of the target distributed along the azimuth axis at time sample 211. There are two pulses (or targets) appearing in Fig. 8.42a, one at azimuth line 563 and the other at azimuth line 939. Only one pulse (or target) appears in Fig. 8.42b, located at azimuth line 818. The two targets, appearing at azimuth lines 939 and 563, are located at range distance R_{0a}. They are separated by 376 samples in the azimuth direction and correspond to 376×0.4 m $=$ 150.4 m in distance. This number matches closely to the target location specification ($y_3 - y_1 = 150$ m). The third target is located at range distance R_{0b} and appears at azimuth line 818. This target is 255 samples away from the first target (818–563) in the azimuth direction, which corresponds to 255×0.4 m $= 102$ m. Since two azimuth matched filters were used at targets located at R_{0a} and R_{0b}; therefore, adjustment of the range ratio, $R_{0a}/R_{0b} = (7500/7650) = 0.98$, should be made. The number 102 m should be adjusted to become 102 m $\times$ 0.98 $=$ 100 m, which matches well with the target specification ($y_2 - y_1 = 100$ m).

Figure 8.43 displays a cross-sectional view of Fig. 8.41 at three azimuth lines. Figure 8.43a shows the magnitude of the target distributed along the time sample axis at azimuth line 563; Fig. 8.43b, at azimuth line 818; and Fig. 8.43c, at azimuth line 939. In Figs 8.43a, 8.43b, and 8.43c one pulse (or target) is located at range samples 181, 211, and 181, respectively. The sample length of the transmitting pulse is 181, which is where the targets in Figs. 8.43a and 8.43c are located. The target of Fig. 8.43b is located at time sample 211, which is 30 time samples away from targets

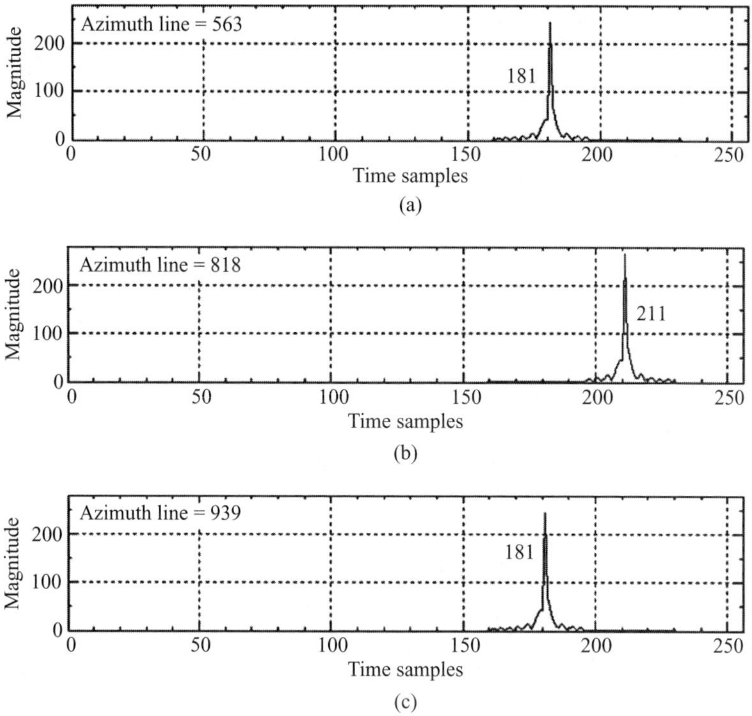

FIGURE 8.43 Cross-sectional view of Fig. 8.41 at azimuth lines 563, 818, and 939.

1 and 3, and corresponds to $10 \times (30/2) = 150$ m. It matches well with specification $[R_{0b} - R_{0a} = (7650\text{–}7500) = 150 \text{ m}]$.

8.5.3 Squint SAR with Single Target

The signal array of the single-target squint SAR described in Section 8.3.1 will be used to reconstruct the target. The baseband signal and the matched filter used in the single-target broadside SAR system will also be used in this case, and they are expressed as

$$p_b(t) = \exp(j\pi\alpha t^2),$$

$$h_r(t) = \exp(-j\pi\alpha t^2).$$

Here $\alpha = 4 \times 10^2$ Hz/s and $T_p = 6 \times 10^{-6}$ s.

The azimuth matched filter with squint angle θ_q can be designed as

$$h_{az}(s) = \exp(-j2\pi f_{Dc}s - j\pi\beta s^2),$$

where f_{Dc} is the centroid of Doppler frequency, and s refers to slow time. The parameters f_{Dc}, T_a, and β can be computed as follows:

$$\begin{aligned} f_{Dc} &= \frac{2V \sin\theta_q}{\lambda} \\ &= 1393.7\,\text{Hz}, \\ Ls &= R_0\left[\tan\left(\theta_q + 0.5\theta_H\right) - \tan\left(\theta_q - 0.5\theta_H\right)\right] \\ &= 227.5\,\text{m}, \\ T_a &= \frac{Ls}{V} \\ &= 1.1375\,\text{s}, \\ \beta &= -\frac{2V^2}{\lambda R_0} \\ &= -355.56\,\text{Hz/s}. \end{aligned}$$

The waveform of azimuth reference function $h_{az}(s)$ is shown in Fig. 8.44. The waveform was plotted with 5 times the sampling frequency f_{PRF} for a better view. The real part of the signal is shown in Fig. 8.44a; the imaginary part, in Fig. 8.44b. No windowing function is used in generating the reference function.

From the LFM waveforms described above, one can observe that the signal is no longer symmetric about the center point as it is in the broadside case. The

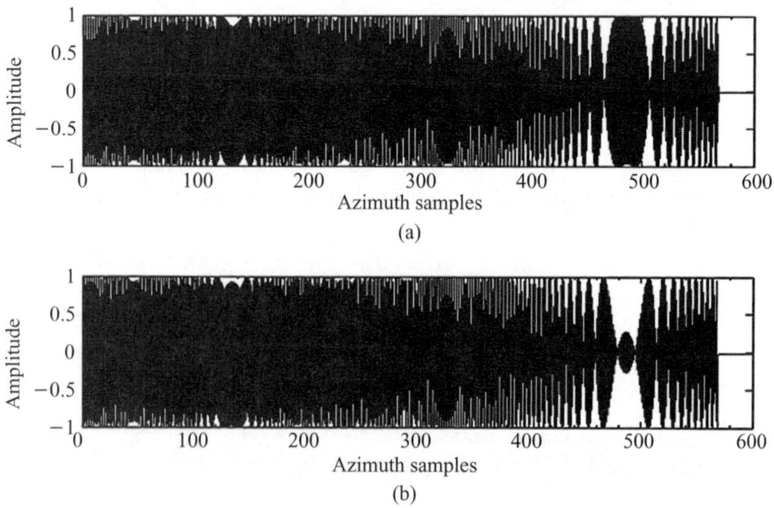

FIGURE 8.44 Waveforms of the real and imaginary parts of azimuth reference function.

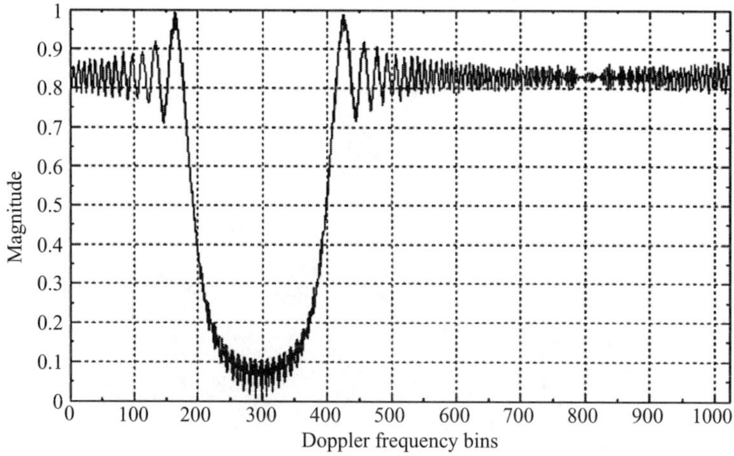

FIGURE 8.45 Frequency spectrum of azimuth reference function.

azimuth reference function has a total of $Naz = Ls/A_s = 569$ azimuth samples, and the azimuth pulse duration time is $T_a = 1.1375$ s.

The Doppler frequency spectrum of the azimuth reference function $h_{az}(s)$ is shown in Fig. 8.45. Since the azimuth signal is naturally digitized at pulse repetition frequency f_{PRF}, only the digitized spectrum is observable, and its replicas are duplicated at the rate of f_{PRF} (=500 Hz). The Doppler frequency spectrum shown in Fig. 8.45 is the observed spectrum. It is the fallback version of the true spectrum (located at passband) and is obtained by applying a 1024-point DFT on $h_{az}(s)$ with origin at bin 1. As described in Section 7.3.1.2, and shown as an example in Fig. 7.15, the true spectrum folds from passband to baseband and splits into two portions to become the observable spectrum. The spectrum shown in Fig. 8.45 consists of two parts; one extends from frequency bin 1 to about bin 190 and represents the upper portion of the frequency band, while the other extends from bin 410 to bin 1024 and represents the lower portion of the frequency band.

The synthesized data array with a single target shown in Fig. 8.17 will be used for range compression. Every row of the time sample data is first transformed into the frequency domain using a 256-point DFT. The frequency-domain data are then multiplied by the Fourier-transformed range reference function. The result is then inverse-Fourier-transformed back to the time domain. A 3D view of the time–spatial-domain range-compressed data array is shown in Fig. 8.46.

The range-compressed data shown in Fig. 8.46 differ significantly from those of the broadside SAR shown in Fig. 8.31. The compressed range data array no longer lines up at time sample 181, but migrates to several time samples around it. In addition, they are no longer equal in magnitude as those shown in Fig. 8.31. The differences are due primarily to the range migration in the squint SAR system.

Figure 8.47 shows a different view of the range-compressed data array. Here, the horizontal and vertical axes are again defined in the same way as in Fig. 8.32. However, the data array migrates to several time samples away from time sample 181

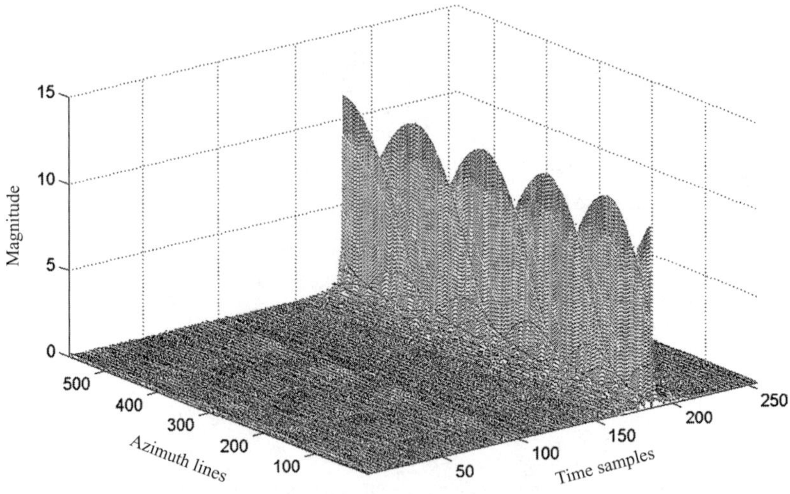

FIGURE 8.46 3D view of a range-compressed signal based on Fig. 8.14.

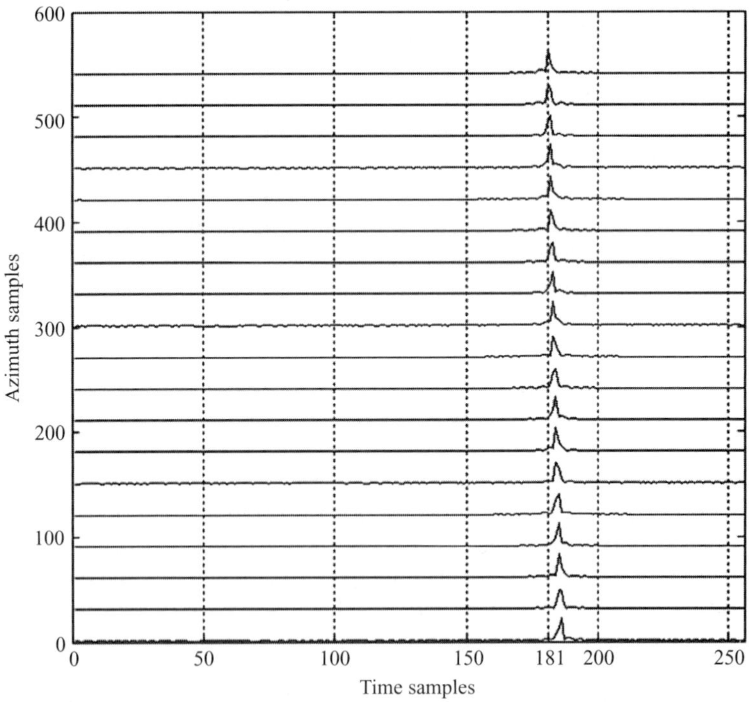

FIGURE 8.47 2D view of a range-compressed signal from Fig. 8.14.

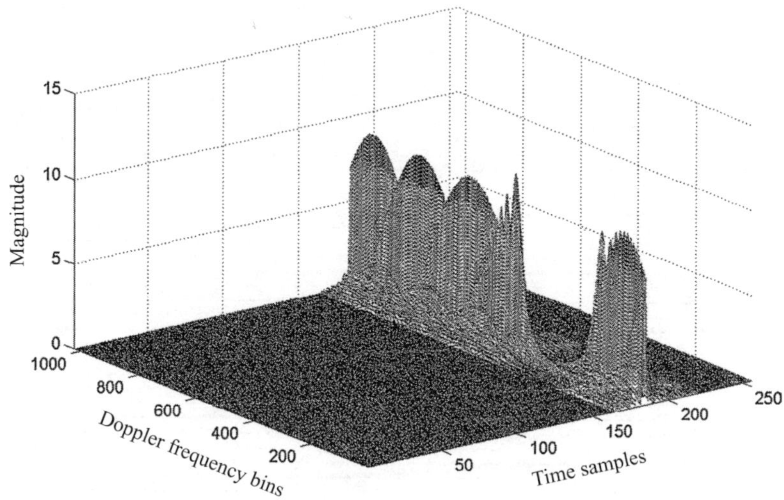

FIGURE 8.48 3D view of a spatial Fourier transformed signal from Fig. 8.46.

for the same reason of range migration. Range cell migration correction is therefore required in this squint SAR case.

To correct the range cell migration problem, the range-compressed data array in the range–azimuth domain is transformed into the range–Doppler frequency domain. Since the number of azimuth samples is 569, the 1024-point FFT is chosen and applied to every column of the range–azimuth data array.

Figure 8.48 is a 3D view of the range-compressed data array in the range–Doppler frequency domain with origin at bin 1. The x axis represents the time sample axis with sample numbers 1–256, the y axis represents the Doppler frequency axis with frequency bins 1–1024, and the z axis represents the magnitude of the data array. For squint SAR system, the data array has a Doppler frequency bandwidth of $B_D < f_{\mathrm{PRF}}$ and a center frequency of f_{Dc}. Similar to the previous discussion when azimuth reference filter was used for squint SAR, the Doppler frequency spectrum is folded into two parts in the baseband display. The upper frequency part, starting from bin ~400 to bin 1024, is the lower portion of the true frequency spectrum, while the lower part, starting from bin 1 to bin ~200, is in fact the upper portion of the true frequency spectrum. The center frequency f_{Dc} is located at bin ~800.

A different view of the range-compressed signal in the range–Doppler frequency domain is shown in Fig. 8.49, where 1 out of 30 frequency bins are displayed. The range-compressed data array along the Doppler frequency direction is no longer aligned at time sample 181; instead, it migrates from time samples 181 to 185. A *range cell migration correction* is therefore required.

For the squint SAR system, the range migration correction with respect to slant range R_3 in the time sample domain equivalently involves aligning the frequency bins with respect to f_{DL} in the Doppler frequency domain, where the maximum amount of range migration occurs at f_{DU}. Since the squint angle θ_q is greater than $0.5\theta_H$,

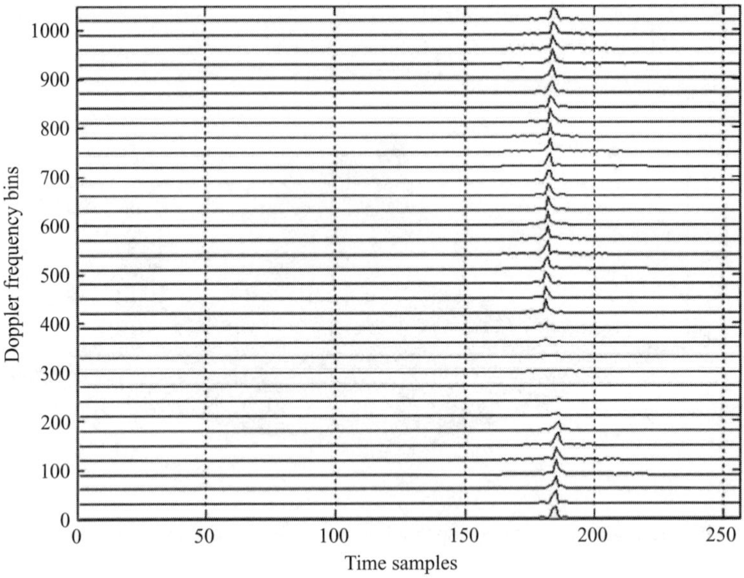

FIGURE 8.49 2D view of a spatial Fourier-transformed signal from Fig. 8.46.

Eq. (6.16d) can be applied to compute the true or unfolded f_{DU} and f_{DL}:

$$\begin{aligned} f_{DU} &= \frac{2V \sin\left(\theta_q + 0.5\theta_H\right)}{\lambda} \\ &= 1593\,\text{Hz}, \\ f_{DL} &= \frac{2V \sin\left(\theta_q - 0.5\theta_H\right)}{\lambda} \\ &= 1195\,\text{Hz}. \end{aligned}$$

Given a 1024-point DFT and sampling frequency $f_{\text{PRF}} = 500$ Hz, the Doppler frequency bin spacing can be computed as $\Delta f_D = (500/1024) = 0.488$ Hz. Therefore the folded f_{DL} corresponds to bin number $1 + (1195 - 2 \times f_{\text{PRF}})/0.488 = 401$, while the folded f_{DU} corresponds to bin number $1 + (1593 - 3 \times f_{\text{PRF}})/0.488 = 191$. The bandwidth of Doppler frequency spectrum then has a total number of frequency bins $= (1024 - 401 + 191) = 814$.

The maximum amount of range cell migration can be computed from Eq. (8.8b) as

$$\begin{aligned} \Delta N_k &\approx \frac{R_0\lambda^2}{4V^2R_s}\left(f_{DL} + \frac{kf_{\text{PRF}}}{N}\right)^2 - \frac{R_0\lambda^2 f_{DL}^2}{4V^2R_s} \\ &= 4.67. \end{aligned}$$

where $k = 814, f_{DL} = 1195$, and $R_s = 10$ were used.

Alternatively, one can also compute the maximum amount of range migration by computing the difference between slant ranges R_1 and R_3, which corresponds to f_{DU} and f_{DL}, respectively; that is, from Eq. (6.11b)

$$\begin{aligned} R_1 &= \frac{R_0}{\cos(0.5\,\theta_H + \theta_q)} \\ &= 7554.07 \\ R_3 &= \frac{R_0}{\cos(0.5\,\theta_H - \theta_q)} \\ &= 7530.28. \end{aligned}$$

The maximum number of range cells that need to be corrected becomes

$$\begin{aligned} \Delta N_{\max} &= \frac{2(R_1 - R_3)}{R_s} \\ &= 4.76. \end{aligned}$$

This matches closely with the result based on Eq. (8.8b).

The fractional range sample interpolation and range sample shift are then applied on the signal array of Fig. 8.48. An 8-tap interpolation sinc filter with $\frac{1}{16}$ sample resolution as described in Section 1.6 (of Chapter 1) is applied here for fractional sample interpolation. A 3D view of the range cell migration corrected Doppler frequency spectrum is shown in Fig. 8.50.

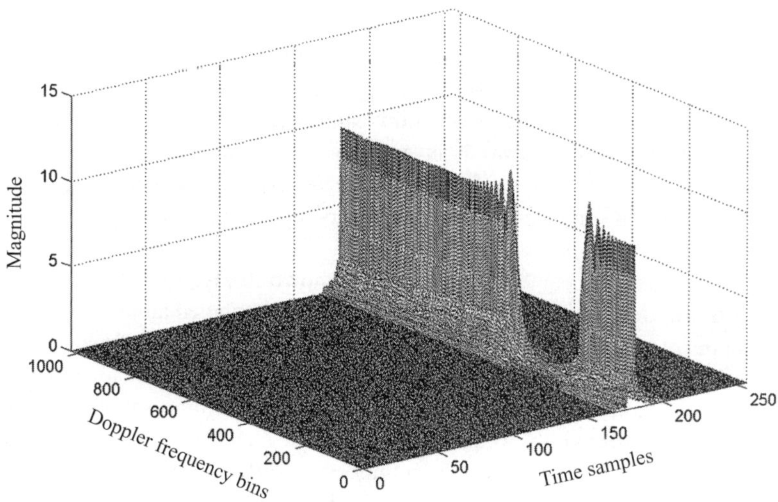

FIGURE 8.50 3D view of Fig. 8.46 after range cell migration correction.

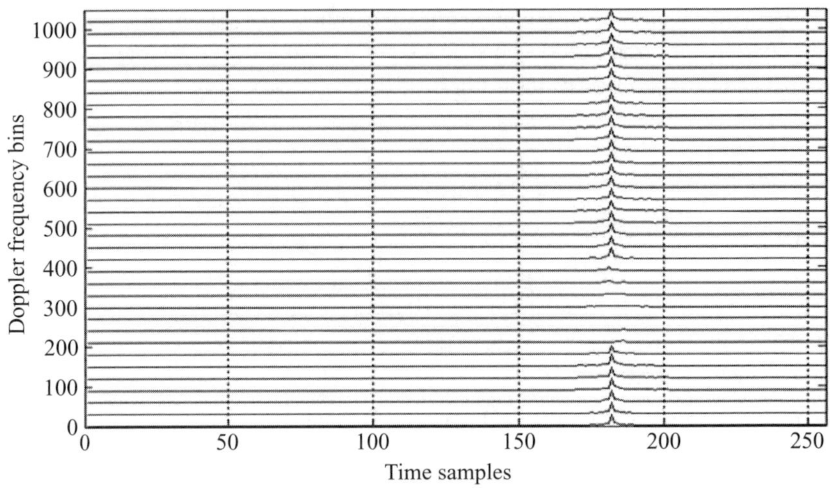

FIGURE 8.51 2D view of Fig. 8.46 after range cell migration correction.

As can be seen, the magnitudes of the spectrum are about equal, and the spectrum consists of two parts: one extending from frequency bins 1 to ~190 and the other, from frequency bins 390 to ~1024. The range cell migration corrected Doppler frequency spectrum now lines up at time sample 181.

A different view of the range migration corrected signal spectrum is shown in Fig. 8.51. Here the horizontal axis represents time samples, and the vertical axis represents Doppler frequency bins with 1 out of 30 bins displayed. Compared to Fig. 8.49, the new plot shows that all Doppler frequency bins are aligned along time sample 181.

The range cell migration corrected data array in the range–Doppler frequency domain is then processed for azimuth compression. Different range-dependent azimuth matched filters $h_{az}(s)$ should be applied to every column of the signal array. However, only one filter is used in this simulation. Once the azimuth compression is completed, the data array is inverse-Fourier-transformed back to the range–azimuth domain. A 3D view of the final range–azimuth data array is shown in Fig. 8.52, where an impulse-like signal appears at time sample ~181 and azimuth sample ~571.

Two cross-sectional views of Fig. 8.52 are shown in Fig. 8.53. Figure 8.53a is a cross-sectional view at time sample 181. The signal distributed along the azimuth direction appears as an impulse at azimuth sample 571, with a very narrow pulsewidth. Figure 8.53b is a cross-sectional view at azimuth sample 571. The signal distributed along the time sample direction appears as an impulse at time sample 181, with a similarly narrow pulsewidth. Again, the number 181 matches the time sample length of the range matched filter, while the number 571 differs slightly from the sample length used in the azimuth matched filter, which has 569 samples. The reconstructed signal appears quite accurately as expected.

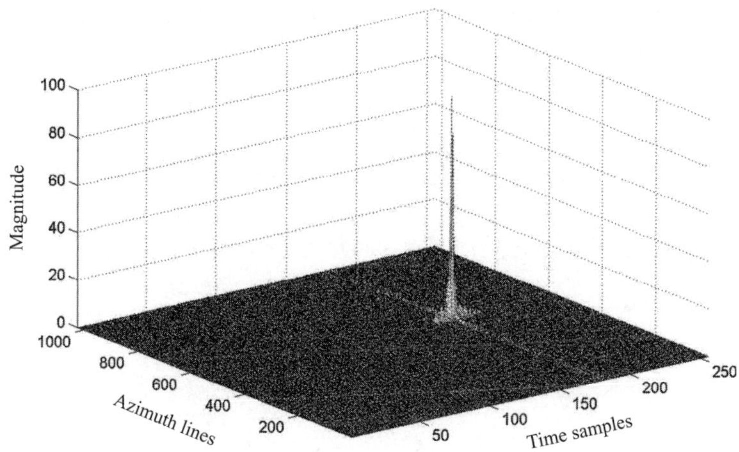

FIGURE 8.52 3D view of a reconstructed target function from Fig. 8.14.

8.5.4 Squint SAR with Multiple Targets

The system model shown in Fig. 8.18 for the multiple-targets squint SAR system, together with the synthesized data array shown in Fig. 8.22, will be used to reconstruct the targets.

Again, the range matched filter remains the same as before:

$$h_r(t) = \exp(-j\pi\alpha t^2).$$

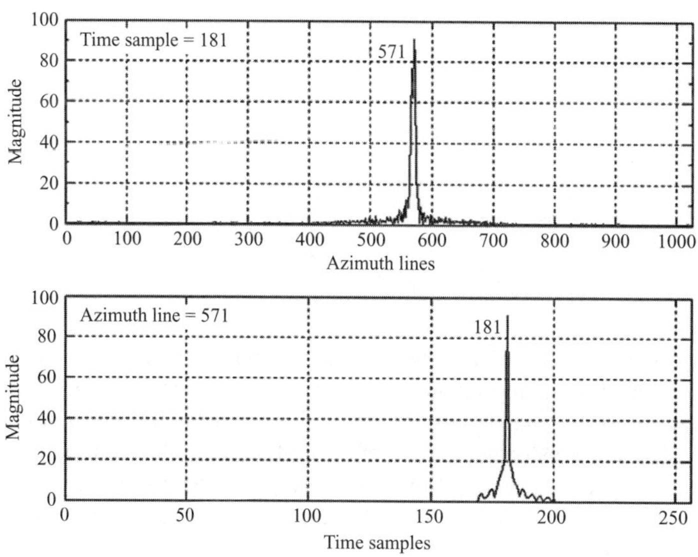

FIGURE 8.53 Cross-sectional view of Fig. 8.52 at range sample 181 and azimuth line 571, respectively.

Similar to the multiple targets of the broadside SAR system, two azimuth matched filters will be used in this example:

$$h_{az1}(s) = \exp\left(-j2\pi f_{Dc}s - j\pi\beta_1 s^2\right)$$

$$h_{az2}(s) = \exp\left(-j2\pi f_{Dc}s - j\pi\beta_2 s^2\right).$$

The parameters f_{Dc}, T_{a1}, T_{a2} β_1, and β_2 can be computed as follows:

$$\begin{aligned} f_{Dc} &= \frac{2V\ \sin\theta_q}{\lambda} \\ &= 1393.7\,\text{Hz}, \\ T_{a1} &= \frac{Lsa}{V} \\ &= \frac{R_{0a}\left[\tan(\theta_q + 0.5\theta_H) - \tan(\theta_q - 0.5\theta_H)\right]}{V} \\ &= 1.1375\,\text{s}, \\ T_{a2} &= \frac{Lsb}{V} \\ &= \frac{R_{0b}\left[\tan(\theta_q + 0.5\theta_H) - \tan(\theta_q - 0.5\theta_H)\right]}{V} \\ &= 1.16\,\text{s}, \\ \beta_1 &\approx -\frac{2V^2}{\lambda R_{0a}} \\ &= -355.56\,\text{Hz/s}, \\ \beta_2 &\approx -\frac{2V^2}{\lambda R_{0b}} \\ &= -348.58\,\text{Hz/s}. \end{aligned}$$

The synthesized data array shown in Fig. 8.22 serves as raw data for range compression. Every row of the range data is first transformed into the frequency domain using a 256-point DFT. The frequency-domain data are then multiplied by the 256-point Fourier-transformed range reference function. The result is then inverse-Fourier-transformed back to the time domain. Figure 8.54 displays a 3D view of the range-compressed data in the time–spatial (range–azimuth) domain. There are two column-like data arrays along the azimuth direction. The left array corresponds to two targets located at range R_{0a}, while the right array corresponds to one target located at range R_{0b}. The magnitudes of the two data arrays appear to be unequal and

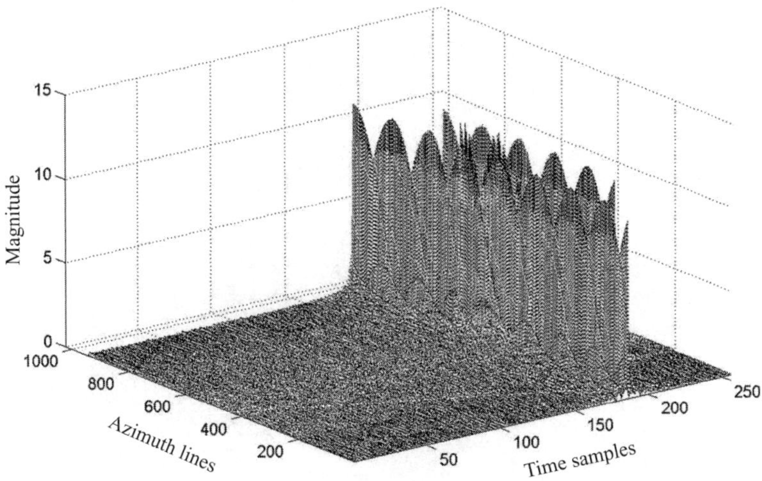

FIGURE 8.54 3D view of a range-compressed signal from Fig. 8.18.

show several peaks. This is due to the range migration that occurs in the squint SAR system, and range migration correction is required.

A different view of the range-compressed data array is shown in Fig. 8.55, where 1 out of 30 azimuth lines are displayed. There are two columns of data arrays in the plot: one located at time sample ~181 and the other, at time sample ~211. The first column of the data array corresponds to two targets located at

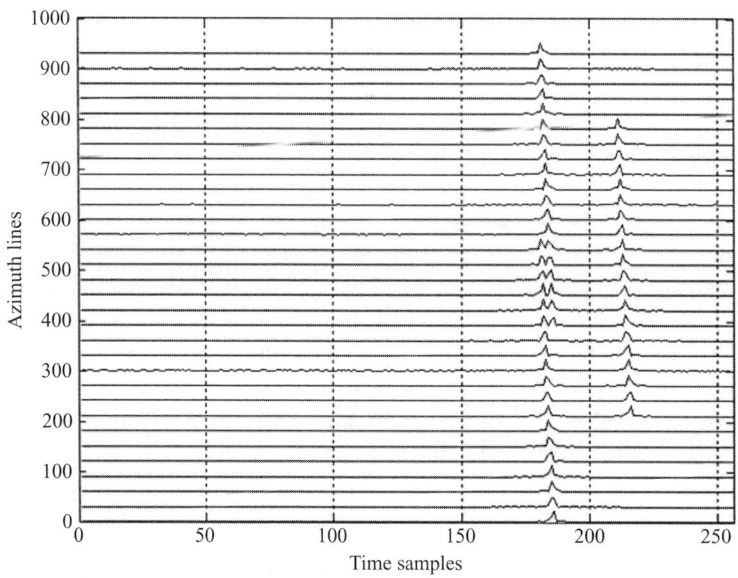

FIGURE 8.55 2D view of a range-compressed signal from Fig. 8.18.

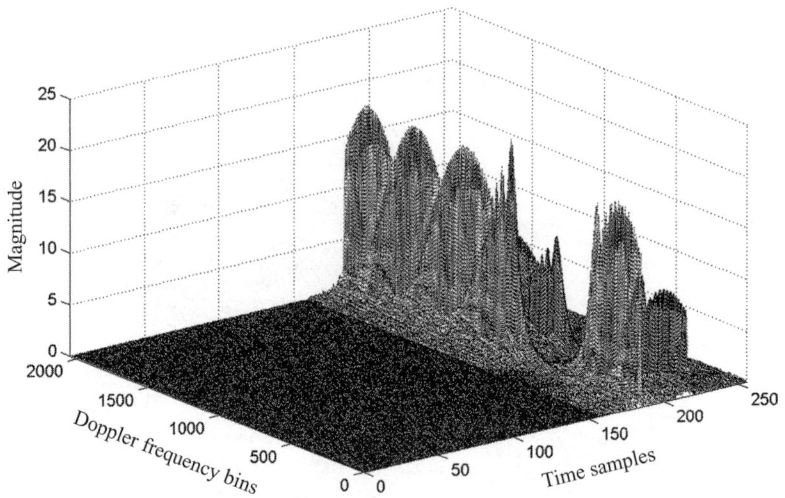

FIGURE 8.56 3D view of spatial Fourier-transformed signal from Fig. 8.54.

range R_{0a}. The dual-magnitude portion of the array is the beamwidth-overlapped region from the two targets. The second column of the data array corresponds to the target located at range R_{0b}. The first column of the data array extends in the azimuth direction for a total of $2Naza - D_1 = 945$ azimuth samples;. the second one extends in the azimuth direction for a total of $Nazb = 581$ azimuth samples. Range migration is clearly visible from either the first or second column of the data array.

To correct the range migration problem, the data array in the range–Azimuth (time–spatial) domain is first transformed into the range–Doppler frequency domain. Since the azimuth sample length of the first column data array is $2Naza - D_1 = 945$ and the sample length of the azimuth matched filter is 569, a 2048-point DFT is chosen to transform every column of the range-compressed data array into the Doppler frequency domain. A 3D view of the range–Doppler frequency spectrum is displayed in Fig. 8.56 with Doppler frequency origin at bin 1.

There are two arrays of the Doppler frequency spectrum: one along time sample ~181 and the other, along time sample ~211. Both spectrum arrays are folded into two parts as a result of the natural digitization at the pulse repetition frequency f_{PRF}, and their magnitudes are unequal because of the range migration problem.

A different view of the Doppler frequency spectrum described above is shown in Fig. 8.57. As can be seen, the two spectrum arrays have the same size along the Doppler frequency direction. Because of phase cancellation from the two targets located at the same range distance R_{0a}, some portions of the spectrum array in the left column appear to have smaller magnitude than do those in the rest of the spectrum array. The magnitude of the left column array is higher than that of the right one because there are two targets.

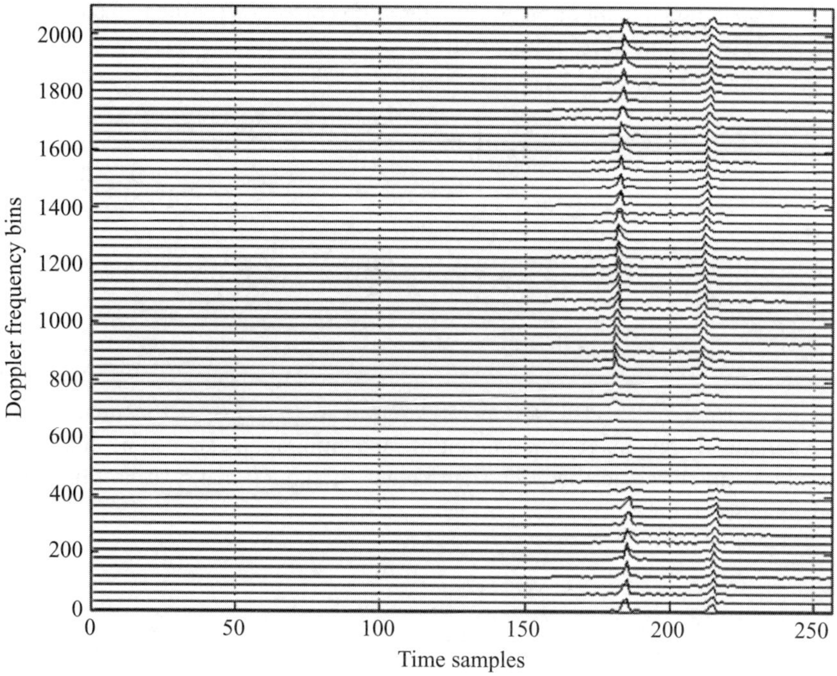

FIGURE 8.57 2D view of a spatial Fourier-transformed signal from Fig. 8.54.

Range cell migration correction is then performed on the Doppler frequency spectrum arrays described above. For a squint SAR system, the amount of range migration will be corrected with respect to the slant range R_3. The maximum amount of range migration occurs at f_{DU}. Both f_{DU} and f_{DL} can be computed as follows, from Eq. (6.16d):

$$f_{DU} = \frac{2V \sin\left(\theta_q + 0.5\theta_H\right)}{\lambda}$$

$$= 1593\,\text{Hz},$$

$$f_{DL} = \frac{2V \sin\left(\theta_q - 0.5\theta_H\right)}{\lambda}$$

$$= 1195\,\text{Hz}.$$

The frequency f_{DU} corresponds to bin number $1 + (1593 - 3 \times f_{\text{PRF}}) \times (2048/500) = 382$. The frequency f_{DL} corresponds to bin number $1 + (1195 - 2 \times f_{\text{PRF}}) \times (2048/500) = 802$. The Doppler frequency spectrum then has a total number of frequency bins equal to $(2048 - 802 + 382) = 1628$, and the bandwidth is $(1628 \times 500)/2048 = 397.5$ Hz.

The maximum amount of range cell migration correction for the two column data arrays can be computed from Eq. (8.8b) as

$$\Delta N_{k1} \approx \frac{R_{0a}\lambda^2}{4V^2R_s}\left(f_{DL} + \frac{kf_{\text{PRF}}}{N}\right)^2 - \frac{R_{0a}\lambda^2 f_{DL}^2}{4V^2R_s}$$
$$= 4.67,$$

$$\Delta N_{k2} \approx \frac{R_{0b}\lambda^2}{4V^2R_s}\left(f_{DL} + \frac{kf_{\text{PRF}}}{N}\right)^2 - \frac{R_{0b}\lambda^2 f_{DL}^2}{4V^2R_s}$$
$$= 4.76,$$

where $k = 1628, f_{DL} = 1195, R_{0a} = 7500, R_{0b} = 7650$, and $R_s = 10$ were used.

Alternatively, the maximum amount of range migration can also be obtained by computing the difference between slant ranges R_1 and R_3; that is, from Eq. (6.11b), one obtains

$$R_{1a} = \frac{R_{0a}}{\cos(0.5\theta_H + \theta_q)}$$
$$= 7554.07,$$
$$R_{3a} = \frac{R_{0a}}{\cos(0.5\theta_H - \theta_q)}$$
$$= 7530.28,$$
$$R_{1b} = \frac{R_{0b}}{\cos(0.5\theta_H + \theta_q)}$$
$$= 7705.15.$$
$$R_{3b} = \frac{R_{0b}}{\cos(0.5\theta_H - \theta_q)}$$
$$= 7680.89.$$

The maximum number of range cells that need to be corrected becomes

$$\Delta N_{a\max} = \frac{2(R_{1a} - R_{3a})}{R_s}$$
$$= 4.76,$$

$$\Delta N_{b\max} = \frac{2(R_{1b} - R_{3b})}{R_s}$$
$$= 4.85.$$

The results closely match those based on Eq. (8.8b).

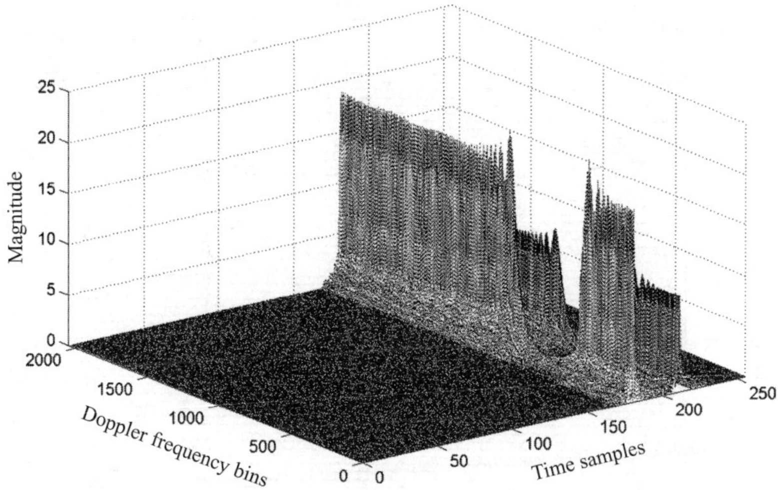

FIGURE 8.58 3D view of Fig. 8.56 after range cell migration correction.

Both fractional range cell interpolation and range sample shift are then applied on the signal array of Fig. 8.56. For simplicity, only ΔN_{k1} was used in simulation and the result is shown in Fig. 8.58.

As can be seen, the magnitudes of the range migration corrected spectrum array become equal for both columns of the spectrum array. The magnitude of the left column array doubles that of the right one because there are two targets.

A different view of the range migration corrected spectrum is shown in Fig. 8.59. The horizontal axis shows the time samples, and the vertical axis represents the Doppler frequency bins with 1 of 30 bins shown. The new plot shows that the two spectrum arrays are aligned at two time samples, one at sample 181 and the other at sample 211. The bandwidth of both arrays is about equal, yet the magnitude of the left column is greater than that of the right column because there are two targets. Again, some portions of the left column appear smaller than the rest of the same array because of phase cancellation from the two targets at the same range distance R_{0a}.

Azimuth compression is then performed on the range migration corrected spectrum array in the range–Doppler frequency domain. The range-dependent azimuth matched filter $h_{az1}(s)$ is applied to columns of time samples 1–190 of the spectrum array, and $h_{az2}(s)$ is applied to columns (time sample) 191–256 of the spectrum array. After the azimuth compression, the Doppler frequency spectrum array is inverse-Fourier-transformed back to the range–azimuth (or time–spatial) domain. A 3D view of the final data array in the spatiotemporal domain is shown in Fig. 8.60 with two impulse-like signals appearing at (181,571) and (181,947), respectively, and the third one at (211,786). All three targets appear to have the same magnitude.

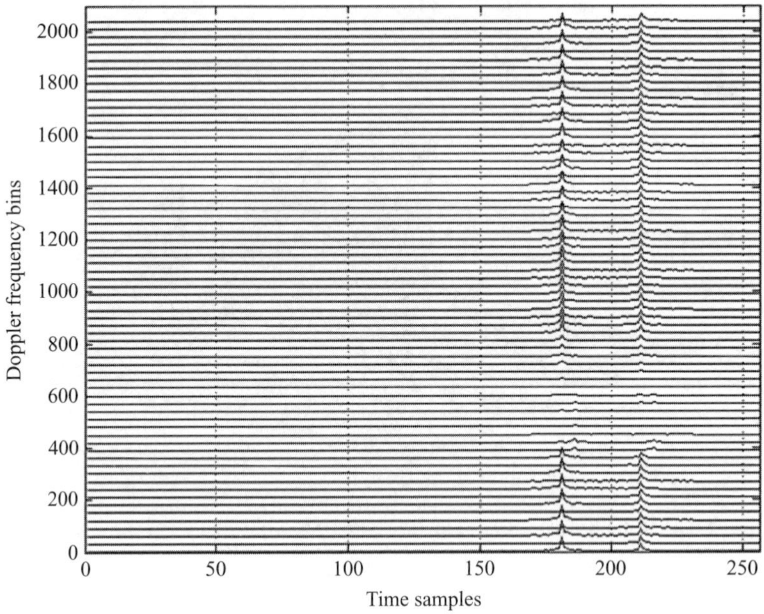

FIGURE 8.59 2D view of Fig. 8.56 after range cell migration correction.

Figure 8.61 displays a cross-sectional view of Fig. 8.60 at two time samples: Fig. 8.61a shows the magnitude of the targets distributed along the azimuth axis at time sample 181; Fig. 8.61b, at time sample 211. Two pulses (or targets) appear in Fig. 8.61a: one at azimuth sample 571 and the other at azimuth sample 947. Only one pulse (or target) appears in Fig. 8.61b, located at azimuth sample 786. The

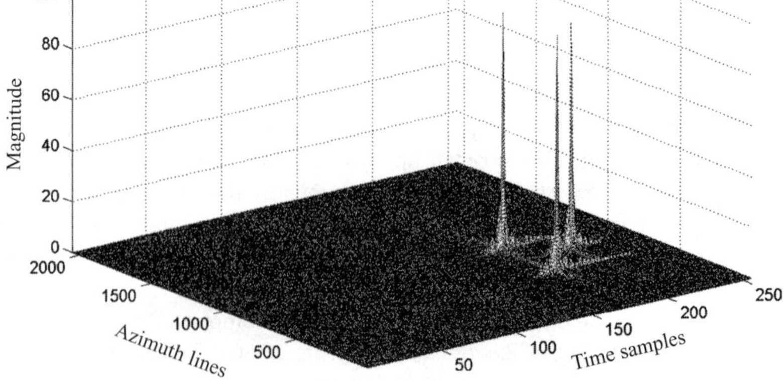

FIGURE 8.60 3D view of a reconstructed target function from Fig. 8.18.

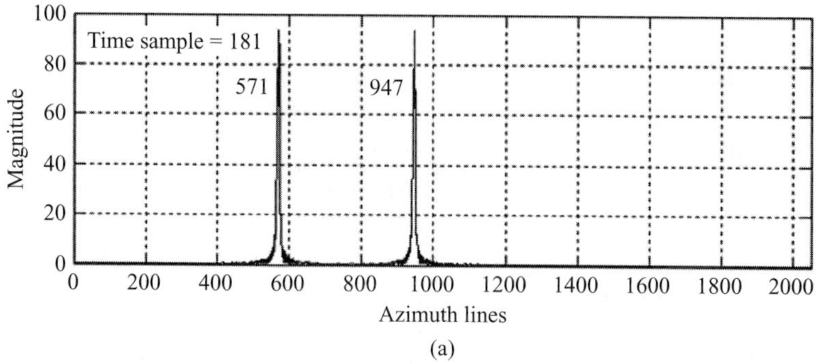

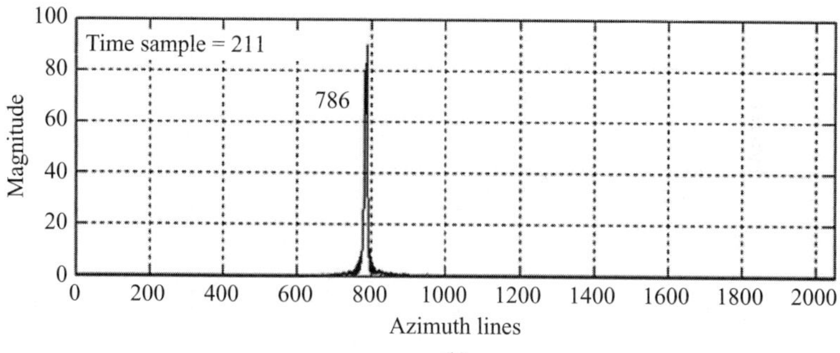

FIGURE 8.61 Cross-sectional view of Fig. 8.60 at range samples 181 and 211.

difference between samples 947 and 571 along the azimuth direction is 376, which corresponds to 376 × 0.4 m = 150.4 m. This number closely matches the specification ($y_3 - y_1 = 150$ m). The difference between samples 786 and 571 along the azimuth direction is 215, which corresponds to 215 × 0.4 m = 86 m. This number is 14 m shorter than the specification, $y_2 - y_1 = 100$ m. This discrepancy is due to two factors: (1) the two targets are located at different range distances, with one located at range R_{0a} and the other at range R_{0b} (range differences require the use of different azimuth matched filters and adjustment of the range ratio); and (2) the nonzero squint angle is $\theta_q = 6°$, which shortens the azimuth distance by an amount $\delta y = \Delta x \tan \theta_q$.

Therefore, adjustment of the range ratio $R_{0a}/R_{0b} = (7500/7650) = 0.98$ should be applied to the number 86 to become 86 × 0.98 m = 84.28 m. In addition, with $\Delta x = 150$ m in the range direction and $\theta_q = 6°$, the azimuth distance Δy is reduced by an amount $\delta y = \Delta x \tan \theta_q = 15.76$ m. Therefore, the theoretical result should be $(100 - 15.76) = 84.24$ m. The simulation result, 84.28 m, matches the specification well in this case.

Figure 8.62 shows a cross-sectional view of Fig. 8.60 at three azimuth lines: Fig. 8.62a shows the magnitude of the target distributed along the time sample axis at azimuth line 571; Fig. 8.62b, at azimuth line 786; and Fig. 8.62c, at azimuth line

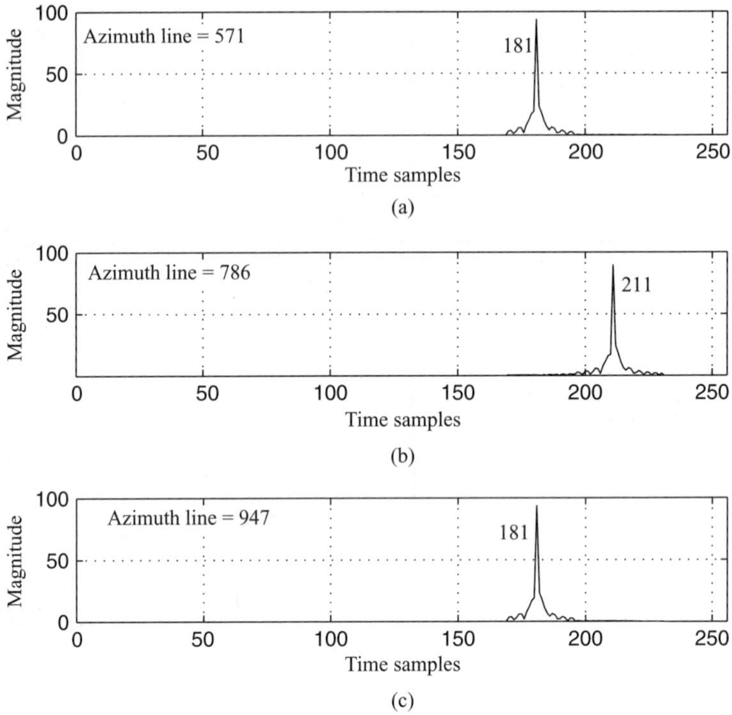

FIGURE 8.62 Cross-sectional view of Fig. 8.60 at azimuth lines 571, 786, and 947.

947. A single target is located at the same time sample (181) in both Figs. 8.62a and 8.62c. The single target in Fig. 8.62b is located at time sample 211. The time sample difference between these two sets of targets is (211−181) = 30, which corresponds to range difference $10 \times (30/2) = 150$ m. The simulation result closely matches the target specification ($R_{0b} - R_{0a} = 150$ m).

9

STOLT INTERPOLATION PROCESSING ON SAR IMAGES

This chapter discusses an alternative approach for processing radar images from the wavenumber (or spatial frequency) domain. Section 9.1 reviews the general background of wavenumber theory. Section 9.2 describes the direct interpolation based on unevenly spaced samples. Section 9.3 reviews the algorithm of Stolt interpolation. Computer simulation for six-target broadside SAR system based on the Stolt interpolation technique is covered in Section 9.4. Computer simulation for six-target squint SAR system is described in Section 9.5. Section 9.6 shows the results of a satellite image file processed by both range–Doppler and Stolt interpolation algorithms. Comparison between range–Doppler and Stolt interpolation algorithms is discussed in Section 9.7.

9.1 WAVENUMBER DOMAIN PROCESSING OF SAR DATA

Several algorithms are used to process SAR image data in wavenumber domains. Examples are direct interpolation from unevenly spaced data, Stolt interpolation (or mapping), time-domain correlation and backprojection, and range stacking.

Consider a group of N targets randomly distributed on the ground with each target located at $(x,y) = (x_n,y_n)$, $n = 1,2,\ldots,N$. Assume that the radar is located at $(x,y) = (0,u)$, and moves at speed V along the y axis. Let $p(t)$ be the transmitting signal; then the total received signal $s(t,u)$ can be expressed as

$$s(t,u) = \sum_n \sigma_n p\left(t - \frac{2\sqrt{x_n^2 + (y_n - u)^2}}{c}\right), \tag{9.1}$$

Digital Signal Processing Techniques and Applications in Radar Image Processing, by Bu-Chin Wang.

where $s(t,u)$ is a function of both t and u, and $p(t)$ is the transmitting signal. The Fourier transform of $s(t,u)$ with respect to time variable t becomes

$$S(\omega,u) = P(\omega) \sum_n \sigma_n \exp\left[-j\frac{2\omega}{c}\sqrt{x_n^2 + (y_n - u)^2}\right], \tag{9.2}$$

where $P(\omega)$ is the Fourier transform of $p(t)$.

By taking the spatial Fourier transform of $S(\omega,u)$ with respect to spatial variable u, as discussed in Section 7.2.1.3 (of Chapter 7), one obtains

$$\begin{aligned} S(\omega, \omega_D) &= P(\omega) \sum_n \sigma_n \exp\left[-j\sqrt{4\frac{\omega^2}{c^2} - \frac{\omega_D^2}{V^2}}x_n - j\frac{\omega_D}{V}y_n\right] \\ &= P(\omega) \sum_n \sigma_n \exp\left[-j\sqrt{4k^2 - k_u^2}x_n - jk_u y_n\right], \end{aligned} \tag{9.3}$$

Here ω_D is the Doppler angular frequency with respect to the radar position u and $k_u = \omega_D/V$ is the spatial wavenumber in the Doppler frequency domain.

Letting k_x be defined as

$$k_x = \sqrt{4k^2 - k_u^2} \tag{9.4}$$

and $k_y = k_u$, one can then rewrite Eq. (9.3) as follows:

$$S(k, k_y) = P(k) \sum_n \sigma_n \exp\left[-jk_x x_n - jk_y y_n\right]. \tag{9.5}$$

Consider an ideal target function in the spatial domain defined as

$$f_0(x,y) = \sum_n \sigma_n \delta(x - x_n, y - y_n). \tag{9.6}$$

Here $\delta(x - x_n, y - y_n)$ corresponds to an ideal impulse-like target located at $(x,y) = (x_n,y_n)$. By applying the 2D spatial Fourier transform on $f_0(x, y)$, one obtains

$$F_0(k_x,k_y) = \sum_n \sigma_n \exp\left(-jk_x x_n - jk_y y_n\right). \tag{9.7}$$

Here k_x and k_y are the wavenumber in the spatial frequency domain, and correspond to x and y in the spatial domain, respectively.

Since we are dealing with a moving spatial variable u, which represents a moving radar position, the corresponding spatial frequency will be considered as Doppler frequency in the following discussion.

Comparing Eqs. (9.5) and (9.7), one can see that if the signal $S(k,k_y)$ can be mapped, or transformed into a spatial frequency k_x as shown in Eq. (9.4), then the target function can be obtained by taking the 2D inverse spatial Fourier transform on $S(k_x,k_y)$. However, the measured signal $S(k,k_y)$ is evenly spaced in the time–frequency (ω) and spatial wavenumber ($k_y = \omega_D/V$) domains, but the transformation from k to k_x, as shown in Eq. (9.4), causes the data to become unevenly spaced in the k_x domain. The inverse 2D spatial transform on $S(k_x,k_y)$ cannot work correctly with a 2D data array, where the data are evenly spaced in the k_y domain but unevenly spaced in the k_x domain. This issue must be resolved to correctly reconstruct the radar image.

Figure 9.1 illustrates the data relationship before and after the square root transformation defined in Eq. (9.4). The black dots in Fig. 9.1a represent the data $S(k,k_y)$ distributed in the (k,k_y) or (ω,ω_D) domain. The white circles in Fig. 9.1b represent the same but transformed data $S(k_x,k_y)$ distributed in the (k_x,k_y) domain. As can be seen, the evenly spaced black dots become unevenly spaced white circles.

In order to use the inverse 2D spatial transform to obtain the target function, the unevenly spaced data in the k_x domain must be interpolated to become evenly spaced data. Figure 9.2 displays the ideal relationship of the data distribution before and after interpolation. The white circles shown in Fig. 9.2 are the unevenly spaced data $S(k_x,k_y)$ after transformation. The black dots are the evenly spaced data $S'(k_x,k_y)$ after interpolation.

The conversion of unevenly spaced data to evenly spaced data in the (k_x,k_y) domain, based on the received data in the (k,k_y) domain, is addressed next.

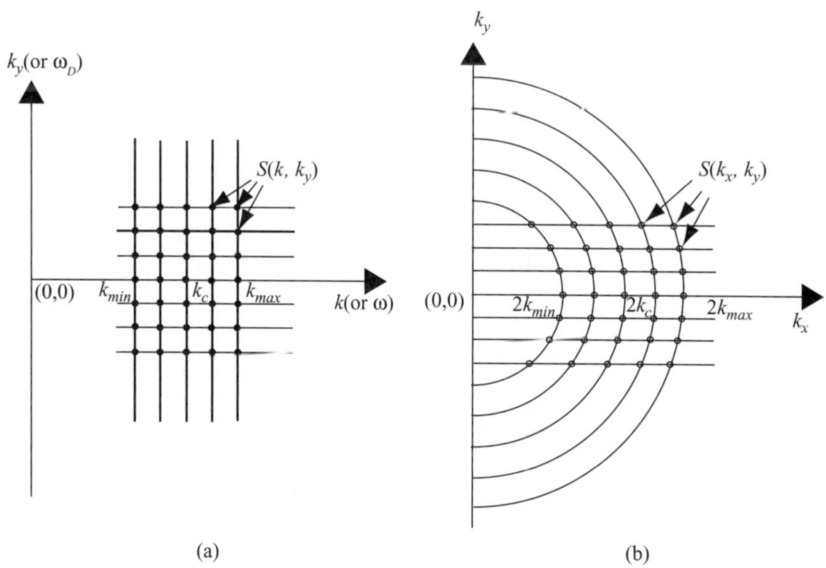

FIGURE 9.1 Data distribution before (a) and after (b) transformation.

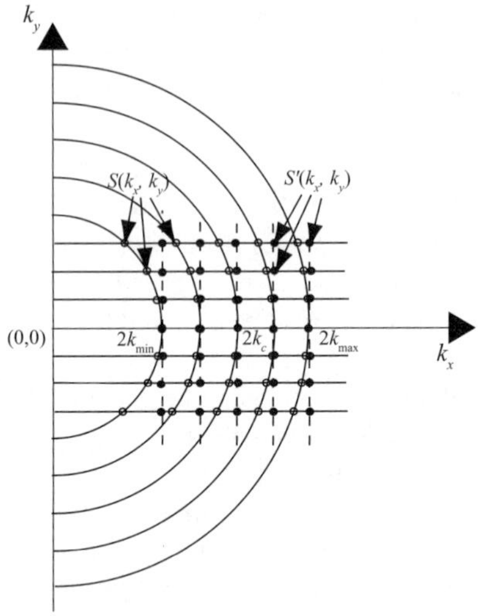

FIGURE 9.2 Data distribution before (○) and after (•) interpolation.

9.2 DIRECT INTERPOLATION FROM UNEVENLY SPACED SAMPLES

As discussed in Section 1.7.5 (of Chapter 1), the unevenly spaced data can be interpolated to obtain the evenly spaced data. Let $S_c'(k_x,k_y)$ and $S_c(k_x,k_y)$ be the evenly spaced and unevenly spaced 2D data in the (k_x,k_y) domain, respectively. Let $S_c(k,k_y)$ be the linearly measured samples in the (k,k_y) domain, with the unevenly spaced data $S_c(k_x, k_y)$ as the direct transformation, through Eq. (9.4), from $S_c(k,k_y)$.

In order to obtain the 2D spatial domain (x,y) signal from the inverse 2D transform on signals in the (k_x,k_y) domain, one method is to interpolate the unevenly spaced samples of $S_c(k_x,k_y)$ to obtain the evenly spaced data $S_c'(k_x,k_y)$, as illustrated in Fig. 9.2. Let the received signal be an array of $(M \times N)$ data. With respect to the mth row of data array $\{S_c(k_{xmn},k_{ym})\}$, $n = 1,2,\ldots, N$, the direct interpolated $S_c'(k_{xm},k_{ym})$ can be expressed as

$$S_c'\left(k_{xm}, k_{ym}\right) \approx \sum_n J_m\left(k\right) S_c\left(k_{xmn}, k_{ym}\right) h\left(k_{xm} - k_{xmn}\right), \tag{9.8}$$

where

$$\begin{aligned} J_m\left(k\right) &= \frac{d}{dk}\sqrt{4k^2 - k_{ym}^2} \\ &= \frac{4k}{c\sqrt{4k^2 - k_{ym}^2}}, \\ h\left(k_x\right) &= \sin\mathrm{c}\left(\frac{k_x}{\Delta k_x}\right). \end{aligned}$$

Here $J_m(k)$ is the Jacobian of transformation from k to k_x, and index n must satisfy $|k_{xm} - k_{xmn}| \leq N_s \Delta k_x$, with N_s as half the number of sinc lobes used for interpolation. The parameters Δk and Δk_x are described below.

Let f_s be the sampling frequency, with $2X_0$ as the swath of target area; then Δk and Δk_x can be expressed as follows:

$$\begin{aligned}\Delta k &= \frac{k}{N},\\ \Delta k_x &= \frac{\pi}{X_0}.\end{aligned} \tag{9.9a}$$

Let the carrier frequency and the bandwidth of the transmitting signal be f_c and $2f_0$; then

$$\begin{aligned}k &\in [k_c - k_0, k_c + k_0],\\ k_y &\in \left[\frac{\omega_{D\min}}{V}, \frac{\omega_{D\max}}{V}\right].\end{aligned} \tag{9.9b}$$

The corresponding k_x is therefore band-limited as

$$k_x \in \left[\sqrt{4(k_c - k_0)^2 - \left(\frac{\omega_{D\max}}{V}\right)^2}, \sqrt{4(k_c + k_0)^2 - \left(\frac{\omega_{D\min}}{V}\right)^2}\right]. \tag{9.9c}$$

For broadside SAR, the minimum value of Doppler frequency is zero and $2(k_c - k_0) \gg \omega_{D\max}/V$; therefore, the range of k_x becomes

$$\begin{aligned}k_x &\in \left[\sqrt{4(k_c - k_0)^2 - \left(\frac{\omega_{D\max}}{V}\right)^2}, 2(k_c + k_0)\right]\\ &\approx [2(k_c - k_0), 2(k_c + k_0)].\end{aligned}$$

The number of samples in the new k_x domain is then

$$N_{kx} = \frac{4k_0}{\Delta k_x}$$

The interpolation function $h(k_x)$ is similar to the one used in range cell migration correction (RCMC). In practical applications, only a finite number of resolution and sidelobes are adopted for interpolation filtering. In the RCMC case, the interpolation filter is chosen with the number of fractional sample shifts equal to 16, and the number of sidelobes equal to 7; therefore the variable n is chosen to be $n = 1,2,\ldots,8$. The direct implementation of Eq. (9.8) is time-consuming. An alternative wavenumber domain processing technique, namely, the Stolt interpolation (or transformation), is discussed next.

9.3 STOLT INTERPOLATION PROCESSING OF SAR DATA

To simplify the discussion, only the targets with ideal reflection coefficients, that is, $\sigma_n = 1$ for $n = 1,2,\ldots, N$, are considered here. Let $s_0(t,u)$ be the received signal from a reference target located at $(x,y) = (X_0,0)$. Then the 2D Fourier transform of $s_0(t,u)$ can be expressed as follows:

$$S_0(\omega,\omega_D) = P(\omega)\exp\left[-j\sqrt{4\frac{\omega^2}{c^2} - \frac{\omega_D^2}{V^2}}X_0\right]. \qquad (9.10)$$

The function $S_0(\omega,\omega_D)$ serves as a 2D reference function for matched filtering on both the ω and ω_D domains. Here $P^*(\omega)$ is considered as the range matched filter, while

$$\exp\left[j\sqrt{4\frac{\omega^2}{c^2} - \frac{\omega_D^2}{V^2}}X_0\right]$$

is the azimuth matched filter. The 2D compressed function $S_c(\omega,\omega_D)$ can be represented as

$$\begin{aligned} S_c(\omega,\omega_D) &= S(\omega,\omega_D)\, S_0^*(\omega,\omega_D) \\ &= P(\omega)P^*(\omega)\sum_n \exp\left[-j\sqrt{\frac{4\omega^2}{c^2} - \frac{\omega_D^2}{V^2}}(x_n - X_0) - j\frac{\omega_D}{V}y_n\right], \end{aligned}$$

or

$$\begin{aligned} S_c(k,k_u) &= |P(k)|^2\sum_n \exp\left[-j\sqrt{4k^2 - k_u^2}\,(x_n - X_0) - jk_u y_n\right] \\ &= |P(k)|^2\sum_n \exp\left[-jk_x(x_n - X_0) - jk_y y_n\right], \end{aligned} \qquad (9.11)$$

where $k = \omega/c$, $k_x = \sqrt{4k^2 - k_u^2}$, and $k_y = k_u = \omega_D/V$.

The 2D compressed signal $S_c(k,k_u)$ shown in Eq. (9.11) serves as the basis for wavenumber domain image processing. Since the azimuth reference function was chosen at $(X_0,0)$, only the target located at $(X_0,0)$ can be correctly focused. Other targets away from $(X_0,0)$ will be unfocused and require more accurate compression or correction.

Let f_b, f_c and $2f_0$ be the baseband frequency, carrier frequency, and bandwidth of transmitting signals, which satisfy the relationship $\omega_c - \omega_0 \le \omega = \omega_c + \omega_b \le \omega_c + \omega_0$, or, in terms of wavenumber representation $k_c - k_0 \le k = k_c + k_b \le k_c + k_0$.

The square root term in Eq. (9.11) can then be approximated as

$$\sqrt{4k^2 - k_u^2}\,(x_n - X_0) = 2k_c \Delta x_n \left[\frac{4(k_c + k_b)^2 - k_u^2}{4k_c^2}\right]^{1/2}$$

$$= 2k_c \Delta x_n \left[D^2 + \frac{2k_b}{k_c} + \frac{k_b^2}{k_c^2}\right]^{1/2}$$

$$= 2k_c \Delta x_n \left[D + \frac{k_b}{Dk_c} + \frac{k_b^2}{2Dk_c^2} - \cdots\right], \qquad (9.12)$$

where

$$D^2 = 1 - \frac{k_u^2}{4k_c^2},$$

$$\Delta x_n = x_n - X_0.$$

Since $k_c \gg k_u$ and $k_c > k_b$, Eq. (9.12) can be represented as follows:

$$\sqrt{4k^2 - k_u^2}\,(x_n - X_0) \approx 2k_c \Delta x_n \left[D + \frac{k_b}{Dk_c}\right]$$

$$= 2\Delta x_n \left[k_c D + \frac{k_b}{D}\right].$$

Equation (9.11) therefore becomes

$$S_c(k, k_u) = |P(k)|^2 \sum_n \exp\left[-j2\Delta x_n \left(k_c D + \frac{k_b}{D}\right) - jk_u y_n\right]. \qquad (9.13)$$

Letting $k_x = k_c + k_{xb}$, where k_{xb} is the baseband of k_x, and from Eq. (9.13), one obtains

$$k_x = 2\left(k_c D + \frac{k_b}{D}\right)$$
$$= k_c + k_{xb}. \qquad (9.14a)$$

Therefore

$$k_{xb} = k_c(2D - 1) + \frac{2k_b}{D}. \qquad (9.14b)$$

The function $D = \sqrt{1 - (k_y^2/4k_c^2)}$ can be approximated as

$$D \approx 1 - \frac{k_y^2}{8k_c^2}$$

and

$$\frac{1}{D} \approx 1 + \frac{k_y^2}{8k_c^2}.$$

Equation (9.14b) then becomes

$$k_{xb} = k_c - \frac{k_y^2}{4k_c} + 2k_b\left(1 + \frac{k_y^2}{8k_c^2}\right) \tag{9.14c}$$

This equation illustrates the transformation from k to k_x in the baseband situation. The Stolt interpolation simplifies the nonlinear transformation of $k_x = k_c + k_{xb} = \sqrt{4k^2 - k_y^2}$ into the form shown in Eq. (9.14c). The baseband component k_{xb} consists of three items: (1) a constant k_c that is independent of k or k_y; (2) a shift that varies for different k_y in the k domain, namely, $-(k_y^2/4k_c)$; and (3) a stretch of k_b in the k domain with stretch factor $2 + (k_y^2/4k_c^2)$. In other words, to perform the transformation of received baseband data from (k_b,k_y) domain to (k_{xb},k_y) domain, the received baseband data must be adjusted as shown in Eq. (9.14c).

By assuming that $|P(k)| = 1$, and substituting Eq. (9.14c) in Eq. (9.11), one obtains

$$\begin{aligned} S_c\left(k_b,k_y\right) &= \sum_n \exp\left[-jk_{xb}\Delta x_n - jk_y y_n\right] \\ &= \sum_n \exp\left\{j\left\{\left[-k_c + \frac{k_y^2}{4k_c} - \left(2 + \frac{k_y^2}{4k_c^2}\right)k_b\right]\right\}\Delta x_n - jk_y y_n\right\} \\ &= \sum_n \exp\left[-j\left(k_b' - k_0\right)\Delta x_n - jk_y y_n\right], \end{aligned} \tag{9.15}$$

where $k_b' = [2 + (k_y^2/4k_c^2)]k_b$ and $k_0 = (k_y^2/4k_c) - k_c$.

The term

$$\exp\left(-j\left(2 + \frac{k_y^2}{4k_c^2}\right)k_b\Delta x_n\right)$$

in Eq. (9.15) represents a stretched and new spatial frequency $k_b' = [2 + (k_y^2/4k_c^2)]k_b$ in spatial domain signal $s_c(x,\omega_D)$. The amount of stretched frequency is constant for a given value of k_y. It can be implemented by varying the sampling frequency f_s. This item can be neglected if only the magnitude of the radar image reconstruction is concerned.

Given $s(x)$ and $S(k)$ as the Fourier transform pair, the well-known modulation and shift property shows that

$$S(k - k_0) \Leftrightarrow s(x) \exp(jk_0 x).$$

The term k_0 in Eq. (9.15) states that in the wavenumber (k_b) domain, $S_c(k_b,k_y)$ has a k_y-dependent shift of $k_0 = (k_y^2/4k_c) - k_c$. This implies that the corresponding spatial domain function $s_c(x, k_u)$, which is in the range–Doppler domain, must be multiplied by a factor

$$\exp\left[-j\left(\frac{k_y^2}{4k_c} - k_c\right)\Delta x_n\right]$$

to correct or compensate the constant frequency shift:

$$s_c'(x,k_y) = s_c(x,k_y)\exp\left[-j\left(\frac{k_y^2}{4k_c} - k_c\right)\Delta x_n\right]. \tag{9.16a}$$

Since k_c is a constant wavenumber corresponding to the carrier frequency of the radar-transmitting signal, it will not cause any difference in the reconstructed image quality. Therefore, Eq. (9.16a) can be simplified as follows:

$$s_c'(x,k_y) = s_c(x,k_y)\exp\left[-j\frac{k_y^2}{4k_c}\Delta x_n\right]. \tag{9.16b}$$

Equation (9.16) is considered as *differential azimuth compression* (DAC) because it not only performs the spatial shift of Δx_n on $s_c(x,k_y)$ but also provides a shift quantity of $k_y^2/4k_c$ that depends on the azimuth value of k_y. This technique, originally developed by Stolt for seismic data processing, is called *Stolt interpolation* (or *approximate Stolt interpolation* in other textbooks).

Alternatively, the Stolt interpolation can be derived in a simple way as

$$\begin{aligned}\sqrt{4k^2 - k_u^2}\,(x_n - X_0) &= 2k\left(1 - \frac{k_u^2}{4k^2}\right)^{1/2}\Delta x_n\\ &\approx 2k\left(1 - \frac{k_u^2}{8k^2}\right)\Delta x_n.\\ &= \left(2k - \frac{k_u^2}{4k}\right)\Delta x_n.\end{aligned}$$

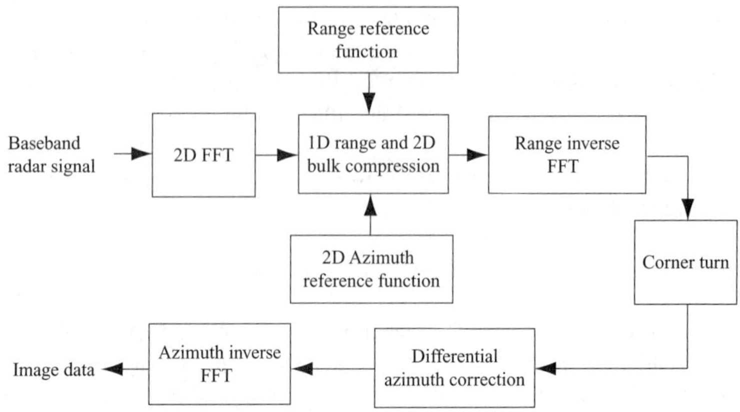

FIGURE 9.3 Block diagram of Stolt interpolation algorithm.

Equation (9.15) can then be expressed as, assuming $|P(k)| = 1$.

$$
\begin{aligned}
S_c(k, k_u) &= \sum_n \exp[-jk_x(x_n - X_0) - jk_y y_n] \\
&= \sum_n \exp[-j(k' - k_0)\Delta x_n - jk_y y_n],
\end{aligned}
$$

where $k' = 2k$ and $k_0 = k_u^2/4k$. Again, the parameter k' can be considered as a stretch of the k-axis; while k_0 is a constant phase and depends on both k_u and k. Since $k = k_c + k_b$ and $k_c \gg k_b$; therefore $k \simeq k_c$. This is the same results as derived by Stolt interpolation.

Figure 9.3 shows the block diagram representation of the Stolt interpolation processing of SAR data.

The Stolt interpolation technique will be used to process some synthesized image data, employing the same radar and signal parameters as listed in Section 8.2. The broadside SAR with six targets is described first, followed by the squint SAR system.

9.3.1 System Model of Broadside SAR with Six Targets

A 2D system model of a broadside SAR with six targets is depicted in Fig. 9.4, where the six targets are located at (R_{0a}, y_1), (R_{0b}, y_2), (R_{0c}, y_2), (R_{0d}, y_2), (R_{0e}, y_2), and (R_{0a}, y_3), respectively. The vertical axis represents the azimuth lines (or cross-range samples), and the horizontal axis shows the range that is perpendicular to the flight path. Lsa, Lsb, Lsc, Lsd, and Lse are the synthetic aperture lengths corresponding to ranges R_{0a}, R_{0b}, R_{0c}, R_{0d}, and R_{0e}, respectively. The total synthetic aperture length that covers the six-target region is $L_{\text{tot}} = Lsa + (y_3 - y_1)$.

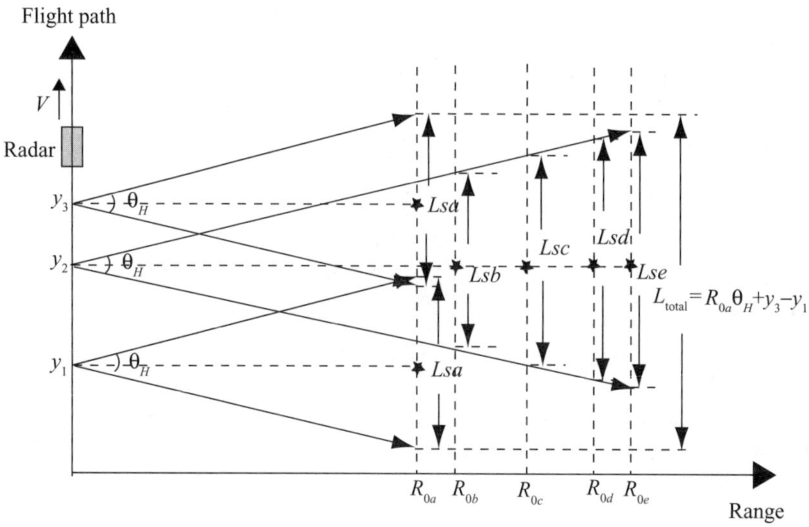

FIGURE 9.4 System model of a six-target broadside SAR.

The received echo signal array based on Fig. 9.4 is displayed in Fig. 9.5. The horizontal axis is changed from time samples, as discussed in Chapter 8, to slant range samples. The slant range sample is shown with index n, where $n = r/R_s$ with r and R_s as the slant range and slant range sample spacing, respectively. Normally, the curved line will occur at the beginning and end of the received signal array due to slant range differences at various radar positions. The differences are negligible for broadside SAR and are replaced with straight lines instead. The slant range is therefore considered approximately equal to range in the broadside case.

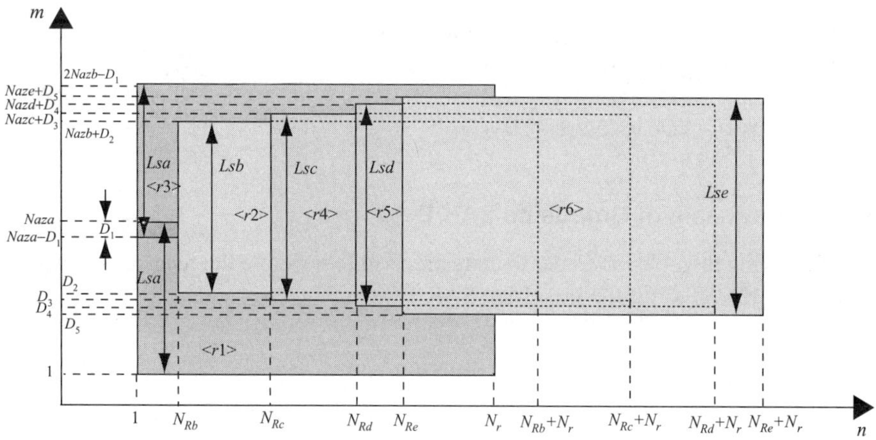

FIGURE 9.5 Received signal array based on Fig. 9.4.

The symbols <$r1$>, <$r2$> ... <$r6$> are the received signal arrays corresponding to targets 1,2,...,6, respectively. *Naza*, *Nazb*, *Nazc*, *Nazd*, and *Naze* are the number of samples within the associated synthetic aperture length, which corresponds to ranges R_{0a}, R_{0b}, R_{0c}, R_{0d}, and R_{0e}, respectively. The parameters $D_1, D_2, \ldots, D_5$ are defined as follows:

$$D_1 = \text{integer of } \{[Lsa - (y_3 - y_1)]/A_s\}$$
$$D_2 = \text{integer of } \{[(y_2 - y_1) - (Lsb2 - Lsa2)]/A_s\}$$
$$D_3 = \text{integer of } \{[(y_2 - y_1) - (Lsc2 - Lsa2)]/A_s\}$$
$$D_4 = \text{integer of } \{[(y_2 - y_1) - (Lsd2 - Lsa2)]/A_s\}$$
$$D_5 = \text{integer of } \{[(y_2 - y_1) - (Lse2 - Lsa2)]/A_s\}$$

Here A_s is the sample spacing along the azimuth axis (or m axis), and $Lsx2$ equals half of Lsx with x equal to a, b, c, d, and e, respectively. The size of total array samples along the vertical axis (or m axis) is 1 to $2Naza - D_1$. N_{Rb}, N_{Rc}, N_{Rd}, and N_{Re} are the sample differences between R_{0a} and R_{0i} for $i = b, c, d, e$. They are defined as follows:

$$N_{Rb} = \text{integer of } ((R_{0b} - R_{0a})/R_s)$$
$$N_{Rc} = \text{integer of } ((R_{0c} - R_{0a})/R_s)$$
$$N_{Rd} = \text{integer of } ((R_{0d} - R_{0a})/R_s)$$
$$N_{Re} = \text{integer of } ((R_{0e} - R_{0a})/R_s)$$

The array size along the horizontal axis (or n axis) is 1 to $N_{Re} + N_r$. The sizes of signal arrays for <$r1$> to <$r6$> are listed below:

Signal	Azimuth	Slant range
<$r1$>	1 to *Naza*	1 to N_r
<$r2$>	D_2 to $(Nazb + D_2)$	$(1 + N_{Rb})$ to $(N_r + N_{Rb})$
<$r3$>	$(Naza - D_1)$ to $(2Naza - D_1)$	1 to N_r
<$r4$>	D_3 to $(Nazc + D_3)$	$(1 + N_{Rc})$ to $(N_r + N_{Rc})$
<$r5$>	D_4 to $(Nazd + D_4)$	$(1 + N_{Rd})$ to $(N_r + N_{Rd})$
<$r6$>	D_5 to $(Naze + D_5)$	N_{Re} to $(N_r + N_{Re})$

9.3.2 Synthesis of Broadside SAR Data Array

The following target location parameters are used to generate the received data array <ri>, $i = 1, 2, \ldots, 6$ for the broadside SAR system.

$$y_1 = 0 \text{ m}$$
$$y_2 = 100 \text{ m}$$
$$y_3 = 150 \text{ m}$$
$$R_{0a} = 7500 \text{ m}$$

$R_{0b} = 7650$ m
$R_{0c} = 8000$ m
$R_{0d} = 8350$ m
$R_{0e} = 8500$ m

In terms of the given parameters of target locations and radar signals, the following data can be computed:

$Lsa = 225$ m
$Lsb = 229.5$ m
$Lsc = 240$ m
$Lsd = 250.5$ m
$Lse = 255$ m
$Naza = Lsa/A_s = 563$
$Nazb = Lsb/A_s = 573$
$Nazc = Lsc/A_s = 601$
$Nazd = Lsd/A_s = 627$
$Naze = Lse/A_s = 637$
$D_1 = \text{integer of } ((Lsa - (y_3 - y_1))/A_s) = 187$
$D_2 = \text{integer of } ((y_2 - y_1) - (Lsb2 - Lsa2))/A_s = 244$
$D_3 = \text{integer of } ((y_2 - y_1) - (Lsc2 - Lsa2))/A_s = 231$
$D_4 = \text{integer of } ((y_2 - y_1) - (Lsd2 - Lsa2))/A_s = 218$
$D_5 = \text{integer of } ((y_2 - y_1) - (Lse2 - Lsa2))/A_s = 212$
Number of azimuth lines $M = 2 \times Naza - D_1 = 939$
Number of range samples $N = N_r + N_{Re} = 281$

The waveforms of the real part of the synthesized signal arrays <r1>, <r2>, . . . ,<r6> are shown in Fig. 9.6. From left to right, Figs. 9.6a and 9.6b are <r1> and <r2>, Figs. 9.6c and 9.6d are <r3 > and <r4>, and Figs. 9.6e and 9.6f are <r5> and <r6>. The vertical axis represents the number of azimuth lines, where 1 out of every 30 azimuth lines is sequentially displayed throughout the total synthetic aperture length. The horizontal axis represents the range samples, with sample spacing of 10 m each and corresponding to a 30-MHz sampling rate.

Let $\{N_{ai} : N_{af}, N_{ri} : N_{rf}\}$ be the size of a signal array, where N_{ai} is the starting sample number and N_{af} is the ending sample number along the azimuth axis (or y axis), while N_{ri} is the starting sample number and N_{rf} is the ending sample number along the range axis (or x axis).

The sizes of signal arrays for <r1> to <r6> can be computed as follows:

<r1>	{1:563, 1:181}	<r4>	{231:832, 51:231}
<r2>	{244:817, 16:196}	<r5>	{218:845, 86:266}
<r3>	{376:939, 1:181}	<r6>	{212:849, 101:281}

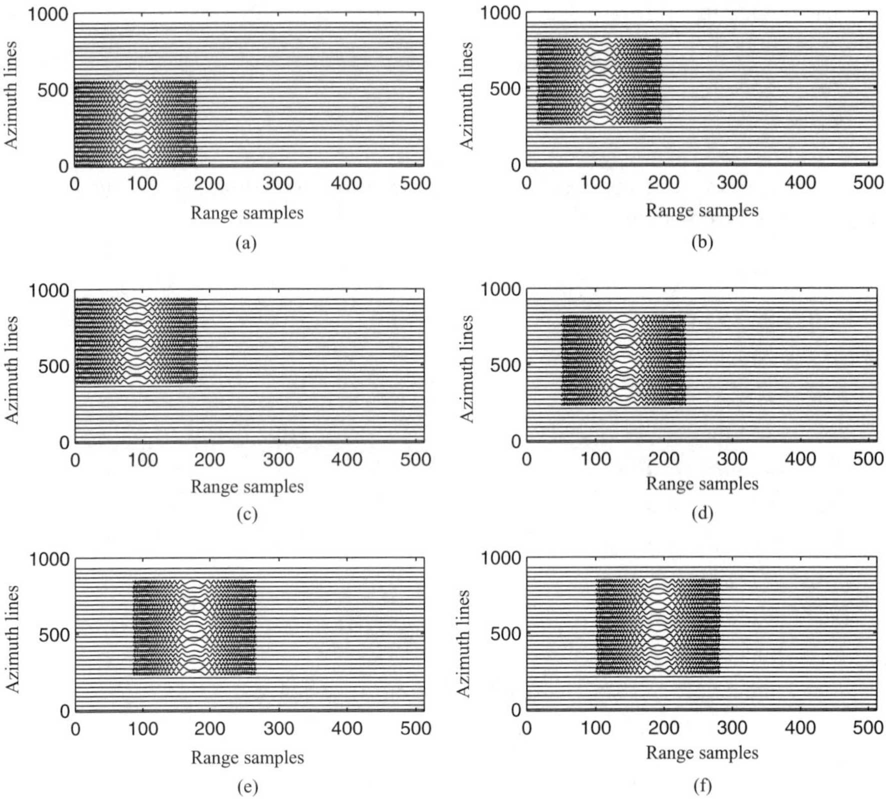

FIGURE 9.6 Waveforms of the real part of individual echo signal based on Fig. 9.4.

The overall received signal array $s(t,u)$, which is the combination (or sum) of six signal arrays <r1>, <r2>, . . . ,<r6>, is depicted in Fig. 9.7. The real part is in Fig. 9.7a; the imaginary part, in Fig. 9.7b. The size of the overall received signal array is $\{1{:}939,\ 1{:}281\}$. Accordingly, the FFT size used to process this time-domain signal array will be chosen as $N_y = 2048$ and $N_x = 512$.

9.3.3 Simulation Results

The data array $s(t,u)$, represented by Eq. (9.1) and shown in Fig. 9.7, is first 2D Fourier-transformed to become $S(\omega,\omega_D)$ as shown in Eq. (9.3). The range reference function $P^*(\omega)$, which is the complex conjugate of the transmitting function $P(\omega)$, is then multiplied by the data array $S(\omega,\omega_D)$ to become the range-compressed data array $S_{1c}(\omega,\omega_D)$. The range-compressed data are then inverse-Fourier-transformed to become $s_{1c}(t,\omega_D)$. Figure 9.8 displays the range-compressed data array $s_{1c}(t,\omega_D)$ in the range–Doppler frequency domain. The origin of the Doppler frequency spectrum

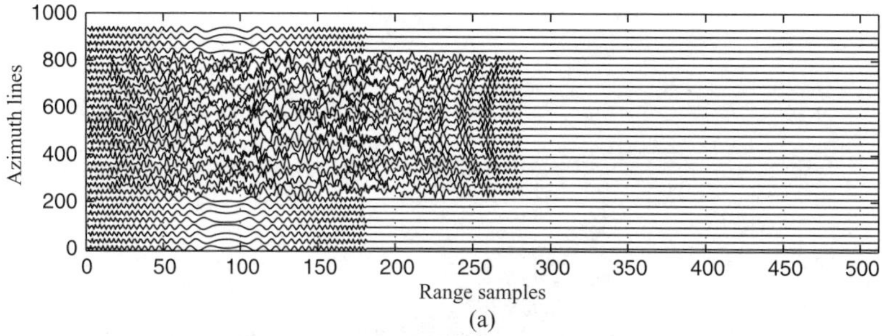

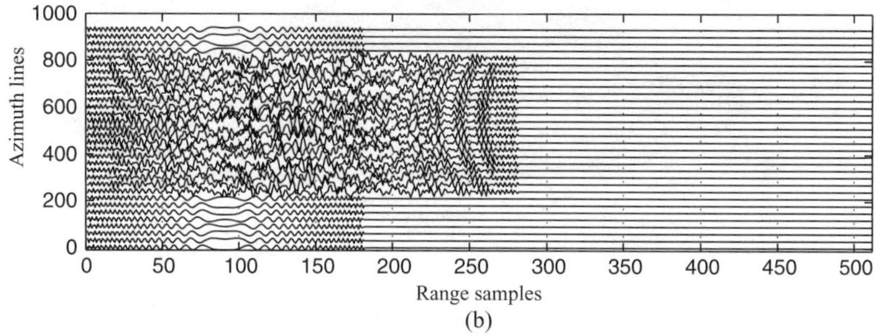

FIGURE 9.7 Waveforms of received signal based on Fig. 9.4.

is at bin 1024. The pulse repetition frequency (or sampling frequency) along the azimuth axis is f_{PRF}, and equals 500 Hz in this example.

Figure 9.9 is the top view of $s_{1c}(t,\omega_D)$ in the range–Doppler frequency domain. Only 1 out of every 50 rows along the Doppler frequency bins is displayed. As can be seen, the range-compressed signal has five columns of spectrum array along the range axis. Each spectrum array extends uniformly (about 80%) along the Doppler frequency axis. The first column array is located at range sample 181, which corresponds to the number of samples within the LFM pulse duration. The second column array appears at range sample 196 and is 15 range samples away from the first column array. Given the range sample spacing $R_s = 10$ m, the 15 range samples equal 150 m. The third column array appears at range sample 231, and is 50 range samples away from the first column array, which corresponds to a range distance of 500 m. The fourth column array appears at range sample 236, has a range sample number difference of 85 from the first column array, and corresponds to a range distance of 850 m from the first column array. The fifth column array appears at range sample 281 and is 1000 m away from the first column array with a range sample number difference of 100. Notice that the Doppler frequency spectrum appears similar for column arrays 2–5, but column array 1 seems different from the remaining four column arrays. This is because two targets contribute to the Doppler frequency spectrum on column array 1, and their phases cancel each other at some frequency bins.

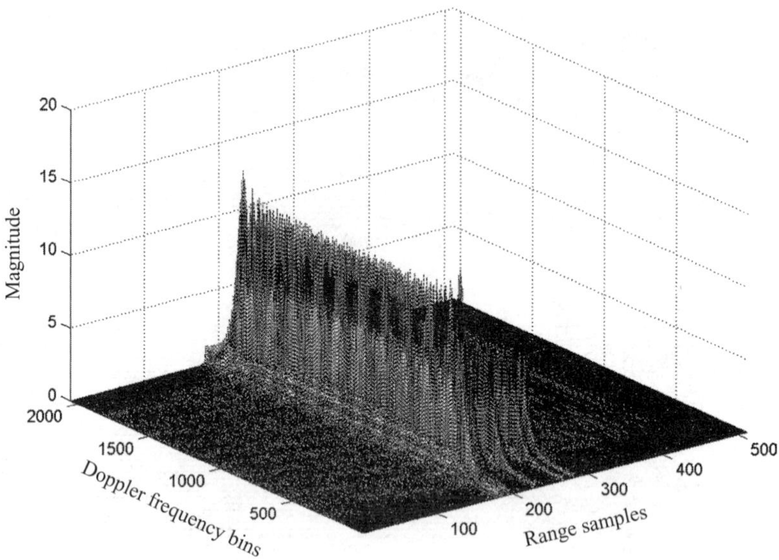

FIGURE 9.8 3D view of $s_{1c}(t,\omega_D)$ in range–Doppler frequency domain.

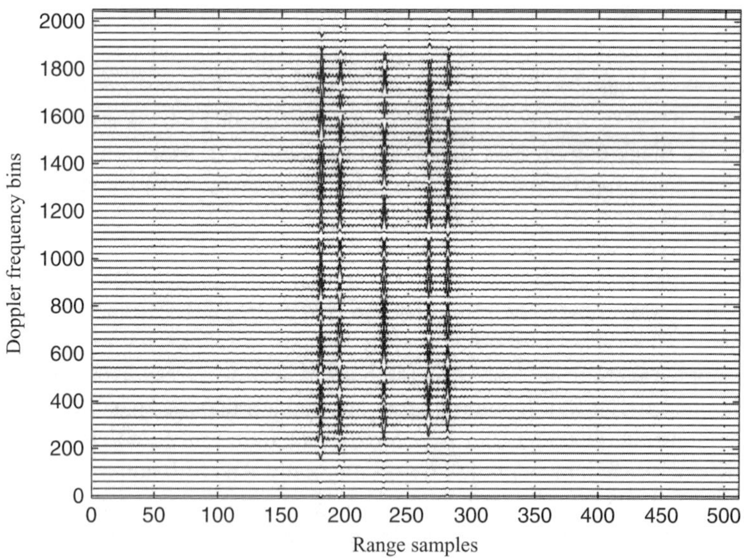

FIGURE 9.9 2D view of $s_{1c}(t,\omega_D)$ in range–Doppler frequency domain.

Since the simulated targets are located between $x = 7500$ m and $x = 8500$ m, the center of the target locations is chosen at $X_0 = R_{0c} = 8000$ m. The ideal target location used by the reference function (or matched filter) for azimuth compression is then chosen at (8000, 0). The corresponding azimuth reference function can be synthesized in two ways.

1. The 1D azimuth reference function in s (slow time) domain, given the f_{Dc} (= 0 for broadside case) and frequency changing rate β, can be designed as

$$\beta = -\frac{2V^2}{\lambda R_{0c}},$$

$$h_{az}(m) = \exp\left[j\pi\beta(m\Delta s)^2\right],$$

where $\Delta s = 1/f_{\text{PRF}}$ is the pulse repetition interval and $-Nazc2 \le m \le Nazc2$. The sample number $Nazc$ corresponds to the reference point located at $X_0 = R_{0c} = 8000$ m. The reference function $H_{az}(\omega_D)$ can be obtained by taking FFT on $h_{az}(s)$, with $s = m\Delta s$.

2. The 2D azimuth reference function can be obtained by substituting $X_0 = 8000$ in Eq. (9.10) and discarding the range matched filter function $P^*(\omega)$:

$$H_{az}(\omega, \omega_D) = \exp\left[j\sqrt{4\frac{\omega^2}{c^2} - \frac{\omega_D^2}{V^2}}8000\right].$$

Notice that the frequency ω is in passband; that is, $\omega_c - \omega_0 \le \omega \le \omega_c + \omega_0$ with ω_c as the carrier frequency and $2\omega_0$ as the bandwidth of the transmitting signal. The frequency ω_D corresponds to the true Doppler frequency.

Since no range migration correction is needed in broadside case, the 2D azimuth reference function $H_{\text{az}}(\omega, \omega_D)$ will be approximated by 1D function $H_{\text{az}}(\omega_D)$, while both 1D and 2D reference functions will be applied to the squint SAR case later.

The azimuth compression is performed by multiplying every column of the range-compressed data array $s_{1c}(t,\omega_D)$ by the reference function $H_{az}(\omega_D)$ to become the 2D compressed signal $s_c(t,\omega_D)$. Since only one azimuth reference function is used to perform the azimuth compression on the whole data array, the result is roughly compressed along the azimuth direction. Figure 9.10 is a 3D view of the target function after taking the inverse FFT on the roughly compressed signal $s_c(t,\omega_D)$.

Figures 9.11 and 9.12 display two side views of Fig. 9.10, one from the range direction into the targets and the other from the azimuth direction into the targets, respectively.

Figure 9.11 shows that the targets appear as pulses located at five different ranges. The center one is located at range sample 231, and corresponds to the reference point located at $x = X_0 = 8000$ m. The large magnitude of this narrow pulse proves that it is accurately compressed in both range and azimuth directions. The other four pulses,

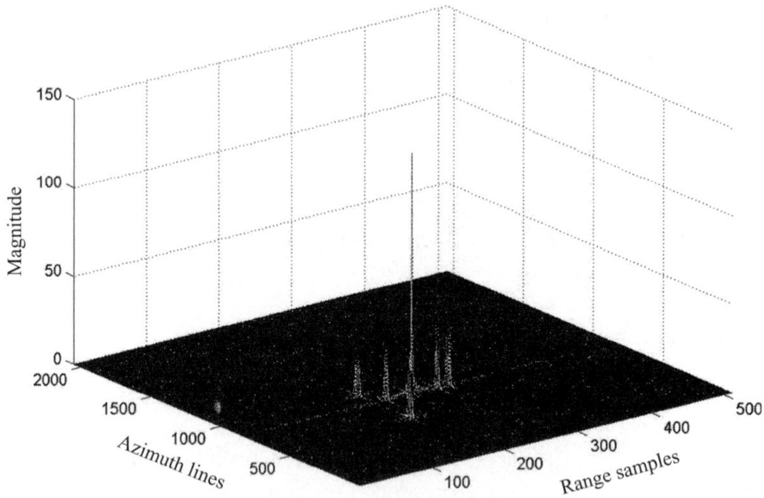

FIGURE 9.10 3D view of the roughly compressed six-target function.

appearing at range samples 181, 196, 266, and 281, have smaller magnitudes and are the targets roughly compressed in the azimuth direction.

Figure 9.12 shows three pulses located at different positions along the azimuth axis. The center pulse is at azimuth sample 832, and corresponds to the reference target for azimuth compression. This position corresponds to the reference function

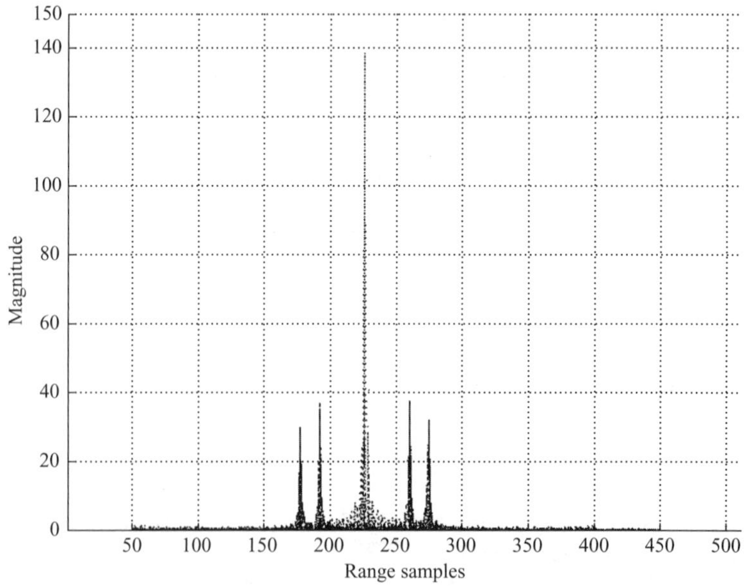

FIGURE 9.11 Side view, from the range direction, of Fig. 9.10.

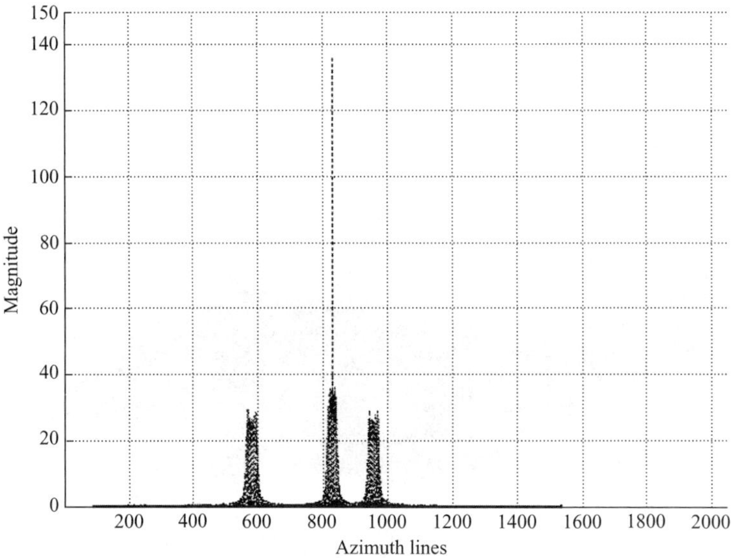

FIGURE 9.12 Side view, from the azimuth direction, of Fig. 9.10.

located at $y = 0$. The other two pulses appear at azimuth samples 582 and 958, and correspond to targets located at $y = -100$ m [(582 – 832) = −250 samples away] and $y = 50.4$ m [(958 − 832 = 126 samples away], respectively.

The roughly compressed data array $s_c(t,\omega_D)$ will now be processed by the Stolt interpolation, which improves the image quality through the *differential azimuth compression* (DAC) technique. The DAC technique is based on the implementation of Eq. (9.16b), which requires computation of two items. The first item is the range difference $\Delta x_n = x_n - X_0$, where x_n is the range value along the range axis and $X_0 = 8000$ m corresponds to range sample 231 as shown Fig. 9.11. For $n = 1,2,\ldots,N_x$ along the range axis with $N_x = 512$ in this example, the range difference can be computed as

$$\Delta x_n = (n - 231)R_s.$$

The second item is $k_y^2/4k_c$, where $k_c = \omega_c/c$ and $k_y = (2\pi m f_{\mathrm{PRF}}/VN_y)$. Here $m = 1,2,\ldots,N_y$ and $N_y = 2048$ in this example. For a given value of m along the Doppler frequency axis, one can compute k_y first, then multiply every range sample of $s_c(t,\omega_D)$ by $e^{-j(k_y^2/4k_c)\Delta x_n}$.

After the Stolt interpolation process, the final target locations can be obtained by taking the IFFT with respect to ω_D. The more accurately compressed and reconstructed target function is obtained and shown in Fig. 9.13.

By comparing the two plots in Figs. 9.10 and 9.13, the difference is obvious. The new plot with DAC improves greatly in both magnitude and sidelobe compression of the target pulses.

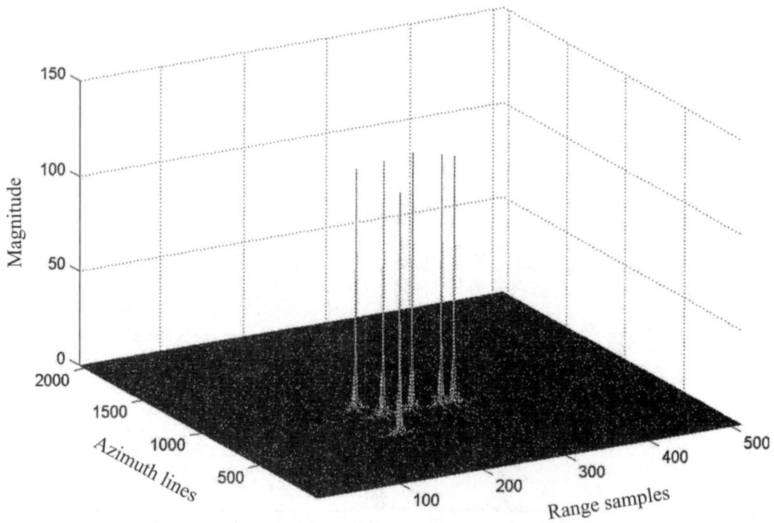

FIGURE 9.13 3D view of refocused six-target function.

Figures 9.14 and 9.15 display two side views of Fig. 9.13. The first view is from the range direction into the targets, and the second is from the azimuth direction into the targets. The magnitudes of the compressed pulses, located at the two sides of the center one, reach about 130 in Figs. 9.14 and 9.15. In Figs. 9.11 and 9.12, their magnitudes are less than 40.

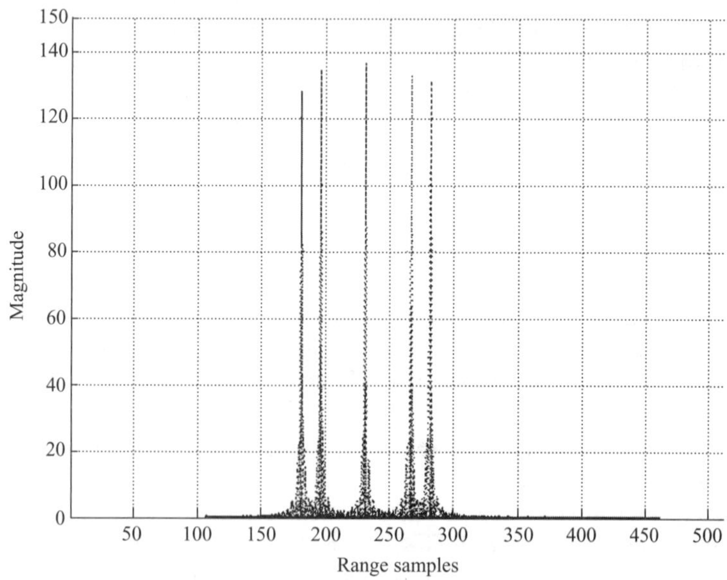

FIGURE 9.14 Side view, from the range direction, of Fig. 9.13.

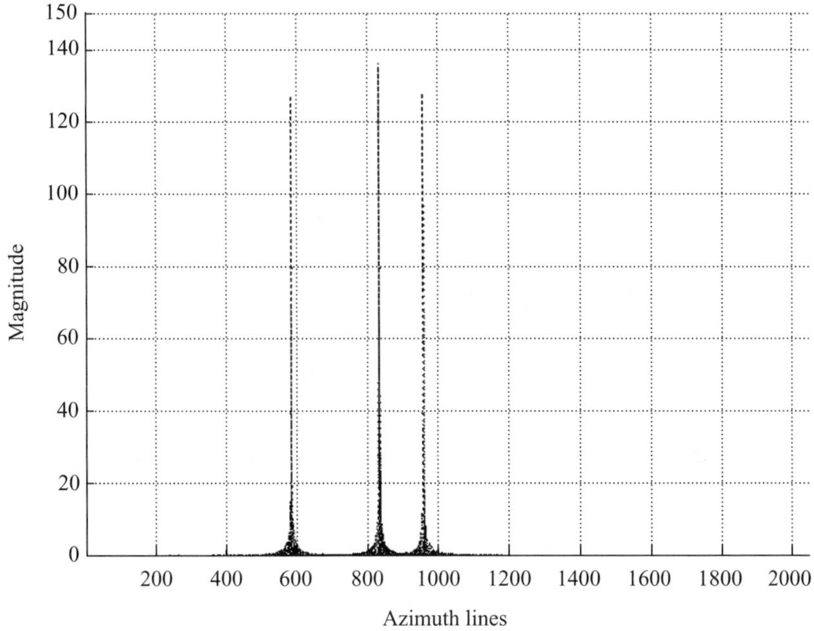

FIGURE 9.15 Side view, from the azimuth direction, of Fig. 9.13.

The principles of synthesizing radar image data and the radar signal parameters used in broadside SAR will be applied to the squint SAR system. In addition, a nonzero squint angle θ_q and different specification of target locations will be incorporated into the squint SAR system. These are described next.

9.3.4 System Model of Squint SAR with Six Targets

The 2D system model of a squint SAR with six targets and a squint angle θ_q is shown in Fig. 9.16. Here the vertical axis represents azimuth samples (or lines), and the horizontal axis represents range. The six targets are located at (R_{0a},y_1), (R_{0b},y_2), (R_{0c},y_3), (R_{0d},y_4), (R_{0e},y_5), and (R_{0a},y_6), respectively. The synthetic aperture length labeled as *Lsa*, *Lsb*, *Lsc*, *Lsd*, and *Lse* corresponds to targets located at ranges R_{0a}, R_{0b}, R_{0c}, R_{0d}, and R_{0e}, respectively.

The radar signal array based on the system model described above is displayed in Fig. 9.17. The horizontal axis is changed from the range to slant range sample index n with $n = r/R_s$, where r is the slant range. The vertical axis is changed from the azimuth position to azimuth lines (or cross-range samples). The curved lines, which appear at the beginning and end of the radar pulse duration, are caused by the slant range difference during the time period when the target is illuminated by the radar beam. The slant range sample difference at two ends of the curved line is represented

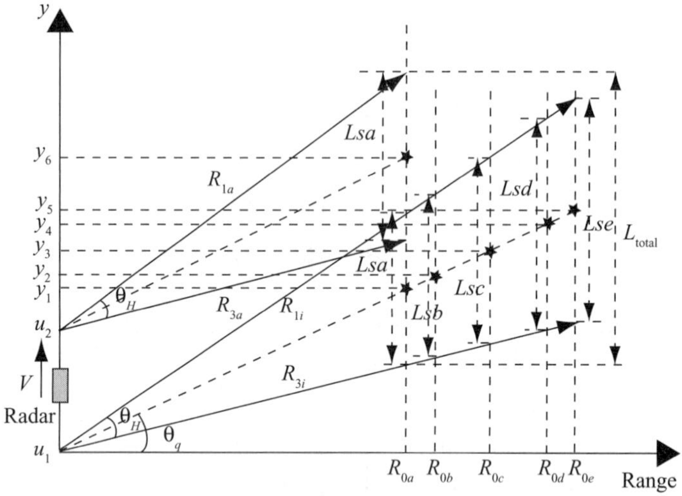

FIGURE 9.16 System model of a 6-target Squint SAR.

as Δ_x with $x = a,b,\ldots,e$. The symbol Δ_a is defined as,

$$\Delta_a = \text{integer of}\left\{\frac{\left[\frac{R_{0a}}{\cos(\theta_q+0.5\theta_H)} - \frac{R_{0a}}{\cos(\theta_q-0.5\theta_H)}\right]}{R_s}\right\}.$$

Other symbols, Δ_b, Δ_c, Δ_d, and Δ_e, can be computed accordingly.

The symbols *<r1>*,*<r2>*,...,*<r6>* are the received signal array corresponding to targets 1,2,...,6, respectively. *Naza*, *Nazb*, *Nazc*, *Nazd*, and *Naze* are the number

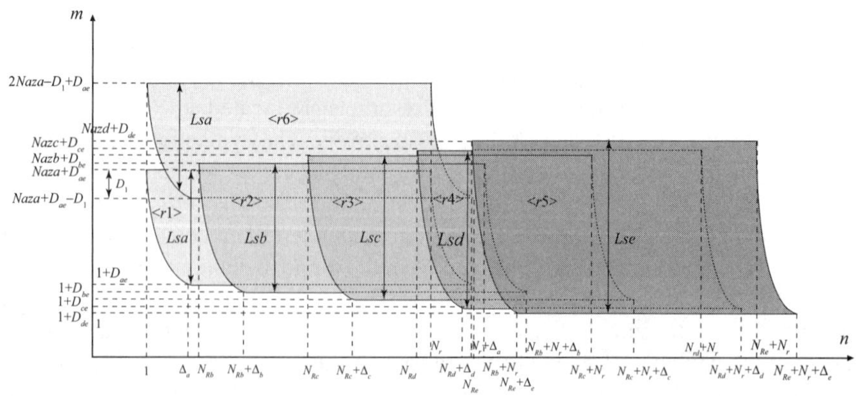

FIGURE 9.17 Received signal array derived from Fig. 9.16.

of samples within the associated synthetic aperture length. The parameters D_1, D_{ae}, D_{be}, D_{ce}, and D_{de} are defined as follows:

$$
\begin{aligned}
D_1 &= \text{integer of } \{[Lsa - (y_6 - y_1)]/A_s\} \\
D_{ae} &= \text{integer of } [(Lse2 - Lsa2)/A_s] \\
D_{be} &= \text{integer of } [(Lse2 - Lsb2)/A_s] \\
D_{ce} &= \text{integer of } [(Lse2 - Lsc2)/A_s] \\
D_{de} &= \text{integer of } [(Lse2 - Lsd2)/A_s]
\end{aligned}
$$

The parameters N_{Rb}, N_{Rc}, N_{Rd}, and N_{Re} are defined in the same form as in the broadside SAR case.

9.3.5 Synthesis of Squint SAR Data Array

The following are the specifications of target locations used to generate the received data array:

$$
\begin{aligned}
R_{0a} &= 7500 \text{ m} \\
R_{0b} &= 7650 \text{ m} \\
R_{0c} &= 8000 \text{ m} \\
R_{0d} &= 8350 \text{ m} \\
R_{0e} &= 8500 \text{ m} \\
\theta_q &= 6^\circ \\
y_1 &= R_{0a} \times \tan\theta_q = 788.3 \text{ m} \\
y_2 &= R_{0b} \times \tan\theta_q = 804 \text{ m} \\
y_3 &= R_{0c} \times \tan\theta_q = 840.8 \text{ m} \\
y_4 &= R_{0d} \times \tan\theta_q = 877.6 \text{ m} \\
y_5 &= R_{0e} \times \tan\theta_q = 893.4 \text{ m} \\
y_6 &= y_1 + 150 = 938.3 \text{ m}
\end{aligned}
$$

Given the target locations and the radar signal parameters, the following data can be obtained:

$$
\begin{aligned}
Lsa &= R_{0a} \times [\tan(\theta_q + 0.5\,\theta_H) - \tan(\theta_q - 0.5\,\theta_H)] \\
&= 227.5 \text{ m} \\
Lsb &= 232 \text{ m} \\
Lsc &= 242.6 \text{ m} \\
Lsd &= 253.3 \text{ m} \\
Lse &= 257.8 \text{ m} \\
Naza &= Lsa/A_s = 569 \\
Nazb &= 581
\end{aligned}
$$

$Nazc = 607$
$Nazd = 633$
$Naze = 644$
$D_1 = 193$
$D_{ae} = 37$
$D_{be} = 32$
$D_{ce} = 18$
$D_{de} = 5$
$N_{Rb} = 15$
$N_{Rc} = 50$
$N_{Rd} = 85$
$N_{Re} = 100$
$\Delta_a = 2$
$\Delta_b = 2$
$\Delta_c = 3$
$\Delta_d = 3$
$\Delta_e = 3$
Number of azimuth lines $M = 2 \times Naza - D_1 + D_{ae} = 982$
Number of slant range samples $N = c \times T_p + (R_{0e} - R_{0a})/R_s = 281$

The real parts of the synthesized signal waveform corresponding to targets 1–6 are shown in Fig. 9.18. The vertical axis shows the number of azimuth lines, where 1 out of every 30 azimuth lines is sequentially displayed throughout the total synthetic aperture length. The horizontal axis shows the slant range samples, with a sample spacing of 10 m and corresponding to a 30 MHz sampling rate.

The sizes of signal arrays for $<r1>$ to $<r6>$ are shown below.

Signal	Azimuth	Slant range
$<r1>$	$(1 + D_{ae})$ to $(Naza + D_{ae})$ (38:606, 1:183)	1 to $N_r + \Delta_a$
$<r2>$	$(1 + D_{be})$ to $(Nazb + D_{be})$ (33:613, 16:198)	$(1 + N_{Rb})$ to $(N_r + N_{Rb} + \Delta_b)$
$<r3>$	$(1 + D_{ce})$ to $(Nazc + D_{ce})$ (19:625, 51:234)	$(1 + N_{Rc})$ to $(N_r + N_{Rc} + \Delta_c)$
$<r4>$	$(1 + D_{de})$ to $(Nazd + D_{de})$ (6:638, 86:269)	$(1 + N_{Rd})$ to $(N_r + N_{Rd} + \Delta_d)$
$<r5>$	1 to $Naze$ (1:644, 101:284)	$(1 + N_{Re})$ to $(N_r + N_{Re} + \Delta_e)$
$<r6>$	$(Naza + D_{ae} - D_1)$ to $(2Naza + D_{ae} - D_1)$ (413:982, 1:183)	1 to $N_r + \Delta_a$

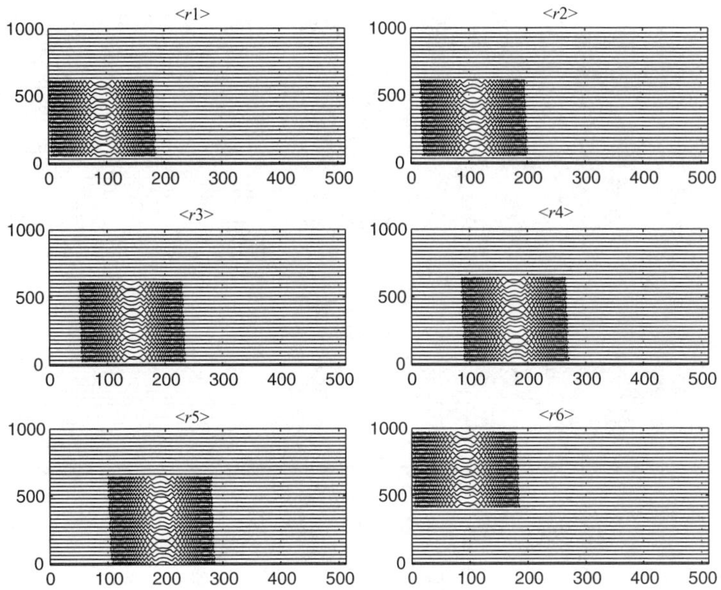

FIGURE 9.18 Waveforms of the real part of individual echo signal from Fig. 9.16.

The overall synthesized signal array $s(t,u)$, which is the combination (or sum) of the six signal arrays <*r*1> to <*r*6>, is displayed in Fig. 9.19. The real part is shown in Fig. 9.19a; the imaginary part, in Fig. 9.19b. The size of the overall synthesized signal array is $\{1{:}982,\ 1{:}283\}$. Accordingly, the FFT size used to process this signal array remains the same as in the broadside SAR case; that is, $N_y = 2048$ and $N_x = 512$.

9.3.6 Simulation Results

The data array $s(t,u)$, represented by Eq. (9.1) and shown in Fig. 9.19, is first 2D Fourier-transformed to become $S(\omega,\omega_D)$. The range reference function $P^*(\omega)$ is multiplied by the received baseband signal $S(\omega,\omega_D)$ to become the range-compressed signal $S_{1c}(\omega,\omega_D)$. The result is then inverse-Fourier-transformed to become $s_{1c}(t,\omega_D)$. Figure 9.20 is a 3D view of the range-compressed function $s_{1c}(t,\omega_D)$ in the range–Doppler frequency domain with frequency origin at bin 1. Comparing the spectrum with that in the broadside case, one can see that (1) the magnitude of the spectrum varies significantly along the Doppler frequency axis, and (2), the spectrum splits into two portions. As discussed previously, this is because the squint angle θ_q causes significant range migration, and the observed spectrum is the fallback version of the true spectrum of $s_{1c}(t,\omega_D)$.

Figure 9.21 is a 2D view of $s_{1c}(t,\omega_D)$ in the range–Doppler frequency domain. Only 1 out of every 30 rows is displayed. There are five columns of spectrum array

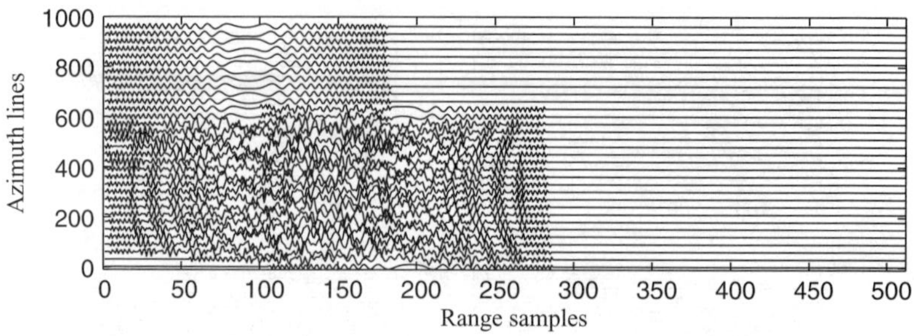

(a)

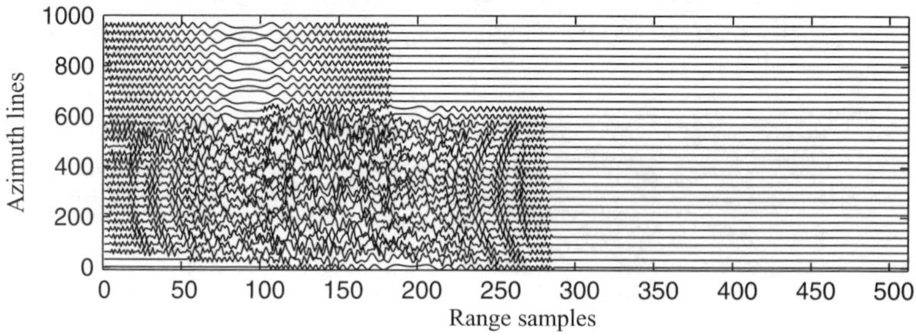

(b)

FIGURE 9.19 Waveforms of received signal from Fig. 9.16.

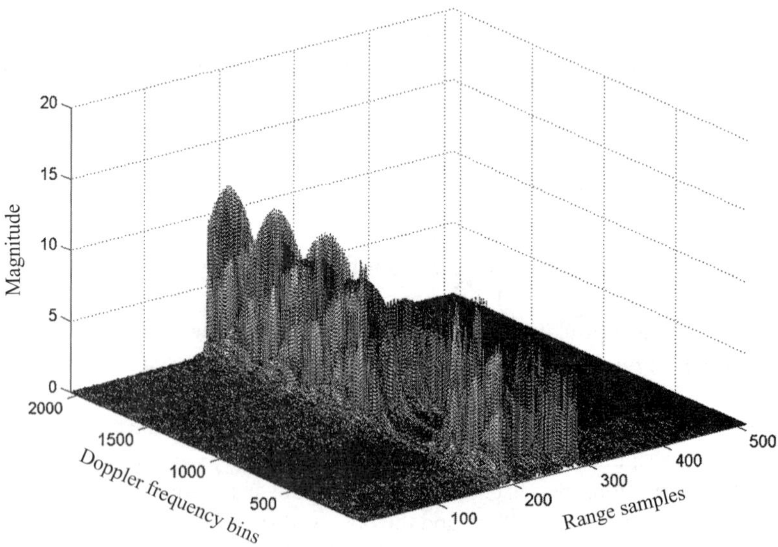

FIGURE 9.20 3D view of $s_{1c}(t,\omega_D)$ in range–Doppler frequency domain.

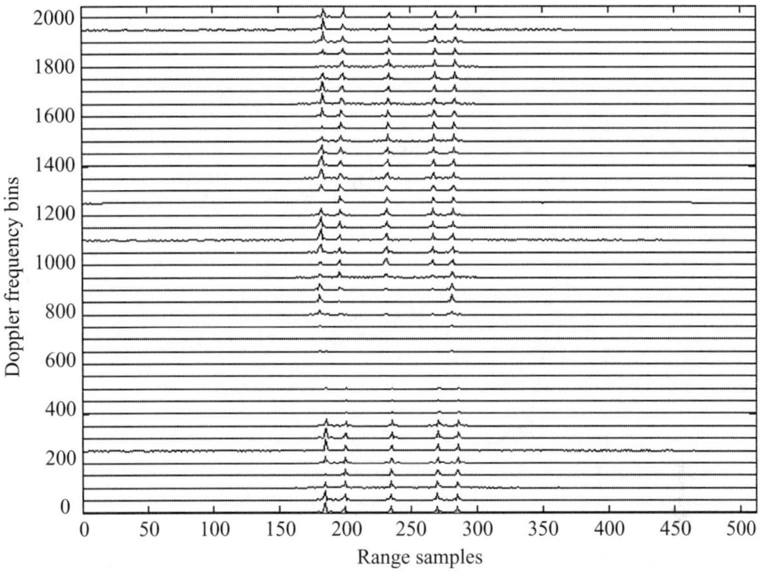

FIGURE 9.21 2D view of $s_{1c}(t,\omega_D)$ in range–Doppler frequency domain.

in the display, and each array corresponds to one of the targets located at ranges R_{0a},R_{0b}, ... ,R_{0e} (there are two targets contribute to the spectrum at location R_{0a}). In addition, every column array splits in two parts and occupies about 80% of the whole Doppler frequency range.

The first column array of the Doppler spectrum corresponds to targets 1 and 2 and tilts from range samples 181 to 185. The number 181 corresponds to the number of samples in the LFM pulse duration. The second column array of the Doppler spectrum tilts from range samples 196 to 200, which corresponds to a target that is 150 m away from the first column array. The third column array tilts from range samples 231 to 235, and corresponds to the target located 500 m away from the first column array. The fourth column array tilts from range samples 266 to 270, and corresponds to the target that is 850 m away from the first column array. The fifth column array tilts from range samples 281 to 285, and corresponds to a target that is 1000 m away from the first column array. The Doppler frequency spectrum looks similar for column arrays 2 to column 5, but column array 1 seems different from the other four column arrays. This is because two targets contribute to the Doppler spectrum on column array 1, and their phases cancel each other out at some frequency bins.

The matched filter for azimuth compression is chosen with a target located at the center of the target area, namely, $(x,y) = (R_{0c}, 0) = (8000, 0)$. For image quality comparison purpose, both the 1D reference function $H_{az}(\omega_D)$ and 2D reference function $H_{az}(\omega,\omega_D)$ will be designed as the azimuth matched filter to reconstruct the targets. The 1D azimuth filter $H_{az}(\omega_D)$ will first be used to reconstruct targets, and the 2D azimuth filter $H_{az}(\omega,\omega_D)$ will be used later.

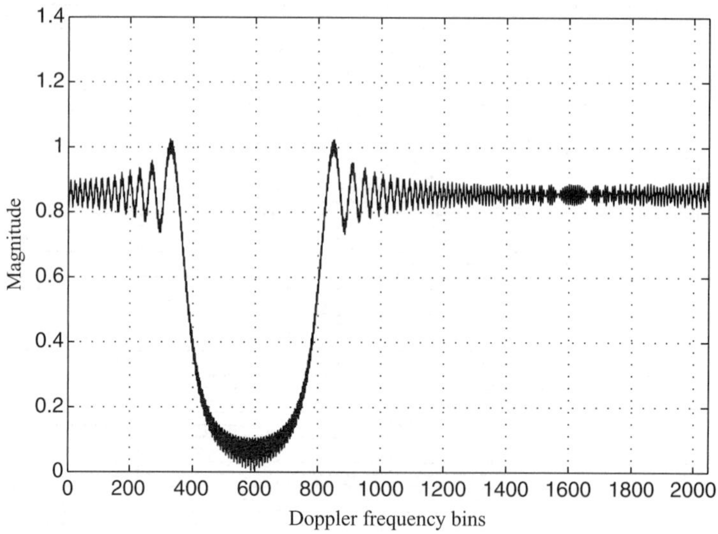

FIGURE 9.22 Synthesized 1D azimuth reference function for squint SAR system.

The azimuth matched filter $H_{az}(\omega_D)$, used in Chapter 8 for squint SAR processing based on the range–Doppler algorithm, is adopted here to serve as the 1D azimuth reference function. The frequency spectrum of $H_{az}(\omega_D)$ is shown again in Fig. 9.22 for convenience.

Azimuth compression is performed by multiplying every column of $s_{1c}(t,\omega_D)$ by the reference function $H_{az}(\omega_D)$ to become the 2D compressed signal $s_c(t,\omega_D)$. Figure 9.23 is a 3D view of the reconstructed target function after taking the inverse spatial DFT on the roughly compressed signal $s_c(t,\omega_D)$. The roughly reconstructed signal array has five targets located at the same y value along the azimuth direction as expected from the slant range viewpoint. The relative positions between the targets are not reconstructed correctly and will be corrected through the Stolt interpolation.

Two side views of Fig. 9.23, one from the range direction into the targets and the other from the azimuth direction into the targets, are displayed in Figs. 9.24 and 9.25. Figure 9.24 shows that the targets appear as pulses located at five different ranges. The center pulse is located at range sample 233, and corresponds to the azimuth reference function with range sample chosen to be $x = X_0 = 8000$. It appears to be accurately compressed in both range and azimuth directions. The other four pulses appear at range samples 183, 198, 233, and 283 and have smaller magnitudes. These four pulses are roughly compressed in the azimuth direction.

Figure 9.25 shows that the targets appear as two pulses located along the azimuth axis. The left pulse is at azimuth sample 626 and corresponds to the azimuth reference function located at $y = 0$. The right pulse appears at azimuth sample 1002. This pulse corresponds to 376 samples, or 150.4 m, away from the reference target.

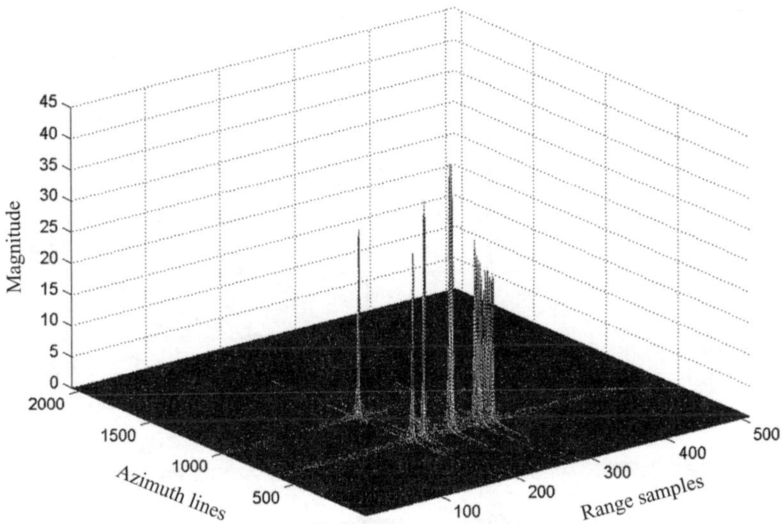

FIGURE 9.23 3D view of roughly compressed target function.

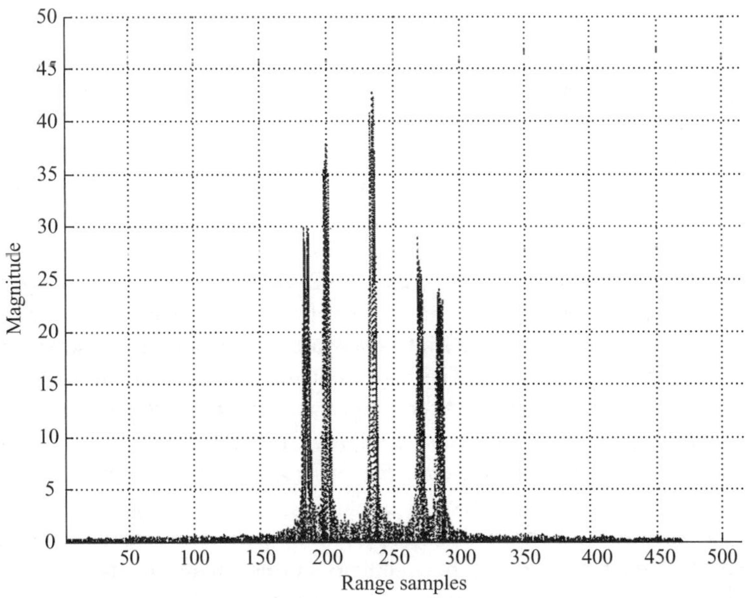

FIGURE 9.24 Side view, from the range direction, of Fig. 9.23.

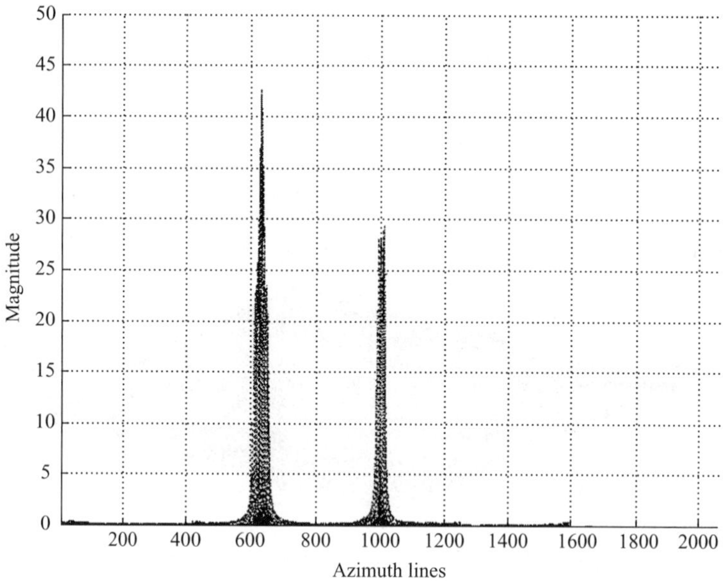

FIGURE 9.25 Side view, from the azimuth direction, of Fig. 9.23.

The Stolt interpolation algorithm requires computation of two parameters:

1. The range difference $\Delta x_n = x_n - X_0$. Given that X_0 corresponds to a range sample 233, the range difference between any range sample n and the reference target can be computed as $\Delta x_n = (n - 233)\, R_s$.
2. The ratio $k_y^2/4k_c$. Here, $k_c = (\omega_c/c)$ and k_y is more complicated than in the case of the broadside SAR. As shown in Fig. 9.20, the Doppler frequency spectrum of the squint SAR system has two characteristics: (1) the true or original Doppler spectrum is a passband signal having a lower frequency edge f_{DL} and upper frequency edge f_{DU}; and (2) because of the natural digitization at sampling frequency f_{PRF}, the Doppler frequency spectrum folds from passband to baseband and splits into two parts to become the observable baseband frequency spectrum. The high-frequency part appears at the lower side of the baseband spectrum. Accordingly, the observed Doppler baseband frequencies f_{DL} and f_{DU} need to be adjusted to obtain their true values in the passband. Given squint angle θ_q, wavelength λ, and radar speed V, as shown in Eq. (6.16), the true values of f_{DL} and f_{DU} can be obtained. Let N_y be the FFT size used to transform the Doppler frequency spectrum, and N_{yL} and N_{yU} be the Doppler frequency bin numbers corresponding to the true (or passband) f_{DL} and f_{DU}. The corresponding true value of k_y can be computed as $k_y = m\,\Delta k_y$, where $\Delta k_y = (2\pi f_{\mathrm{PRF}}/VN_y)$ and $m = N_{yL}$ to N_{yU}. Differential azimuth compression is then implemented by multiplying every range sample of $s_c(t, \omega_D)$ by $\exp[-j(k_y^2/4k_c)\Delta x_n]$.

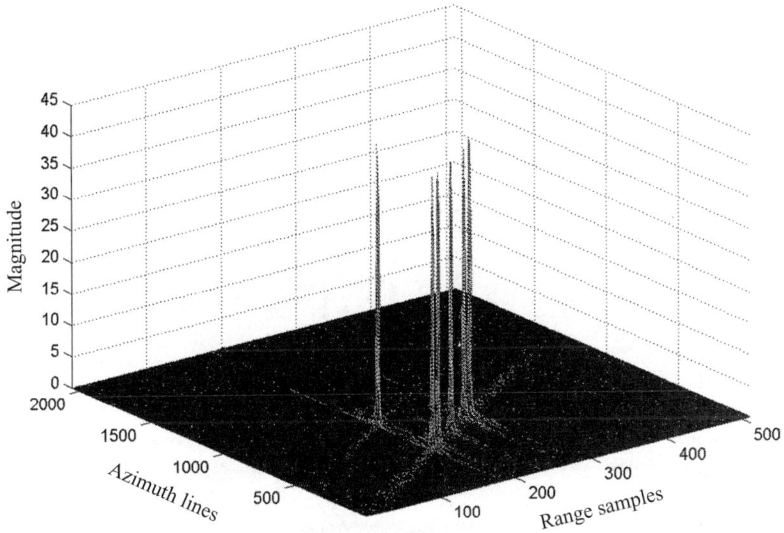

FIGURE 9.26 3D view of refocused target function.

After the Stolt interpolation, the final target function can be obtained by taking the inverse FFT with respect to the Doppler frequency. The result is shown in Figure 9.26.

By comparing Fig. 9.26 with Fig. 9.23, it can be seen that the new target function with differential azimuth compression improves greatly in magnitude with the target pulses. It also displays the correct target positions along the azimuth axis.

Two side views of Fig. 9.26, one from the range direction into the targets and the other from the azimuth direction into the targets, are depicted in Figs. 9.27 and 9.28, respectively.

Figure 9.27 shows that there are five pulse-like signals along the range axis, with the center signal located at range sample 233 and corresponding to the azimuth reference function. The magnitude of this center signal remains the same as the one in Fig. 9.24, while the magnitudes of the other four signals are improved from less than 30 in Fig. 9.24 to over 40 in Fig. 9.27.

Figure 9.28 shows that there are six pulses located at different azimuth lines (or y axes), which differ from the two pulses shown in Fig. 9.25. This demonstrates that the differential azimuth compression accurately compressed the targets and reconstructed the target locations correctly. The six pulses, corresponding to targets 1–6, are located at azimuth samples 495, 535, 626, 718, 758, and 871 respectively.

By setting target 1 at the position of $y = 0$, the sample number differences between target 1 and targets 2–6 are 40, 131, 223, 263, and 376, respectively. Given the azimuth sample spacing $A_s = 0.4$ m, the positions of y_2 to y_6 (corresponding

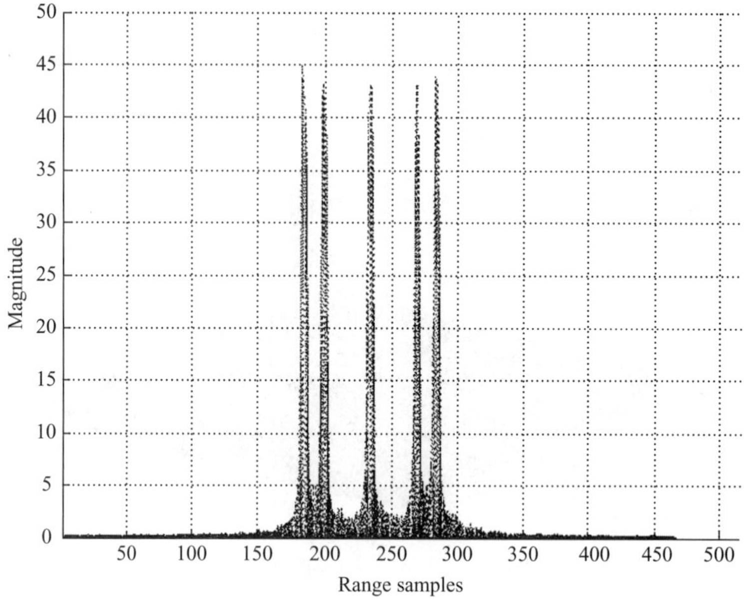

FIGURE 9.27 Side view, from the range direction, of Fig. 9.26.

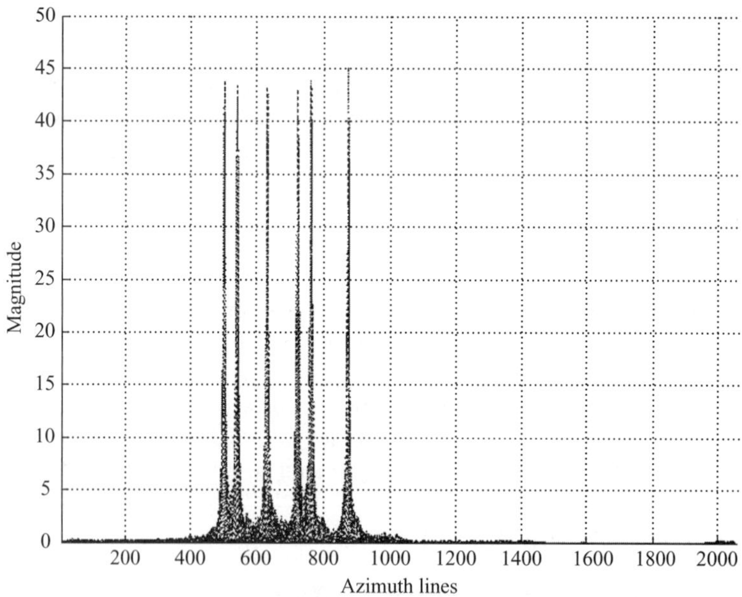

FIGURE 9.28 Side view, from the azimuth direction, of Fig. 9.26.

to targets 2–6) can be computed and compared against the specifications listed below:

Simulated	Specification
$y_1 = 0$	0
$y_2 = 16$ m	15.7 m
$y_3 = 52.4$ m	52.5 m
$y_4 = 89.2$ m	89.3 m
$y_5 = 105.2$ m	105.1 m
$y_6 = 150.4$ m	150 m

The second azimuth reference function $H_{az}(\omega,\omega_D)$, based on Eq. (9.10) with $X_0 = 8000$ m, is now applied to process the same image data. By taking the inverse DFT on the 2D azimuth reference function $H_{az}(\omega,\omega_D)$ with respect to the frequency domain, the range–Doppler frequency-domain view of $H_{az}(t,\omega_D)$ is shown in Fig. 9.29 with bin 1 as the Doppler frequency origin. As can be seen, the Doppler frequency spectrum of $H_{az}(t,\omega_D)$ is folded from passband to baseband and divided in two portions, the upper one representing the low-frequency band and the lower one, the high-frequency band. The thin-wall-like Doppler frequency spectrum is not equal in magnitude and aligns around range sample 440.

A different view of Fig. 9.29 is shown in Fig. 9.30, where 1 out of 50 rows along the Doppler frequency axis is displayed. As can be seen, the column-like Doppler frequency spectrum has two portions along the Doppler frequency axis and tilts to the left, opposite to the squint targets spectrum shown in Fig. 9.21. The column

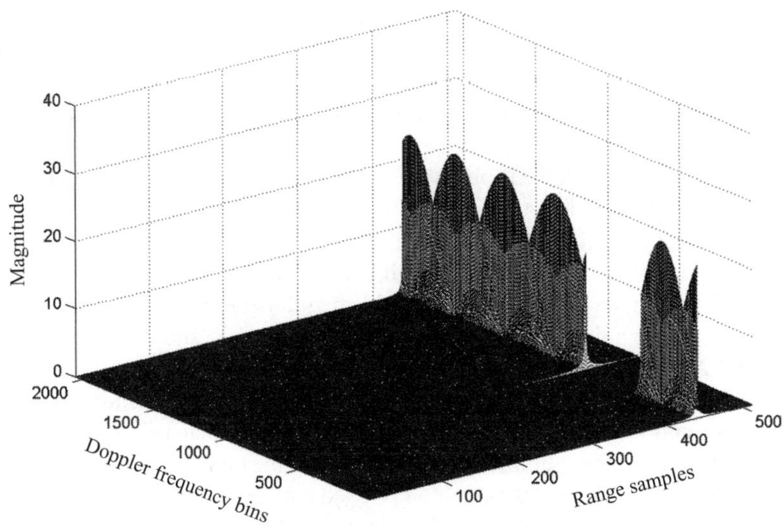

FIGURE 9.29 3D view of $H_{az}(t,\omega_D)$.

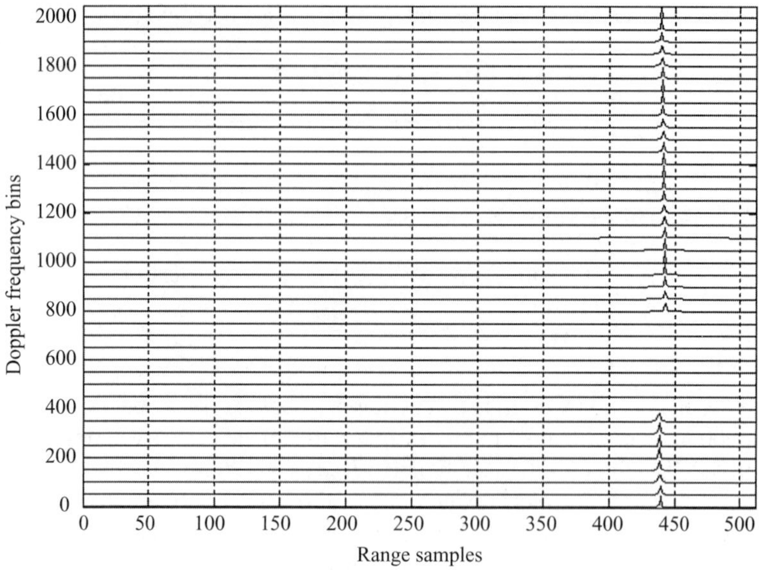

FIGURE 9.30 2D view of $H_{az}(t,\omega_D)$.

spectrum is located at range sample 442 and tilts toward the left at range sample 438 along the Doppler frequency direction.

The 2D azimuth reference filter $H_{az}(\omega,\omega_D)$ is then multiplied by the 2D range compressed data array $S_{1c}(\omega,\omega_D)$. The result becomes the roughly compressed data array. By using the same process of differential azimuth compression as in the 1D azimuth matched filtering case, the final target function can be obtained.

Figure 9.31 shows the reconstructed targets based on the 2D azimuth reference function. Compared with the results from the 1D azimuth matched filter, the 2D azimuth filter appears to have a better pulse compression ratio, and therefore better image quality.

Two side views of Fig. 9.31, one from the range direction into the targets and the other from the azimuth direction into the targets, are depicted in Figs. 9.32 and 9.33, respectively.

By comparing Figs. 9.32 and 9.33 with Figs. 9.27 and 9.28, one can see that the targets appear sharper in Figs. 9.32 and 9.33 than those in Figs. 9.27 and 9.28. The magnitudes of five pulses are near or above 70 in Figs. 9.32 and 9.33, while in Figs. 9.27 and 9.28, the pulse magnitudes are around 43 only.

From Figs. 9.32 and 9.33, one can see that the reference target, located at range sample 161 and azimuth sample 377, is different from that using a 1D azimuth matched filter. The range sample 161 is the result of the convolution of 2D reference function $H_{az}(t,\omega_D)$ and the 2D range-compressed data array $s_{1c}(t,\omega_D)$. The $H_{az}(t,\omega_D)$ peaks around range sample 442, while in the 2D range-compressed data array the reference target is located at range sample 231. The convolution of these two

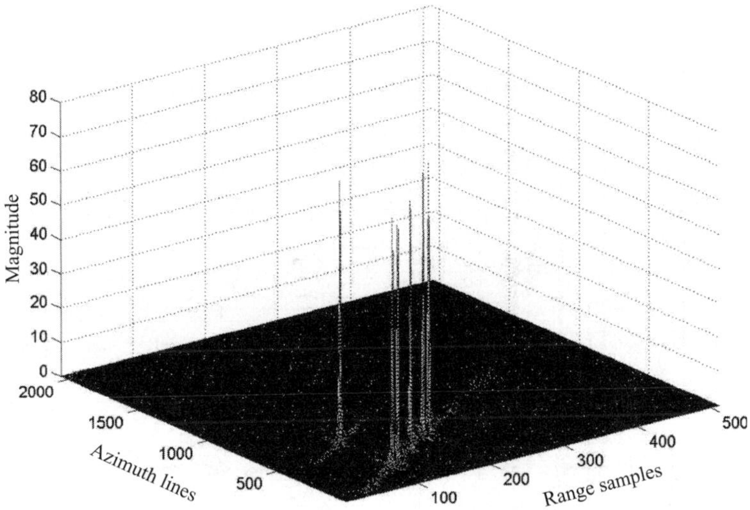

FIGURE 9.31 3D view of a reconstructed target function.

functions will have a peak at Mod[(442 + 231), 512] = 161. Similarly, azimuth sample 377 can be computed accordingly. The relative positions among other targets are quite accurate. The third (center and reference) target, located at range sample 161, is 50 range samples (50 × 10 = 500 m) away from target 1 (located at range 111) and targets 5 (at range 211). The distance between the targets 1 and 2 is 15 samples

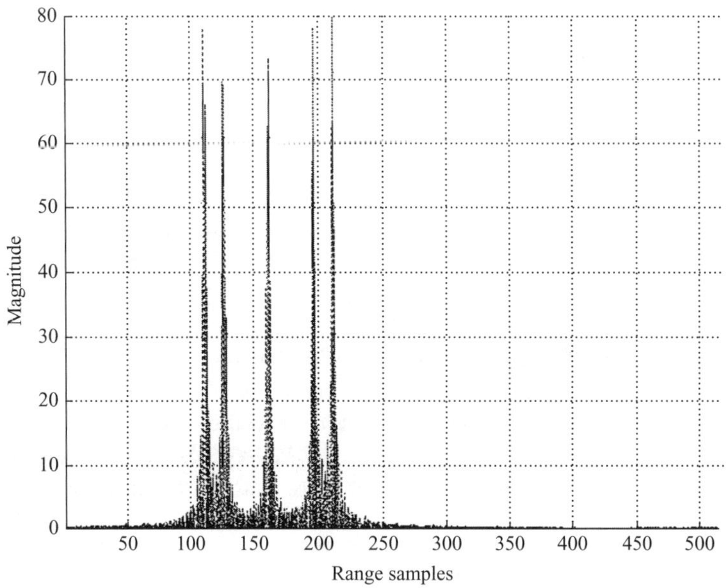

FIGURE 9.32 Side view, from the range direction, of Fig. 9.31.

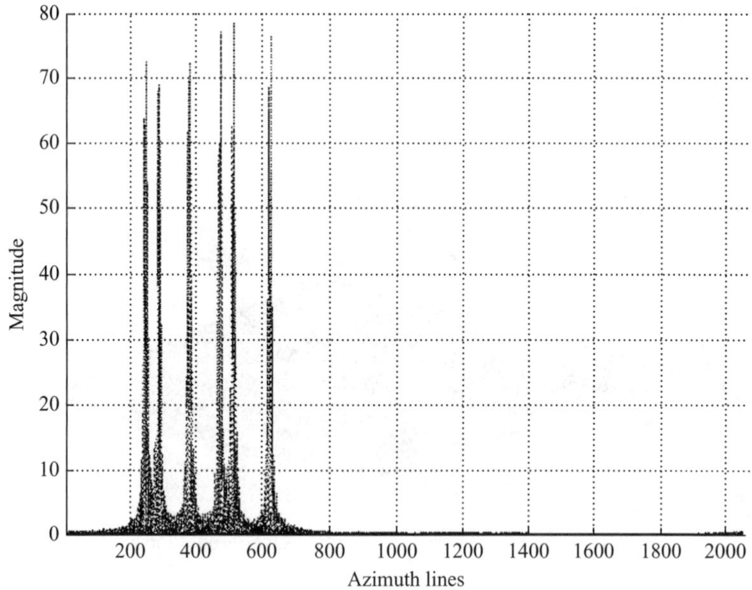

FIGURE 9.33 Side view, from the azimuth direction, of Fig. 9.31.

(15 × 10 = 150 m), which is also true for targets 4 and 5. Similarly, the locations of six targets along the azimuth direction are 244, 284, 377, 469, 509, and 620, respectively. With target 1 be assigned as $y_1 = 0$ and with azimuth sample spacing $A_s =$ 0.4 m, the simulated results and original specifications are listed below:

Simulated	Specification
$y_1 = 0$	0
$y_2 = 16$ m	15.7 m
$y_3 = 53.2$ m	52.5 m
$y_4 = 90$ m	89.3 m
$y_5 = 106$ m	105.1 m
$y_6 = 150.4$ m	150 m

9.4 RECONSTRUCTION OF SATELLITE RADAR IMAGE DATA

A satellite-based (RADARSAT-1) raw radar image file, generated by the Canada Space Agency and processed and distributed by MDA Geospatial Services Inc., will be used to reconstruct the radar image. The key parameters of this radar image data are listed below:

Antenna size ($L \times W$): 15 × 1.5 m

Closest range between target and radar $R_0 = 989{,}340$ m

Radar image data array size:

Number of azimuth lines = 1536

Number of range samples = 2048

Effective radar velocity $V_r = 7062$ m/s

Radar signal:

$f_c = 5.3$ GHz

$\theta_H = \lambda/L = 0.0377$ (radian)

$\theta_q = -1.6°$

Pulse duration time $T_p = 41.74\ \mu s$

Range FM rate $\alpha = 0.72135$ MHz/μs

Pulse bandwidth: 30.111 MHz

Data format: baseband complex data, 4 bits each

A/D sampling frequency $f_s = 32.317$ MHz

Pulse repetition frequency $f_{\text{PRF}} = 1257$ Hz

On the basis of this information, the following data can be computed:

Number of range samples in LFM pulse $= f_s\, T_p$
$= 1349$

Doppler frequency bandwidth $B_{\text{Dop}} = 2V \cos\theta_q\, /L$
$= 940$ Hz

Doppler frequency centroid $f_{Dc} = 2V \sin\theta_q\, /\lambda$
$= -\ 6968$ Hz

Doppler ambiguity number $M = -5$

Azimuth FM rate $(\beta) = 2V^2 \cos^3\theta_q/(\lambda R_0)$
$= 1780$ Hz/s

Synthetic aperture length $Ls = R_0\theta_H$
$= 3733$ m

Range sampling space $R_s = c/f_s$
$= 9.28$ m

Azimuth sampling space $A_s = V/f_{\text{PRF}}$
$= 5.62$ m

Number of azimuth samples within $Ls = 664$.

The complex baseband data, which appear in an in-phase–quadrature-phase format with 4-bit accuracy each, is first decoded and modified with automatic gain control provided with the data set. The real and imaginary part of the received signal, $s_b(m,n)$ with $m = 1,2,\ldots,1536$ and $n = 1,2,\ldots,2048$, is shown in Fig. 9.34: Fig. 9.34a shows the real part and Fig. 9.34b the imaginary part, of the baseband signal. Only 1 out of 50 azimuth lines is shown in the plot.

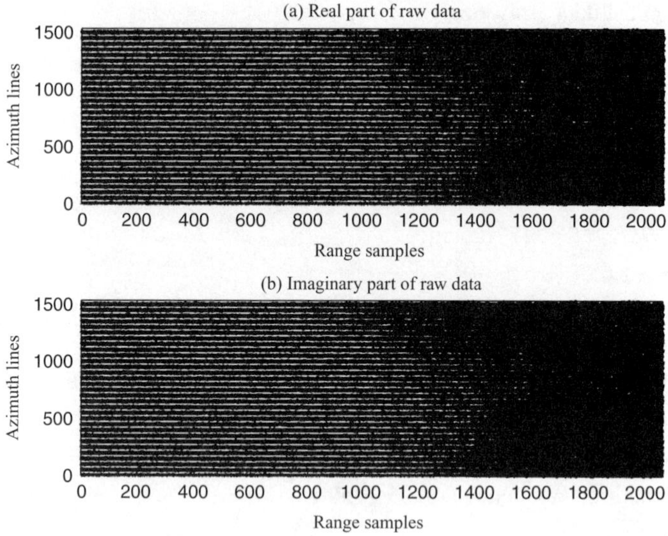

FIGURE 9.34 Waveforms of the real and imaginary parts of a received satellite baseband signal. (With permission from MDA Geospatial Services.)

A range matched filter is required to perform range compression on the received raw data. Given the sampling frequency f_s, the chirp rate α, and the radar beamwidth, the range matched filter can be designed as

$$h_r(n) = \exp\left[j\pi\alpha\,(n\Delta t)^2\right],$$

where $\Delta t = 1/f_s$ is the sampling interval, $\alpha = 0.72135$ MHz/μs, and $-674 \le n \le 674$.

Range compression is then performed by first transforming both the matched filter $h_r(n)$ and every line of the range signal $s_b(m,n)$ into the frequency domain using a 4096-point FFT. The product of the transformed signal and matched filter is then inverse-Fourier-transformed to become a time-domain range-compressed signal $s_{1c}(m,n)$. The range-compressed signal $s_{1c}(m,n)$, which appears in the range-azimuth (or time–spatial) domain, is shown in Fig. 9.35 for $n = 1349$–3396. Here the plot is shown in image format, with a gray level used to represent the magnitude of the signal (the white color corresponds to a large magnitude and the dark color corresponds to a small magnitude).

Since the sample length of the range matched filter is 1349, the range-compressed signal $s_{1c}(m,n)$, which is obtained through 4096-point IFFT, will have range samples 1–4096. Because of the edge effect of circular convolution of IFFT, only range samples 1349–3396, corresponding to $n' = 1,2,\ldots,2048$, are chosen for further image compression.

The range-compressed signal array $s_{1c}(m,n')$, for $n' = 1,2,\ldots,2048$, is then transformed into range–Doppler frequency-domain signal $S_{1c}(m,n)$. This is done by taking

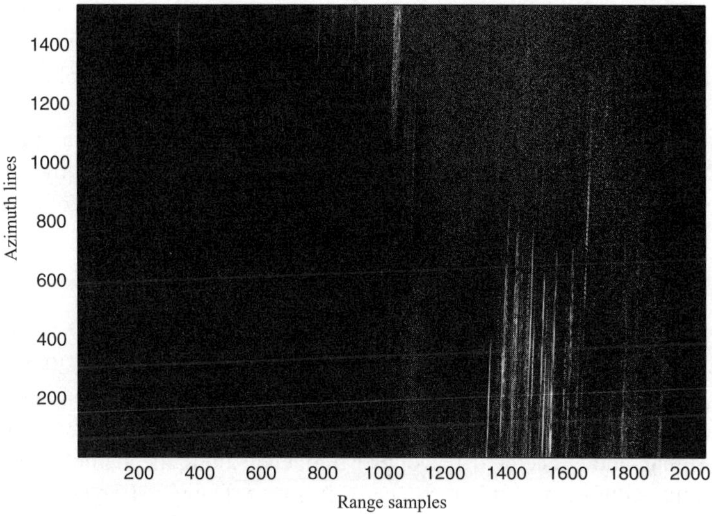

FIGURE 9.35 Image of a received satellite signal after range compression.

a 2048-point FFT on every column of the range–compressed signal. Since the sample length of the azimuth pulse is $Ls/A_s = 664$, and the number of azimuth lines of raw data is 1536, $M_{\mathrm{FFT}} = 2048$ is chosen as the azimuth FFT length. Figure 9.36 displays an image of the range-compressed signal $S_{1c}(m,n)$ in the range–Doppler frequency domain.

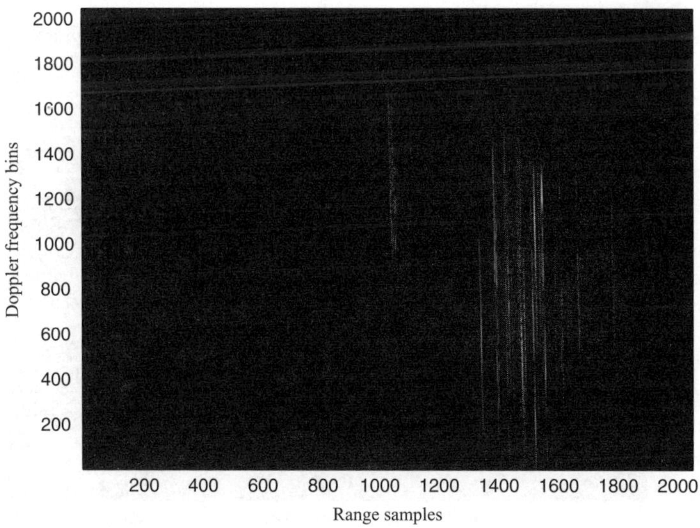

FIGURE 9.36 Image of a range-compressed signal in range–Doppler frequency domain.

The range-compressed data array $S_{1c}(m,n)$ in the range–Doppler frequency domain serves as the basis for image reconstruction. This data array will first be processed in the wavenumber domain based on Stolt interpolation, followed by the range–Doppler algorithm.

The Stolt interpolation starts with a reference function serving as the azimuth matched filter to perform the bulk azimuth compression, followed by differential azimuth compression. The 2D azimuth matched filter is designed as follows:

$$H_{az}(\omega,\omega_D) = \exp\left[j\sqrt{\frac{4\omega^2}{c^2} - \frac{\omega_D^2}{V^2}}X_0 \right].$$

Here $X_0 = 1{,}000{,}000$ m is the reference range, $\omega = 2\pi[f_c + (nf_s/2048)]$ and $\omega_D = 2\pi f_{\text{PRF}}(m + 2048(M - 1))/2048$. The row and column variables $m, n = 1,2, \ldots, 2048$, and $M = -5$. The variable ω refers to the passband frequency of the time-domain signal, and ω_D is the true Doppler frequency along the azimuth direction.

The range-compressed signal $S_{1c}(m,n)$, in range–Doppler frequency (t,ω_D) domain, is then transformed into (ω,ω_D) domain by taking a 2048-point FFT on every row of $S_{1c}(m,n)$ to become $S_2(m,n)$. The 2D matched filter $H_{\text{az}}(\omega,\omega_D)$ is then applied on $S_2(m,n)$ to become the roughly compressed signal $S_{2c}(m,n)$ in the (ω,ω_D) domain. A roughly reconstructed image $u(m,n)$ can then be obtained by taking 2D IFFT on the bulk compressed signal $S_{2c}(m,n)$ with respect to ω and ω_{D}, and the result is shown in Fig. 9.37. As can be seen, the roughly reconstructed image provides quite detail and accurate map of the ground. Since the azimuth matched filter has sample length 664, and the data file along the azimuth direction is 1536, the edge effect of circular convolution will cause distortions on both ends of the image. Only azimuth lines ranging from 332 to 1867 will provide a correct image.

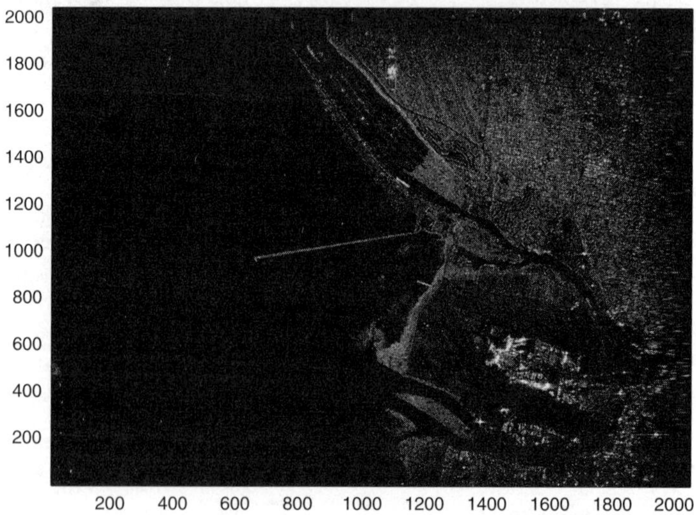

FIGURE 9.37 Radar image after bulk compression.

The bulk compressed signal $S_{2c}(m,n)$ in the (ω,ω_D) domain is then inverse Fourier transformed into (t,ω_D) domain as $S(m,n)$. It is then further processed by the differential azimuth correction, which requires computation of Δx_n and $dk(m)$ and are defined as

$$\Delta x_n = \frac{c\left(n - N_{rf}\right)}{f_s}.$$

$$dk\,(m) = \frac{k_y^2}{4k_{fc}}$$

where

$$k_y = \frac{(m + 2048M)\,(2\pi f_{\mathrm{PRF}})}{2048V},$$

$k_{fc} = (2\pi f_c/c)$, $M = -5$, and $m = N_{DL}, \ldots, N_{DU}$, $n = 1,2,\ldots,2048$.

The reference sample $N_{rf} = 1148$ is chosen and corresponds to the slant range $X_0 = 1{,}000{,}000$ m. On the basis of the estimated Doppler frequency centroid $f_{Dc} = -6968$ Hz and Doppler bandwidth $= 940$ Hz, the true upper and lower frequency band edges can be computed as $f_{DU} = -7338$ Hz and $f_{DL} = -6398$ Hz. The corresponding folded or observed baseband frequencies are $f_{Dc}' = -683$ Hz, $f_{DU}' = -1153$ Hz and $f_{DL}' = -213$ Hz. Given the Doppler sampling frequency $f_{\mathrm{PRF}} = 1257$ Hz and $N_{\mathrm{FFT}} = 2048$, differential azimuth compression is operated on $S(m, n)$ from $N_{DL} = 347$ (corresponding to f_{DL}') to $N_{DU} = 1879$ (corresponding to f_{DU}'), and for the whole range samples from 1 to 2048. That is, for every Doppler frequency column, only Doppler frequency bin numbers from 347 to 1879 will be adjusted with a phase factor of exp $[-jdk(m)\Delta x_n]$. In equation form,

$$U(m,n) = S(m,n)\exp[-jdk(m)\Delta x_n]$$

After the differential azimuth compression, the reconstructed image is obtained by taking IFFT on (m,n). The result is shown in Fig. 9.38.

The image quality after differential azimuth compression is about the same as compared with the bulk compressed image shown in Fig. 9.37. This implies that the Stolt interpolation with bulk compression provides good quality of images for radar with steady moving speed V and stable squint angle θ_q, and for ground targets that cause smooth variation in the Doppler frequency centroid f_{Dc}.

The range–Doppler algorithm will now be applied to the same image file. It operates on data in the range–Doppler frequency domain, namely, $S_{1c}(m,n)$ as shown in Fig. 9.36. The sample length of azimuth matched filter equals $N_{az} = R_0\theta_H/A_s = 674$, where $R_0 = 1000{,}000$ m, $\theta_H = 0.00377$ radian, and $A_s = V/f_{\mathrm{PRF}} = 5.62$ m are used. Given the slant range sample spacing $R_s = 9.28$ m, for a swath of 2048 range samples, the maximum slant range difference is $9.28 \times 1148 = 10654$ m, which is less than 1.1% of R_0. Therefore, the same azimuth matched filter $h_{az}(m)$ can be

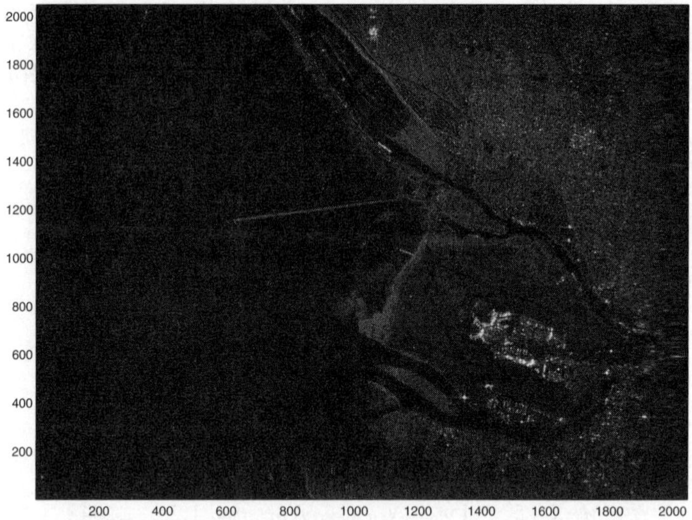

FIGURE 9.38 Radar image after differential azimuth compression.

used throughout the whole image frame without significant errors. The 1D azimuth matched filter is designed as follows:

$$h_{az}(m) = \exp\left[-j2\pi f_{D_C} m\Delta s + j\pi\beta(m\Delta s)^2\right].$$

Here $\Delta s = 1/f_{\text{PRF}}$ is the pulse repetition interval, $\beta = 1780$ Hz/s, $f_{D_c} = -6968$ Hz, and $-332 \le m \le 332$. The Doppler frequency-domain filter $H_{\text{az}}(m)$, for $m = 1,2\ldots,$ 2048, is obtained by taking a 2048-point DFT on $h_{az}(m)$.

The range migration of the squint SAR system is caused by the slant range difference at the edges of the 3-dB radar beamwidth. The amount of range migration can be computed, from Eq. (8.8b) as follows:

$$\begin{aligned}\Delta N_{Rk} &= \frac{2\Delta R_k}{R_s}\\ &\approx \frac{R_0\lambda^2}{4V^2R_s}\left(f_{DL} + \frac{kf_{\text{PRF}}}{N}\right)^2 - \frac{R_0\lambda^2 f_{DL}^2}{4V^2R_s}.\end{aligned}$$

The integer part of ΔN_{RK} will be used for the range sample shift, while the fractional part will be used for interpolation on the range samples. In this example, the 16-set 8-tap sinc filter is adopted again as the interpolation filter. The maximum amount of range migration is 22 samples in this example.

After computing the range migration amount, every row of the range-compressed signal $S_{1c}(m,n)$ along the Doppler frequency axis is then filtered with the corresponding sinc interpolation filter. The sinc-filtered output is then further adjusted by two elements. The first one is 4-sample delays caused by the 8-tap sinc filter. The second one is the range sample shift, which varies along the Doppler frequency axis. Every

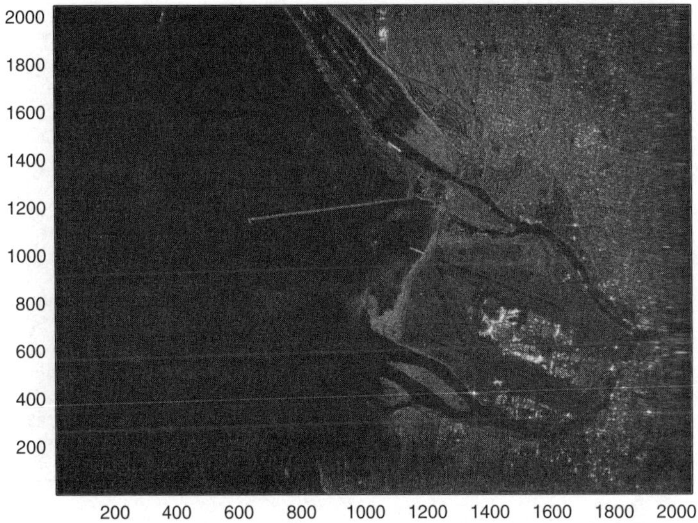

FIGURE 9.39 Radar image processed by range–Doppler algorithm.

column of the adjusted output of the sinc filter is then multiplied by the azimuth reference filter $H_{az}(m)$ to become $U(m,n)$. The inverse DFT is then applied on $U(m,n)$, and the result is shown in Fig. 9.39.

By comparing Figs. 9.38 (or 9.37) and 9.39, it is difficult to tell which one has better image quality. Both the range–Doppler and Stolt interpolation algorithms generate good quality of radar images.

For comparison purposes, a different image data file was used and processed by the two algorithms described above. Figure 9.40 displays the image using the Stolt

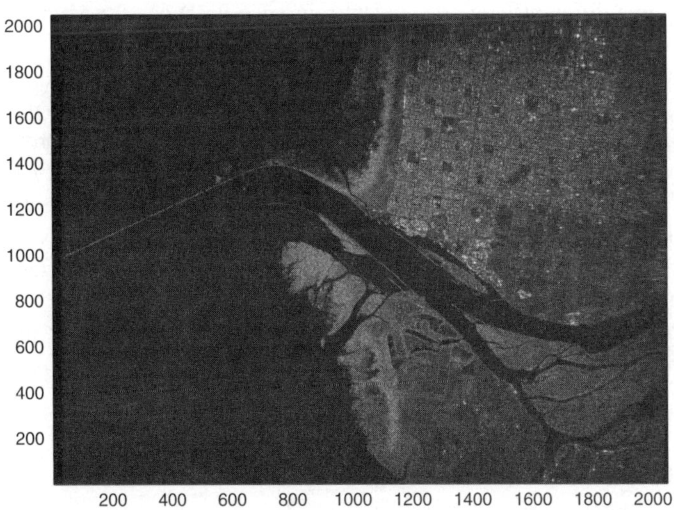

FIGURE 9.40 Radar image processed by Stolt interpolation technique.

FIGURE 9.41 Radar image processed by range–Doppler algorithm.

interpolation technique (without DAC), while Fig. 9.41 presents the image using the range–Doppler technique. Again, both algorithms appear to have similar image quality.

9.5 COMPARISON BETWEEN RANGE–DOPPLER AND STOLT INTERPOLATION ON SAR DATA PROCESSING

The difference between the range–Doppler and Stolt interpolation algorithms on processing SAR data can be summarized as follows: (1) the range–Doppler algorithm requires different range-dependent reference functions for azimuth compression, while the Stolt interpolation uses only one reference function for azimuth compression; (2) the range–Doppler algorithm uses a synthesized 1D azimuth reference function, while the Stolt interpolation utilizes a 2D azimuth reference function; (3) the range–Doppler algorithm corrects the range cell migration problem based on interpolation filtering of range samples for all Doppler frequency bins in the range–Doppler domain, while the Stolt interpolation algorithm corrects the problem through the multiplication of an azimuth and range-dependent phase factor, with respect to a reference point, on Doppler frequency samples for every range column in the range–Doppler domain.

The major computation requirements for range–Doppler and Stolt interpolation algorithms are based on the following general assumptions:

1. The basic computation requirement of a radix-2 FFT (or IFFT) are

   One complex multiplication (four real multiplications and two real additions)

Two complex additions (four real additions)

2. The sample length of the range matched filter is N_{fr}.
3. The range matched filter is predesigned in the frequency domain and stored in memory as a table ROM (read-only memory)
4. The radar raw image data array is $M_i \times N_i$.
5. The FFT (or IFFT) length used along the range or x axis is $N \geq N_i + N_{fr} - 1$, and N is chosen to be a power of 2.
6. The reconstructed radar image data array is $M_i \times N_i$.

In addition, the following assumptions are made for the range–Doppler algorithm:

- The sample length of the azimuth 1D matched filter is M_{fa}.
- The FFT (or IFFT) length used along the y axis is $M \geq M_i + M_{fa} - 1$, and M is a power of 2.
- The interpolation filter coefficients are real numbers, and the filter length is N_{fi}.
- The group of interpolation filter is predesigned in the range domain and stored in memory as a tabular ROM.

The major computation requirements for range–Doppler processing algorithms are

1. Range compression
   a. FFT along x axis for N range samples
      (1) Complex multiplications: $0.5M_iN \log_2 N$
      (2) Complex additions: $M_iN \log_2 N$
   b. Range matched filtering on $M_i \times N$ data array (azimuth–Doppler domain)
      (1) Complex multiplications: M_iN
   c. IFFT along x axis for N range samples
      (1) Complex multiplications: $0.5M_iN \log_2 N$
      (2) Complex additions: $M_iN \log_2 N$

   (After the IFFT operation, the range-compressed output data are in a range–azimuth domain, and the size of the data array is rescaled to $M \times N_i$. The azimuth sample length is zero-padded to render $M \geq M_i + M_{fa} - 1$ sample length for the following range cell migration correction and azimuth compression.)
2. Range cell migration correction
   a. FFT along y axis for M azimuth samples
      (1) Complex multiplications: $0.5N_iM \log_2 M$
      (2) Complex additions: $N_iM \log_2 M$
   b. Fractional interpolation and sample shift
      (1) Complex multiplications:* $0.5MN_i N_{fi}$

(2) Complex additions:* $MN_i(N_{fi} - 1.5)$
(* Since the interpolation filter coefficients are real, only two real multiplications are needed instead of 4 multiplications and two additions.)

3. Azimuth compression
   a. Azimuth matched filtering on $M \times N_i$ data array
      (1) Complex multiplications: MN_i
   b. IFFT along y axis for M azimuth samples
      (1) Complex multiplications: $0.5N_iM \log_2 M$
      (2) Complex additions: $N_iM \log_2 M$

   The data array is rescaled to $M_i \times N_i$.

In summary, the total computation requirements for the range–Doppler algorithm are

Complex multiplications: $M_iN(\log_2 N + 1) + N_iM(\log_2 M + 1 + 0.5N_{fi})$
Complex additions: $2M_iN\log_2 N + N_iM\,(2\log_2 M + N_{fi} - 1.5)$

For the Stolt interpolation algorithm, two additional assumptions are made:

1. The 2D matched filter is in the (ω,ω_D) domain, and its array size is $M \times N$.
2. The FFT (IFFT) length used along the y axis is $M \geq M_i + M_{fa} - 1$, and M is a power of 2. Here M_{fa} is the azimuth sample length within the synthetic aperture length.

The major computation requirements for the Stolt interpolation algorithm are

1. Range compression
   a. FFT along x axis for N range samples
      (1) Complex multiplications: $0.5M_iN \log_2 N$
      (2) Complex additions: $M_iN \log_2 N$
   b. Range matched filtering on $M_i \times N$ data array
      (1) Complex multiplications: $M_i \times N$

   (The range-compressed output data are in the azimuth–frequency domain with array size $M_i \times N$. The data array is rescaled to $M \times N$ by zero-padding the azimuth sample to render $M \geq M_i + M_{fa} - 1$ with M equal to a power of 2.)
2. Rough azimuth compression
   a. FFT along y axis for M azimuth samples
      (1) Complex multiplications: $0.5NM \log_2 M$
      (2) Complex additions: $NM \log_2 M$

b. 2D Matched filtering for $M \times N$ data array
   (a) Complex multiplications: MN
c. IFFT along x axis for N range samples
   (a) Complex multiplications: $0.5MN \log_2 N$
   (b) Complex additions: $MN \log_2 N$

(The 2D bulk-compressed output data are in the range–Doppler domain with array size $M \times N$. The data array is rescaled to $M \times N_i$, which corresponds to the effective values of 1D convolution between the 1D range matched filter and 2D data.)

3. Differential azimuth compression
   a. Complex multiplications: MN_i
   b. IFFT along y axis for M azimuth samples
      (1) Complex multiplications: $0.5N_iM \log_2 M$
      (2) Complex additions: $N_iM \log_2 M$

(The 2D differential azimuth-compressed output data are in a range–azimuth domain with an $M \times N_i$ array. The data array is rescaled to $M_i \times N_i$, which corresponds to the effective values of 2D convolution between the 2D matched filter and 2D data.)

In summary, the total computation requirements for the Stolt interpolation algorithm are

$$\text{Complex multiplications: } 0.5M_iN(\log_2 N + 2) + 0.5MN_i(\log_2 M + 2) + 0.5MN(\log_2 MN + 2)$$

$$\text{Complex additions: } M_iN\log_2 N + MN\log_2 MN + MN_i\log_2 M$$

For comparison purposes, letting the input image array be 1536×2048 ($M_i = 1536$ and $N_i = 2048$), the intermediate array $M \times N$ be 2048×4096, and the interpolation filter be an 8-tap filter, one obtains the following data:

For the range–Doppler algorithm:
  Number of complex multiplications: 1.42×10^8
  Number of complex additions: 2.58×10^8

For the Stolt interpolation without DAC:
  Number of complex multiplications: 1.42×10^8
  Number of complex additions: 2.56×10^8

For the Stolt interpolation algorithm:
  Number of complex multiplications: 1.68×10^8.
  Number of complex additions: 3.0×10^8.

Since the range–Doppler algorithm and the Stolt interpolation algorithm process the $M_i \times N_i$ image data array in the same spatiotemporal and frequency–Doppler frequency domains, both algorithms require roughly the same memory capacity for SAR processing. Taking the double-buffering requirement for the FFT/IFFT process, the range cell migration correction, and the differential azimuth correction, the total memory required for SAR processing can be estimated at about $2.5MN$ complex samples. For a data array of $M \times N = 2048 \times 4096$, the memory requirement is 2×10^7 complex samples.

FURTHER READING

1. V. Oppenheim and A. S. Willsky, *Signals and Systems*, Prentice-Hall, Uppersaddle River, NJ, 1996.
2. B. P. Lathi, *Signal Processing and Linear Systems*, Oxford University Press, New York, 1998.
3. S. J. Orfanidis, *Electromagnetic Waves and Antennas*, 2004 (`www.ece.rutgers.edu/~orfanidi/ewa`).
4. D. R. Wehner, *High Resolution Radar*, 2nd ed., Artech House, Boston, 1995.
5. A. G. Stove, *Linear FMCW Radar Techniques*, IEE Proceedings F. 1992, No. 5.
6. M. Soumekh, *Synthetic Aperture Radar Signal Processing with Matlab Algorithms*, Wiley, New York, 1999.
7. J. C. Curlander and R. N. McDonough, *Synthetic Aperture Radar Systems and Signal Processing*, Wiley, New York, 1991.
8. Ian G. Cumming and Frank H. Wong, *Digital Processing of Synthetic Aperture Radar Data: Algorithms and Implementation*, Artech House, Boston, 2005.
9. A. Meta and P. Hoogeboom, "Time analysis and processing of FM-CW SAR signals," paper presented at International Radar Symposium (IRS 2003), Sept. 30–Oct. 2, 2003.
10. Atlantis Scientific Inc., *Theory of Synthetic Aperture Radar*, 1997.
11. R. Bamler, "A comparison of range-doppler and wavenumber domain SAR focusing algorithms," *IEEE Trans. Geosci. Remote Sensing* **30**: 706–713 (July 1992).
12. T. Einstein, *Realtime Synthetic Aperture Radar Processing on the RACE® Multicomputer,* Application Note 203.0, Mercury Computer Systems, Chelmsford, MA, June 1996.
13. D. T. Sandwell, *SAR Image Formation: ERS SAR Processor Coded in Matlab* (`http://www.topex.ucsd.edu/insar/sar_image_formation.pdf`).

Digital Signal Processing Techniques and Applications in Radar Image Processing, by Bu-Chin Wang.

INDEX